T0362158

Intelligent Warfare

This book examines the future trend toward "intelligent" warfare considering the global environment, the history of warfare, and scientific and technological advancement. It develops a comprehensive set of theoretical frameworks, application concepts, and evaluation criteria for military intelligence.

The volume is packed with theoretical highlights and vivid examples, including the tracking of Osama bin Laden, the decapitation strike against Qasem Soleimani, the remote assassination of Iranian nuclear scientists, the drone war in the Nagorno–Karabakh conflict, modern equipment deployed in the Palestinian–Israeli conflict, and the war between social media groups. In addition, the author envisions a possible future for "intelligent" wars in which adversarial parties engage in combat through virtual and unmanned systems. This nature may help avoid the brutality and high death toll associated with traditional warfare.

The book explores the possibility of future civilized warfare. It will be of interest to researchers, academics, and students in the fields of politics, military intelligence, and military technology, as well as those who are interested in intelligent warfare in general.

Mingxi Wu, a distinguished Chinese research fellow and an expert in military equipment and intelligence, has conducted research for decades on the frontier of defense science and technology, equipment development, and revolutions in military affairs, with a focus on artificial intelligence and intelligentization. He is the deputy director of the Intelligent Weapons and Equipment Technology Committee, China Ordnance Society.

Intelligent Warfare

Prospects of Military Development in the Age of AI

Mingxi Wu

Translated by Qichao Zhu and Chaowei Pang

Routledge
Taylor & Francis Group

LONDON AND NEW YORK

First published 2023
by Routledge
4 Park Square, Milton Park, Abingdon, Oxon OX14 4RN

and by Routledge
605 Third Avenue, New York, NY 10158

Routledge is an imprint of the Taylor & Francis Group, an informa business

© 2023 Mingxi Wu

Translated by Qichao Zhu and Chaowei Pang

British Library Cataloguing-in-Publication Data
A catalogue record for this book is available from the British Library

Library of Congress Cataloging-in-Publication Data
A catalog record has been requested for this book

ISBN: 978-1-032-29863-4 (hbk)
ISBN: 978-1-032-29864-1 (pbk)
ISBN: 978-1-003-30242-1 (ebk)

DOI: 10.4324/b22974

Typeset in Times New Roman
by Newgen Publishing UK

Contents

Figures

Tables

Translator's Preface

Mr. Mingxi Wu is a research fellow who specializes in the application of artificial intelligence (AI) technology. This book, *Intelligent Warfare – Prospects of Military Development in the Age of AI*, was first published in China in January 2020 and has received widespread praise from both professionals and general readers. The book is distinguished by its perceptive ideas, extensive content, and exceptional performance in the Chinese book market.

Our translation team was awestruck by the book. Therefore, when we learned that the Chinese National Defense Industry Press and Taylor & Francis Publishing Group were discussing the publication of an English translation, we agreed immediately. We planned to complete the translation, review, and proofreading work within 6 months, based on the contract agreed upon by the two publishing houses.

In comparison to the commercially available Chinese version, the English version has undergone some structural and content changes. The new edition is divided into the theoretical and practical sections. The first seven chapters examine intelligent Warfare in general and lay the theoretical groundwork for the remainder of the book. The following eight chapters discuss how military AI can be used in various contexts, including unmanned and cyberspace operations. The final chapter discusses military intelligence's evolution and evaluation.

Notable is the author's initial inspiration, which came from the 2016 human–machine Go battle between Google's AlphaGo and world champion Lee Sedol. His determination to write this book was fueled by AlphaGo's eventual victory in the match. The author developed the concepts and viewpoints through academic and technical analysis, predictive thinking, and exploratory research, all of which were motivated by personal interests and an ambition to explore future warfare. They are unrelated to the Chinese government's or military's strategic plans and involve no disclosure of state secrets.

The author attempts to study and analyze the global military revolution brought about by new technology from the standpoint of all humanity. Real events in this book are interpreted based on information provided by the media in their entirety, and do not reflect the author's political leanings or views on civilization superiority, ideological disputes, racism, and religious discrimination. In the concluding remarks, the author expresses his hope and vision for a peaceful world. The book argues that the emergence of new operational capabilities, such as unmanned systems and intelligent high-fidelity simulation systems, which serve as the "amulets" of human warfare, will enable future warfare to be conducted in a peaceful (or less harmful) manner, thereby sparing humanity the great suffering caused by traditional warfare in physical space.

The translation process was as follows: in the first round, the author lectured the translation team to ensure they understood the book's primary contents and viewpoints. In the second round, the team decided on translation strategies based on the readability of English-speaking readers, created a unified termbase and style guide, and assigned tasks to each team member. In the third round, the completed translation draft was reviewed by subject matter experts. We would like to express our appreciation for the efforts of the translators and reviewers. Due to the scope of the book's vision, the depth of its content, and its professionalism, translation errors and flaws are almost unavoidable. We welcome reader feedback and suggestions.

Qichao Zhu
Translation Coordinator and Final Reviewer
May 2022

Preface

Military intelligence seeks to build and optimize decision-making algorithms in operational systems. These algorithms are based on artificial intelligence (AI), and they affect every aspect of warfare, including reconnaissance, command and control, attack, defense, logistics, research, production, mobilization, and procurement. While AI is only a component of the operational system, due to its "brain-like" functions and strong ability to surpass human limitations, it will dominate future warfare, accelerating the transformation of modern warfare from informatization to intelligentization.

Intelligent warfare may take on diversified forms, most notably cognitive confrontations with AI at their core and integrated operations utilizing "intelligence+" and "+intelligence." The intelligent operational system is comparable to the human body, which has numerous interconnected components, including the five sense organs, the nervous system, the heart, the limbs, and, most importantly, the brain as its center.

Human civilization develops due to interactions between politics, economy, technology, military, and social development, all of which are influenced by technology in a long-term, endogenous, disruptive, and revolutionary manner. Intelligent technologies such as AI, big data, cloud computing, interdisciplinary biology, unmanned systems, and parallel training are advancing at a breakneck pace and becoming increasingly integrated with established technologies, altering humans' epistemology, methodology, and operational mechanisms and enhancing humans' ability to transform the world. Following mechanization and informatization, intelligentization will be the third stage of human civilization. Emerging intelligent technologies will also accelerate transformations in machine, bionic, swarm, and human–machine integrated intelligence; intelligent perception, decision-making, action, confrontation, logistics; and intelligent design, research and development (R&D), testing, and manufacturing, which span almost all societal sectors. They will also facilitate intelligent industries characterized by knowledge sharing and ubiquity, the emergence of new economic and social forms, and changes in the military field, including operational patterns, capabilities, formation, theory, and force structure and organization.

Modern warfare will become increasingly technical and complex as new technologies, equipment, and methods are applied to the military. They will encompass multiple operational dimensions such as time, space, sea, and land, as well as new domains such as frequency, energy, information, biology, society, and cognition. These factors and domains interweave and interact with each other, necessitating a higher level of operational capability and efficiency that will enable the seamless execution of military actions such as detection, perception, communication, command, observation, analysis, judgment, decision-making, confrontation, support, and calculation. Intelligent technologies applied to the military sector may help reduce uncertainty and increase winning chances. These technologies have the potential to process massive amounts of data for tasks such as situational awareness, target recognition, mission planning, rapid attack, precision support, cyber operations, and cognitive confrontation, while also addressing issues such as multi-strategy responses, multi-tasking, multi-tactics selection, and multi-modality support in an intense confrontation with complex environments, highly dynamic responses, and fuzzy information.

The intelligent military ecosystem powered by AI will be highly mobile and interconnected, supported by network communication and distributed clouds. Data-processing models and algorithms will serve as the core technology and cognitive confrontation as the primary application. Military operations in the future will take the form of multi-domain integration, cross-domain attack and defense, conflicts between unmanned systems and swarms, and interaction between virtual and physical forces. Either "intelligence+" or "+intelligence" will usher in a new era of unstoppable military development. Whoever controls the advantage of intelligence will have the initiative in future warfare.

This book discusses how the advancement of AI-based technologies may alter the future military landscape and the characteristics of war. The following summarizes the major arguments:

1. **New operational elements embodied by AI, cloud computing, networking, swarms, and terminals will reshape the battlefield ecosystem and alter the rules and laws of victory.** Among these, the AI system built on models and algorithms will serve as the central operational capability, traversing all facets and connections while multiplying, surpassing, and enabling itself. For example, operational vehicles can be controlled by AI, swarms guided by AI, and decisions assisted by AI. Models and algorithms will supplant traditional human-based strategies and tactics, and AI will dominate the operation system of systems (OSoS) and the whole warfare progress. In this sense, algorithmic warfare and intelligence superiority will be critical in future conflicts.

2. **Virtual space will play a growing role in the OSoS and achieve deep convergence with physical space operations.** Physical space will give way to virtual space as operational space expands. Virtual space will unify segmented

operational forces and elements to enhance networked and systemized operational capabilities, while serving as the primary battlefield for cognitive factors such as cyber power, electromagnetic power, intelligence, public opinion, psychology, and consciousness. Virtual battlefields, operational simulations and experiments, virtual–real interactions, and virtual–real parallel operations will become possible, enhancing the military capability to suppress reality through virtual reality.

3. **Unmanned operations will supplant manned operations as the primary mode of warfare.** Technological integrations will accelerate the evolution of war forms to unmanned stages. At the primary stage, humans will play a significant role in wars, assisted by unmanned systems. At the intermediate stage, unmanned systems will take the lead, assisted by humans. At the advanced stage, completely unmanned operations will be carried out based on human-defined rules and programs. The advanced stage can also be summarized as "unmanned control with manned design." Humans will create the framework for unmanned systems, defining autonomous behaviors and game rules and delegating most operational tasks to unmanned vehicles and robotic forces. Unmanned systems powered by AI will expand into new fields such as cyber attack and defense, electronic confrontation, multi-source perception, connection verification, human tracking, and infrastructure control.

4. **Joint all-domain and cross-domain operations will evolve from a physically connected pattern to one that is heterogeneous, incorporating data cross-linking, mutual tactical control, and integrated cross-domain attack and defense.** AI and multi-disciplinary fusion will enable multi-source perception, heterogeneous information integration, data cross-linking, cross-domain joint strike and defense; multi-domain integrated logistics, force deployment, command and control, mission planning and correlation; weaponry and equipment interoperability. AI brain systems and human–machine hybrid intelligence will combine resources from multiple domains to support operations in a single domain.

5. **Traditional human-centered command, control, and decision-making models will be entirely replaced by a hybrid of centralized, weakly centralized, and decentralized models.** In its natural state, the intelligent OSoS is distributed, networked, flattened, and parallelized. The human-centered decision-making model will be phased out in favor of decentralized or weakly centralized models, such as AI-based unmanned systems, autonomous swarms, and manned–unmanned cooperation systems. Decentralized decision-making models can be applied to numerous lower-level operations and simple tasks. However, manned and centralized decision-making models continue to be dominant in solving complex high-level tasks.

6. **Military operations will no longer be characterized by energy release or a linear power superposition, but by a rapid convergence of multiple nonlinear, emergent, self-growing, and self-focusing effects.** When military

intelligence reaches a certain level, operational effects may exhibit non-linearity, asymmetry, self-growth, rapid confrontation, and uncontrollable amplification through the use of advanced AI, quantum computing, an IPv6-based Internet, and hypersonic weapons. These effects are more noticeable in unmanned control, swarm attack and defense, cyber public opinion manipulation, and cognitive confrontations. Under AI control, the interwoven elements – cognition, information, and energy – will rapidly accumulate around the target, allowing for more rapid confrontations and causing a sharp amplification of multiple effects and rapid convergence of results.

7. **The interactions between humans and weapons/equipment will be transformed, with the physical distance between them increasing but the mental distance decreasing.** Human thought and wisdom will be incorporated into the R&D, iterative optimization, and operational testing of weapons and equipment. Unmanned systems will combine human creativity and foresight with machine precision, speed, reliability, and fatigue resistance, and their role in operations may shift from assistance to dominance, eventually supplanting human roles. Weapons and equipment will evolve into cyber-physical systems supported by clouds and human–machine interaction systems. Their mode of development and application will also change. While mechanical equipment degrades over time, intelligent software and algorithms may improve in sophistication and precision with each upgrade. This revolutionizes the existing system for developing managing weapons and equipment.

8. **Intelligent OSoS will develop self-adaptive, self-learning, self-confrontation, self-repairing, and self-evolving capabilities, eventually morphing into an evolvable and ecological self-competition system.** Technological advancements and applications, such as operation simulation, virtual reality, digital twins, parallel systems, intelligent models and algorithms, brain-like chips and systems, bionic systems, natural resource collection, and novel machine learning, will drive future OSoS upgrades. The single-task subsystems will resemble individual living organisms, while the multi-task system will have cyclic functions and evolutionary mechanisms similar to those found in forest species groups, allowing it to confront and compete with itself under complex environmental conditions.

9. **The national defense industry will eventually transition from a closed, self-contained, physical-based, and long-cycle model to an open-source, crowd-sourcing, self-adaptive, and intelligent model capable of rapidly responding to military demands.** Following joint and multi-domain operations requirements, an intelligent manufacturing system with interactive humans, machines, materials, and environments will emerge, characterized by hardware–software integration, virtual–real interaction, vertical industrial chain linkage, horizontal distributed collaboration, and civil–military fusion. The industry will shift its emphasis from developing products for a single military service to developing products

for multiple military services, revolutionizing previous research, design, and manufacturing models. Intelligence construction and combat effectiveness generation will be a collaborative effort between the military and civilian sectors, incorporating continuous optimizations via virtual and real-world operations and training, as well as concurrent R&D, trials, applications, and iterations.

10. **Humans may suffer an unforeseeable disaster without pre-programming a "termination button" for intelligent operational systems.** Intelligent operational systems have the potential to evolve and become superhuman. Thus, it can be dangerous if no "termination button" is designed in advance to control the programs and nodes.

This issue should be discussed in the context of the global community in order to ensure a shared future for humanity and the sustainable development of human civilization. Intelligent operations should be regulated to ensure that they evolve in a predictable, dependable, credible, safe, and civilized manner, with appropriate technical, procedural, moral, and legal safeguards in place, as well as mandatory constraints, inspection, and management.

In warfare, intelligence levels may reflect civilization levels. As technological transparency and economic interdependence increase, conflicts between human forces will gradually be supplanted by clashes between robots and AI.

We hope that in the future, warfare will evolve from intra-human society conflicts that result in widespread devastation in the physical world to a high-simulation war game between AI and robots in the virtual world that limits conflict to deterrence and balance based on countries' military capabilities and overall strength. Throughout the war, humans will transition from planners, designers, participants, dominators, and victims to rational thinkers, organizers, controllers, bystanders, and adjudicators. They will be spared the physical agony, mental anguish, and property destruction that accompany warfare. This may be the final stage of intelligent warfare, the book's ultimate goal, my dream, and humanity's grand vision.

We have arrived at a critical juncture on the road to the age of AI. This book attempts to envision the future of warfare by examining the strategic and technological factors at play. It focuses on a few key issues surrounding intelligent war, including its nature, driving forces, evolutionary process, forms of qualitative changes, supporting science and technology, and other topics such as parallel military and training, reconfiguration of battlefield ecology, the evolution of OSoS, the nine-dimensional evaluation criteria, and the evolution of war. In conclusion, the book provides an overview of intelligent warfare.

Thought determines action, which in turn affects the future. Conceptual innovation may catalyze the emergence of novel technologies, applications, models, operational forms, and demands. We may understand our current situation by looking to the future, obtain local directions by examining the global framework, and comprehend evolutionary laws by exploring the changes. This book is intended to educate readers about AI-powered military competition and inspire them to consider military intelligence. AI is advancing at a breakneck pace, with unpredictable consequences for war. As a result, this book may contain inaccuracies or viewpoints that are debatable. Suggestions and discussions are welcomed.

Mingxi Wu
May 2022, Beijing

Part I

Theoretical Discussions

Intelligent warfare is cognitive-centric, with algorithms at its heart and intelligence as the dominant force, eclipsing fire power, machine power, and information power. The basic pursuit is to harness energy through intelligence, to increase efficiency through intelligence, to control the physical world through the virtual, and to overcome inferiority through superiority. Whichever side of the battlefield has numerically superior and more intelligent artificial intelligences will obtain a greater initiative on the battlefield.

DOI: 10.4324/b22974-1

1 Global Factors that Influence Warfare

Global factors that influence warfare in the twenty-first century are diverse, spanning politics, economy, science and technology, military affairs, and society. By the middle of the twenty-first century, there will be seven strategic factors determining the direction of military reform and warfare development, as well as some inescapable rules and trends that will have a direct or indirect effect on future warfare.

1.1 Reshaping the International Strategic Landscape

For the foreseeable future, the international strategic landscape may shift from "one superpower, multi great powers" to "two superpowers, multi great powers," gradually forming an international landscape in which all countries are autonomous but also interdependent, with China and the US at the center. Unlike the all-out confrontation between the US and the former Soviet Union during the Cold War, contemporary China–US relations are characterized by a balance of competition, cooperation, and conflict, as well as a high degree of economic interdependence. The two powers' trade war has had a significant impact on bilateral and global trade since 2018, sparking a series of frictions and competition in political, economic, scientific, technological, and military fields. However, the global strategic landscape will not be fundamentally reversed in the long run. The trade war may even hasten the emergence of a "two superpowers" landscape.

Regardless of the inevitable conflicts and competition in politics, the military, and high-end scientific and technological fields, the driving force for the two superpowers' cooperation and mutual benefits in developing economy and combating climate change will always exist. It is less likely for China to escalate the China–US trade war into a full-fledged military confrontation, as the country possesses the patience and confidence to confront the issue rationally and wait for a peaceful resolution. By contrast, the US is more aggressive in challenging the status quo because it believes the recent China–US trade agreement is a bad deal for the US. Nevertheless, China and the US have developed intricate ties during the globalization process as a result of their structural contradictions and complementary advantages. Capital and the

DOI: 10.4324/b22974-2

market have shaped this complex relationship, which may not be transformed overnight. Economic and trade imbalances will eventually be resolved as negotiations, compromise, and competition between the two parties intensify. We require additional time to resolve the matter.

Even with the trade war, China's gross domestic product (GDP) exceeded 98 trillion RMB in 2019, a 6.0% increase over 2018 at constant prices [1]. Despite the global pandemic of COVID-19, China's GDP exceeded 101 trillion RMB in 2020, an increase of 2.2% over 2019, making China the only major economy to experience positive growth in 2020 [2]. In 2021, China's GDP reached 114 trillion RMB, an increase of 8.1% year on year [3], incorporating both recovery and inertial growth.

China's imports and exports, according to the General Administration of Customs, totaled 39.1 trillion RMB in 2021, an increase of 21.4% [4]. China's trade value with the US increased 28.7% to 755.6 billion dollars in the same year. China has long been the world's leader in terms of foreign exchange reserves, grain production, and telecommunications services. It has constructed the world's largest expressway, high-speed railway, and mobile broadband networks. While China's economy continues to experience fluctuations, the overall trend is upward. McKinsey forecast in its January 2019 report *Globalization in Transition* that the US and Canada's share of global consumption would decline from 31% in 2017 to 23% in 2030, while this proportion will increase from 10% in 2017 to 16% in 2030 in China, indicating that the country will transition from a global factory to a global market [5]. According to the Brookings Institution, an American think tank, total economic consumption in Europe will decline to 27% by 2030, while total economic consumption in the Asia-Pacific region will increase to 31%, implying that the international market center is shifting to the Asia-Pacific region [6].

China's GDP accounted for 18% of the world's total economy in 2021, which is equivalent to 77% of the US economy and exceeds the combined GDP of the EU's 27 member countries. China contributes approximately 30% to global economic growth and continues to serve as a critical engine for global economic recovery [1]. China's total economic volume will surpass that of the US by 2027–2030, and its per-capita gross national product (GNP) will reach that of developed countries around 2050, according to forecasts from several international economic organizations and prediction models. By the middle of the twenty-first century, it is anticipated that the US will continue to be the world leader in scientific and technological innovation, finance, investment, and services, military strength, economic quality, GDP per capita, and international rule-making. Meanwhile, China is expected to have comparative advantages in economic volume, global trade, outbound investment, new economic development, business, tourism, and cutting-edge science and technology, in addition to country-specific agendas such as China's "the Belt and Road Initiatives" and growing influence in international rule-making. In comparison to China and the US, the European Union's (EU's) solid foundation ensures that it will retain its influence for an extended period. However, given its nature as a loose alliance comprised of independent sovereign countries and subsequent events such as Brexit and the European migrant crisis, its development momentum and global influence may be constrained. As hostilities between Russia and Ukraine erupted on February 24, 2022, Europe's division will deepen, making it more difficult to reconcile Russia's conflicts with the EU, the US, and NATO. The US, China, Russia, and the EU remain the four global quadrupoles, but the latter two will suffer significant developmental setbacks, which will have a negative impact on their influence and unity. Japan, India, Canada, Australia, and countries of the Association of Southeast Asian Nations (ASEAN) all have unique advantages and characteristics. They have strengths and potential, but will be unable to compete for global leadership by 2050 due to geographical, resource, population, institutional, cultural, and social system constraints.

1.2 Far-Reaching Impact of Scientific and Technological Development

Science and technology are advancing rapidly in three directions in the twenty-first century: enormity, miniature, and complexity. Innovative concepts are emerging in advanced and strategic technologies. Breakthroughs are occurring in understanding the evolution of the universe, the structure of matter, the origin of life, and the nature of consciousness. Almost every sector is impacted by information, biology, new materials, new energy, and intelligent manufacturing technology. Artificial intelligence, mobile Internet, satellite navigation, and interdisciplinary biology advancements all contribute to the massive changes associated with green, intelligent, and ubiquitous technologies as well as the emergence of new industries, social forms, and economies characterized by knowledge, sharing, sustainability, and health. Civil–military fusion has been extended, blurring the lines between civilian and military technology, talent, management, and standards. A more flexible chain of communication and innovation allows for faster technology upgrades and seamless transformation and application. Mechanization is giving way to informatization and intelligentization.

Breakthroughs in basic sciences will enrich people's perceptions of the physical and mental world. Basic science research is progressing steadily in macro, micro, and extreme conditions, and the previously unknowable questions are on the verge of major breakthroughs. Observation technologies broaden our understanding of the universe's origin and evolution, dark matter and dark energy, the microscopic structure of matter, rare physical phenomena occurring under extreme conditions, and complex systems, elevating our understanding of the material world to a new level. Synthetic biology advances our understanding of life's activities from macro-systematic and micro-quantal perspectives, paving the way for future research on the origin, evolution, and reconstruction of life while also igniting a new round of biological technology revolution. Neuroscience research also extends our knowledge of human brain's activity and functioning mechanisms, unraveling the mystery of cognition, and furthering artificial intelligence and complex network theory.

Following mechanization and informatization, intelligentization will catalyze a new round of scientific and technological revolutions. Intelligent technologies such as artificial intelligence, big data, cloud computing, interdisciplinary biology, unmanned systems, and parallel systems will accelerate machine, bionic, swarm, and human–machine integrated intelligence, as well as intelligent decision-making, mission execution, confrontation, and logistics. Intelligent products such as service robots, security robots, unmanned vehicles, drones, and intelligent wearable devices will increase the level of intelligence in complex system management and control. Intelligent technology satisfies individuals' unique needs, enhances life quality, and increases productivity. New-generation mobile communications (5G/6G), global satellite network communication, software-defined and cognitive radio communications, mimetic networks, supercomputing, model-based system design and simulation, and digital twin technologies are forming a world based on the Internet of Things (IoT). Intelligentization will become more pervasive, serving as the "brain" and "nervous system" of the digital economy, sharing economy, and potentially the entire society. In the future, we may live in a world in which all information is instantly accessible. A slew of industrial innovations and significant economic and strategic benefits may emerge. Intelligent chips, bionics, neuroscience, supercomputing, nanomaterials, new devices, new energy and power, advanced manufacturing, and other technologies may enable products with higher speed, capacity, performance and lower cost, smaller size, and increased durability that can meet military and commercial requirements.

Disruptive technologies are the tipping point for social productivity enhancement and may trigger another industrial revolution. The fields that receive the most research and development (R&D) investment worldwide, such as intelligent technology, information networks, biotechnology, clean energy, new materials, and high-end manufacturing, are breeding a number of disruptive technologies that have the potential to completely reshape the industry. Quantum computers, material genomes, stem cells, and regenerative medicine, synthetic biology (e.g., artificial chloroplasts), nanotechnology, quantum dot technology, graphene materials, and non-silicon-based information functional materials, among others, have demonstrated a broad range of potential applications. Advanced manufacturing is evolving toward structural–functional integration and material–device integration. Extreme manufacturing is advancing rapidly in both macro (e.g., aircraft carriers, very-large-scale integrated circuits) and micro manufacturing (e.g., microchips and nanochips). Intelligent manufacturing, which combines intelligent materials and three-dimensional (3D) printing, will transform central mass manufacturing into distributed customized manufacturing, forming the "material nature of the digital world" and the "intelligent nature of the material world." These disruptive technologies will continue to generate new products, demands, and business models, act as catalysts for economic and social development, and contribute to the profound reorganization of economic patterns and industrial structures. They are critical to a country's innovation-driven development and competitiveness.

Scientific and technological advancements continue to improve human capacity for resource utilization, and expand its landscape to deep space, deep sea, deep Earth, "Deep Blue,[1]" and new energy. The three primary directions of space technologies are access, utilization, and control. Researchers continue to prioritize the exploration, landing, and even long-term habitation of the Moon, Mars, and other extraterrestrial bodies. The observation system, based on a 3D and global network, e.g., small satellite constellations, and a combination of space- and ground-based technologies, will enhance Earth observation, communication and navigation, deep-space exploration, and information resource utilization. Oceanography breakthroughs will give birth to a new "blue economy," featuring multi-functional underwater cable-controlled robots, high-precision underwater self-propelled vehicles, deep-sea and seabed observation systems, and deep-sea space stations, enabling deep-sea marine monitoring, resource exploration, and higher maritime security. The Earth's subterranean environments become more transparent with geological exploration technology and equipment such as gravity gradiometers and high-sensitivity magnetic probes. A better understanding of the Earth's deep structure and underground resources may encourage the exploration of new resources. Renewable energy sources such as solar, wind, geothermal, and biological energy will be used to boost energy efficiency and economy, alter the existing energy structure, and accelerate the rate of national energy self-sufficiency. In addition, nanothermoelectric energy generation, hydrogen energy, and nuclear fusion may be necessary to meet future energy demands.

The global technological innovation structure, underpinned by broad national strength, is undergoing profound transformation. Since the onset of the international financial crisis and the acceleration of economic globalization, the driving forces of scientific and technological innovation have quietly shifted from developed to developing countries. China, India, Brazil, and Russia are emerging economies that have become active creators of new science and technology, contributing to global scientific advancements and reaping a greater share of the benefits. From the Atlantic ocean, particularly Europe and the US, to the Asia-Pacific region, innovation centers are extending. North America, East Asia, and the EU will continue to be the three major innovation hubs that will dominate the global innovation landscape in the future.

1.3 Global Interconnectivity is Increasingly Rapidly Upgraded

The Internet has amassed an enormous amount of dynamic data in the fields of politics, economics, science, technology, militaries, society, and market, among others, and has archived numerous human life experiences and behavioral patterns. It is all-pervading and all-powerful. At the moment, mobile communications, soft switch networking communications, instant messaging software and social media have removed the geographical constraint on information transmission, and increased its efficiency. Individuals are connecting faster and more frequently. Online shopping, banking, administration, education, and scheduling have become ingrained in many people's daily lives. The Internet has infiltrated all spheres of human activity.

In the future, new technologies including Internet plus, Internet of Things, 5G/6G, virtual reality (VR), augmented reality (AR), and mixed reality will connect everything visible on a network. The concept of smart cities, buildings, farms, factories, hospitals, and families will gain popularity. The virtual and physical worlds will seamlessly interact and integrate. The importance of networks, information, data, algorithms, and models will be on a par with that of air and sunlight. They may even become the new capitals, serving as currency to aid nations and societies in their development.

Global interconnectivity and information fusion will surely enhance daily convenience. However, they may also result in potential risks and cyberspace confrontations. With the proliferation of instant messaging, social media, and network media, sensitive information is likely to be amplified in the public sphere, resulting in unexpected emergence effects and consequences such as public opinion reversal, psychological disorder, and social unrest. As more ways for people to express their opinions become available, those from disparate regions or backgrounds who share common views or dissatisfaction can find and connect more easily, forming groups with divergent viewpoints, which adds to social complexity. Additionally, cyber fraud and crime will develop into malignant "tumors" that will be difficult to eradicate. Cyber attacks and defenses will become a significant component of future warfare. Cyber technologies have the potential to attack and disrupt the information control infrastructure, resulting in large-scale economic and social disasters.

1.4 The Shaping of New Economic Forms

Following the previous round of globalization, deglobalization and trade protectionism have become the global economy's dominant trends. For an extended period of time in the future, the struggle for economic leadership between major powers will become the source of interstate or regional conflict. As a result of global economic restructuring and growing uncertainty, trade leadership and international rule-making have taken on a new character. Globalization's ebb is accompanied by a flood of new regionalism. Due to the complexity of the global economy, the development of globally accepted trade rules is hampered. Regional economic cooperation, on the other hand, is facilitated. While emphasizing bilateral trade and negotiation strategies, the US maintains a limited presence in international and trade organizations such as the North American Free Trade Zone, the Transatlantic Trade Partnership, the Asia-Pacific Economic Cooperation Organization, the

Group of Seven, and the Group of Twenty. In contrast, the countries and organizations represented by China, the EU, and Russia are promoting economic and trade organizations, including the Belt and Road Initiatives, the Shanghai Cooperation Organization, BRICS (Brazil, Russia, India, China, and South Africa), the China–ASEAN Free Trade Agreement, the China–Japan–Korea Free Trade Agreement, the Asian Investment Bank, and the Silk Road Fund. Regional economic and trade agreements have been established by regional organizations such as the League of Arab States, the Gulf Cooperation Council, the African Union, and the Union of South American Nations.

McKinsey Global Research Institute released the report *Globalization in Transition* at the start of 2019, highlighting five structural shifts: trade in goods is contracting, while trade in services is accelerating; labor costs continue to decline in importance, while research and development continue to grow in importance; and regional trade becomes more concentrated. The global economy is exhibiting a regionalization and diversification trend [5].

New economic forms are rapidly emerging as a result of demand, technology, and market forces:

1. **Knowledge economy will guide future global economy.** The growth and maturation of the knowledge economy will accelerate the development of industrial robots, service robots, and intelligent manufacturing. Data, models, experiences, and knowledge will all gain value, giving rise to networked military intelligence and new operational patterns such as cyberspace and cognitive confrontations, cloud operations, distributed killing, cyber attack and defense, and unmanned swarm operations, all of which will evolve into intelligent war forms.

2. **Sharing economy based on the IoT will transform our way of life.** The sharing economy is also expected to evolve into a new operational mode and domain of control, with new operational objectives, methods, and characteristics. The sharing economy also provides an alternative method of acquiring intelligence, controlling and managing critical infrastructure, capturing major public opinion, and ensuring operational forces' mobility and security. Featuring joint contribution and shared benefits, the sharing economy is inextricably linked to most people's interests. It may also contribute to the containment and expansion of warfare.

3. **Green economy will become the future trend in global economic development, garnering increased attention from all countries.** The green economy is more energy-efficient, environmentally friendly, and sustainable than the conventional one. It requires enterprises to safeguard the local ecology and the military to regulate the operational environment, forms, and methods. Military competition in the future may be focused on energy utilization, transformation, and precise control.

4. **Health economy, centered on biomedicine, will be a critical sector of industrial and economic growth in the twenty-first century,** triggering

the development of brain science, biomedicine, gene therapy, bionics, biomanufacturing, life reconstruction, and psychological treatment. The health economy will enhance combatants' body function, bionic equipment, and human–machine intelligent fusion. It may, however, introduce new threats to humanity, including bioweapons, genetic weapons, novel viruses, and epidemic diseases such as severe acute respiratory syndrome (SARS).

1.5 Intensifying Competition for Natural Resources

In today's world, resource competition is becoming increasingly fierce. As of 2012, fewer than a billion people consumed more than three-quarters of the world's resources. By 2030, "new and expanded middle classes in developing countries could generate an additional two billion consumers. An explosion of this magnitude will result in a scramble for raw materials and manufactured goods" [7]. Population growth, migration, and environmental change will all contribute to resource scarcity. Nowadays, the race for water, food, fossil fuels, and rare minerals occurs in many parts of the world, and when it becomes unmanageable, it can even result in war. Prior to the Arab Spring, Syria's water scarcity and Egypt's soaring bread prices were direct manifestations of resource conflicts. Excessive development, utilization, and pollution in the upper reaches of transnational rivers and lakes have resulted in conflict with downstream countries. Certain countries interpret international law in relation to territories below sea level in accordance with their national interests. Changes in the environment, such as grassland desertification and glacier melting, will result in population migration and exacerbate resource competition. Unless and until alternative resources are discovered, human beings will continue to compete for traditional resources. Additionally, countries competing for control of freight and trade ports may result in conflict, especially given the predictable rise in sea level. At the moment, people are at a crossroads where available resources are diminishing and alternative energy sources have yet to be developed, resulting in intense competition for scarce resources.

1.6 The Rapid Growth of Megacities

The movement of populations between countries and regions increases the scale of cities. Numerous cities are likely to grow into megacities with populations exceeding ten million. According to the National Intelligence Council, the year 2030 will be a time of migration, with approximately 60% of the population living in megacities [7]. According to United Nations population projections, the world's population will reach nearly 9.7 billion by 2050, with two-thirds of people living in cities [8].

Since the industrial revolution, urban development has been classified into three distinct types: skyscraper cities, suburban development, and urban sprawl. Skyscraper cities are characterized by the presence of skyscrapers, dense apartment blocks, and subways, as well as highly centralized urban management and clustered local communities based on culture or occupation. Suburban cities have subcenters that radiate out from the central business district and develop into dense, self-governing urbanized areas. Urban sprawl refers to urban development that lacks sufficient planning and resources.

In many countries, the city center has become overrun by slums and has devolved into urban wastelands, ethnic ghettos, and outlawed areas where gangs and criminal groups thrive. Megacities frequently face issues of social inequality, which can result in violent and political riots on a regular basis.

Megacities form as a result of people's natural tendency to congregate in urban areas. Drought, hunger, surface changes caused by climate change, political persecution, and the hope for development opportunities are all reasons for large-scale migration. For example, the Syrian civil war has resulted in a massive influx of refugees, with millions fleeing the country in search of new homes and many settling permanently in Europe, reshaping Europe's population and social patterns.

The majority of futuristic operations will take place in cities. Megacities' growth will have a significant impact on future warfare. Cities are the most complicated battlefields, with densely packed buildings, intricate streets, and, most importantly, a plethora of targets. Military and civilian targets, above-ground and underground targets, fixed and mobile targets, hard targets such as equipment and facilities, and soft targets such as key figures are all included in this category. The urban environment imposes significant constraints on equipment and technology. Simple methods and means are incapable of resolving tactical issues in streets, alleys, indoor locations, and underground operations, let alone preventing and controlling social issues such as marches, demonstrations, riots, and terrorist attacks.

1.7 Long-Standing Religious and Cultural Conflicts

Economic ties, technological advancements, and industrial revolutions, particularly the rise of global interconnectivity, instant messaging, and online social media, will reorganize power and interests among states, transnational groups, non-governmental organizations (NGOs), religious groups, and the general public. To begin, disagreements over ideologies and values will continue to exist to varying degrees. Countries differ in their national systems, development paths, modes, and foundations due to historical and ethnic factors. They will be unable to obtain completely equitable and just treatment as a result of economic globalization and global governance, which will result in conflicts of interest and ideological disagreements in politics and ideology, further fueling nationalism. Second, future conflicts will be fought over differences in religions, beliefs, and cultures. Globally, Islamic extremists and terrorist forces exacerbate cultural clashes and estrangement, triggering social unrest and crisis. Middle Eastern immigrants in Europe have distinct cultures and customs from the natives, and the difficulty of cultural integration will result in a schism within the EU. The "arc zone" – North Africa, the Middle East, West Asia, Central Asia, South Asia, and Southeast Asia – is rife with terrorist activity. They are rapidly spreading throughout Europe, Asia, and other parts of Africa, posing a threat to national security and global stability. Third, unresolved border and territorial disputes are strongly intertwined with religious and cultural conflicts, which may spark conflict, riots, and even war.

Note

1 IBM's supercomputer.

References

[1]   National Bureau of Statistics 国家统计局. (2020, December 30). *Guanyu 2019nian guonei shengchan zongzhi (GDP) zuizhong heshi de gonggao* 关于 2019 年国内生产总值 (GDP) 最终核实的公告. Official Website of National Bureau of Statistics 国家统计局官方网站.www.stats.gov.cn/xxgk/sjfb/zxfb2020/202012/t20201230_1811898.html.

[2]   National Bureau of Statistics 国家统计局. (2021, December 17). *Guanyu 2020nian guonei shengchan zongzhi zuizhong heshi de gonggao* 关于 2020 年国内生产总值最终核实的公告. Official Website of National Bureau of Statistics 国家统计局官方网站. www.stats.gov.cn/xxgk/sjfb/zxfb2020/202112/t20211217_1825447.html.

[3]   National Bureau of Statistics 国家统计局. (2022, January 17). *2021nianguomi njingjiyunxingqingkuang*2021年国民经济运行情况. Official Website of National Bureau of Statistics 国家统计局官方网站. www.stats.gov.cn/xxgk/sjfb/zxfb2020/202107/t20210715_1819477.html.

[4]   General Administration of Customs of the People's Republic of China 中华人民共和国海关总署. (2021, January 14). *Haiguan zongshu 2020nian quannian jinchukou qingkuang xinwen fabuhui* 海关总署2020年全年进出口情况新闻发布会. Official Website of General Administration of Customs of the People's Republic of China 中华人民共和国海关总署官方网站. http://fangtan.customs.gov.cn/tabid/1106/Default.aspx.

[5]   McKinsey Global Institute. (2019). *Globalization in transition: The future of trade and value chains 2019*. New York, NY: McKinsey.

[6]   Dollar, D., Huang, Y., & Yao, Y. (Eds.). (2020). *China 2049: Economic challenges of a rising global power*. Washington, DC: Brookings Institution Press.

[7]   National Intelligence Council 美国国家情报委员会. (2016). *Quanqiu qushi 2030: bianhuan de shijie* 全球趋势2030:变换的世界 (American Studies Center, China Institute of Contemporary International Relations 中国现代国际关系研究院美国研究所, Trans.). Beijing: Shishi Press.

[8]   Department of Economics and Social Affairs, UN. (2019). *World population prospects 2019*. New York, NY: United Nations.

2   The Future Trends of Warfare

Since its onset in history, war in nature has been dynamic, involving both constant and inconstant factors. The constant variables are national security demands, the violent nature of war, and the attributes of confrontation. The outbreak of war originates from geopolitical competition, fights for economic interests, historical and cultural strife, ethnic conflicts, the escalation of domestic political struggles, social contradictions, and other politically defined purposes. The underlying causes of war have remained constant over the past few thousand years and will possibly remain so in the future, regardless of how they manifest themselves. War theories, operational elements, space, means, and patterns are constantly changing. While these are the primary changes, more are reflected in force composition, structure, organization, military training, comprehensive support, armaments development, defense procurement, etc. These facets keep evolving with time and social development, the core of which is exactly the evolution of war forms.

Historically, the evolution of war forms is closely related to scientific and technological progress, economic development, and social patterns, and is especially directly connected to social productivity and production modes. The combined effects of demand-based planning, technology advancement, economic support, scientific management, military practices, and other elements all contribute to the changing forms of war. In the short term, political factors and military demands play a relatively larger role, while the influence of economic development and financial support is more obvious in the middle term. But in the long term, it is the scientific and technological development that has a more lasting, endogenous, disruptive, and revolutionary impact on warfare, while scientific management and practical war experience may accelerate or retard the future shape of war.

2.1 World Peace and Security as the Intrinsic Driving Forces

Demand is an essential condition to drive the evolution of war. World peace and security and national development and security are the dominant factors and the primary driving forces. In the foreseeable future, profound changes are taking place in the balance of powers, global governance structure,

DOI: 10.4324/b22974-3

Asia-Pacific geostrategic landscape, and international competition patterns in the economic, technological, and military fields. The struggle of various forces for the redistribution of power and interests is becoming increasingly intense. Under the current situation and conditions, three major strategic trends in world peace and security deserve consideration in the future.

First, the world is generally peaceful, but military competition in emerging areas is intensifying. From a political and military point of view, due to the gradually balanced and coordinated military development among big powers such as the US, Russia, and China, their mutual constraints and checks and balances may not cause global catastrophic events. In the current Russia-Ukraine conflict, it is precisely due to the effect of nuclear deterrence that it will not quickly escalate into World War III. In addition, the desire and power to maintain peace are still strong in the European Union (EU), Association of Southeast Asian Nations (ASEAN), and other Asian, African, and Latin American regions, which means the restraints on war still exist. The core capabilities that support the strength of military powers and maintain their balances are mainly manifested in two major areas: nuclear deterrence and conventional deterrence. Because nuclear deterrence technology is mature and restricted by international nuclear non-proliferation treaties, arms control remains the focus of global attention. Future military competition among major powers will be more about conventional deterrence, especially in those emerging areas. Among these, the military competition initiated by artificial intelligence (AI) ranks first, improving national economic competitiveness and overall military combat capability and enhancing the deterrence and balance capability among major powers. Therefore, from a technological point of view, to maintain world peace, the world's major powers will accelerate the pace of military intelligence development, which will shape their strategic deterrence so that nuclear and conventional deterrence are equally developed.

Second, the security risks among major powers are more reflected in their development speeds, and the security issues caused by economic factors are noteworthy. In the communication processes of countries in the world, economic factors have always played a very important role. On the one hand, due to the convenience of global investment and the role of the market economy, the global economy continues to develop toward mutual benefit and win–win situations. On the other hand, political and military conflicts caused by the imbalance of economic development always exist and are gradually rising. In the face of the rapid development of rising economies such as China, the traditional conservative powers would surely have a sense of crisis and urgency and thus adopt political, economic, technological, and military means to suppress the rising. In early March 2021, the National Security Commission on Artificial Intelligence (NSCAI) released a research report fully expressing these concerns. The report states,

> For the first time since World War II, America's technological predominance—the backbone of its economic and military power—is

under threat. China possesses the might, talent, and ambition to surpass the United States as the world's leader in artificial intelligence (AI) in the next decade if current trends do not change.

The report iterates, "the United States must prepare to defend against these threats by quickly and responsibly adopting AI for national security and defense purposes." It concludes that "AI is going to reorganize the world" and that "America must lead the charge." [1] The economy is the cause, while the military is the effect. In recent years, a series of conflicts and frictions around China have manifested the intention of suppressing China's development. The economic sanctions imposed by Western countries on Russia, although originating from the Crimea crisis, are also intended to contain Russia's development in the long run.

Third, national interests are intertwined with the clash of civilizations and religious conflicts. Constant skirmishes and conflicts are still common in many parts of the world. Terrorist activities are increasingly active; ethnic and religious conflicts, as well as border and territorial disputes, are complex and volatile. Constant wars and crises such as the Palestinian–Israeli conflict, the Syrian war, the Turkish–Syrian border conflict, the Libyan civil war, the Nagorno-Karabakh conflict, the war between the Houthis and the Saudi coalition in Yemen, the killing of Suleimani by the US military, and the assassination of Iranian nuclear scientists are proof that the world still faces real and potential threats of war.

Under the veils of clashing civilizations or ethnic conflicts, local wars, or precision decapitation strikes, the intrinsic driving forces of war are always world peace and security and national development and security, in which AI and its military applications will become the focus of global military competition.

2.2 The Disruptive Effect of Science and Technology

Throughout the history of warfare, cold weapons have been used for millennia, thermal weapons (firearms) for centuries, mechanized weapons for nearly a century, thermonuclear weapons for more than half a century, and informatized weapons for several decades. Throughout human history, science and technology (S&T) have played an increasing role in winning wars. They have accelerated the evolution of war forms and will become the deciding factor in the outcome of future warfare, which will be shorter in duration, more intense, and more reliant on technology (Figure 2.1).

Since the beginning of the twenty-first century, the rapid development of S&T has led to a new revolution in military affairs (RMA). Breakthroughs are about to happen in basic and frontier science fields, where information technology, the Internet, cognitive science, biotechnology, new energy, new materials, and advanced manufacturing are the leading fields. Transformative innovations are being promoted by interdisciplinary integration. All these changes are brewing a revolutionary impact on war.

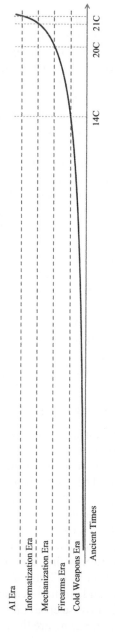

Figure 2.1 Science and technology's roles in War.

1. **Emerging technologies, including AI, mobile Internet, Internet of Things (IoT), space-based information, cloud computing, big data, and cyberattack and defense, which are developing rapidly and integrating creatively, will speed up the development of intelligent warfare capabilities.** A global, distributed, and networked military cloud and AI-based brain system, supported by cloud computing and big data, could be built by developing network information systems (NIS) that are space–ground-coordinated, civil–military-integrated, elastic, self-organizing, jamming-resistant, and destruction-resistant. The system will shorten the "observe–orient–decide–act" (OODA) loop and meet various operational demands, including intelligent perception for multi-dimensional battlefields, confrontations on multi-source information and cognition, integrated command and autonomous decision-making, joint and cross-domain fire strikes, and comprehensive support for diversified military operations (Figure 2.2). Modern warfare concepts, characterized by the interconnection of all things, autonomous action, parallel combat, and "war defined by software," will replace traditional warfare concepts by adapting more to mission requirements and realizing cross-domain self-organization and synchronous cooperation of an army. In addition, cyberspace, the core of intelligent warfare, will become a key part of virtual and physical space-integrated operations. Likewise, cognitive confrontations, including

Figure 2.2 Intelligent military products (2021 China Zhuhai Airshow).

network attack and defense, electronic warfare, intelligence warfare, and public opinion control, will also be strategic points that engage both warring sides.

2. **New materials, new energy, military bionics, advanced power and manufacturing, network communications, and machine learning will foster the development of unmanned, autonomous, bionic, and swarm combat vehicles.** Future weapons and combat vehicles, in addition to traditional mechanized functions such as mobility, protection, and firepower, will be equipped with information capabilities (e.g., reconnaissance, surveillance, tracking, navigation, data link, mobile interconnection, Identification Friend or Foe (IFF), command and control, and system networking), intelligent functions (e.g., multi-source information collection in various networks, data mining, intelligent perception, adaptive planning, autonomous decision-making, health diagnosis, offline memory, and online upgrade), and combat performance (e.g., information security protection, network confrontations, electronic confrontations, and stealth confrontations).

Machine learning, bionics, brain–computer control, new power, and additive manufacturing will accelerate the development of land, space, aerial, offshore, and unmanned underwater vehicles (UUVs), military robots, bionic robots, exoskeletons, individual aerial vehicles, and amphibious combat vehicles. With a continuously increasing autonomous combat level, unmanned equipment will be widely deployed, resulting in new combat modes such as manned–unmanned cooperative combat as well as swarm attack and defense (Figure 2.3).

Figure 2.3 Unmanned aerial vehicle (UAV) series (Zhuhai Airshow).

3. **New probes, intelligent perception, composite guidance, advanced missile power, high-speed flight control, networked cruise flight, and high-efficiency damage technologies will significantly improve long-range and hypersonic precision strike and damage capabilities.** Multi-mode and compound guidance, missile-borne data links, satellite communication, and navigation technology, with their wide and practical use in the future, will improve the information, networking, anti-jamming capabilities, and hit rate of precision-guided munitions (PGMs). PGMs will focus on intelligent, high-speed, general-purpose, modular, and multi-purpose development (Figure 2.4). Guided weapons will inevitably develop the features of high speed, swarm deployment, and precision damage, especially with the breakthrough and maturity of core technologies in this field, including high-altitude and high-speed boost-jump glide, solid ramjet engine, hypersonic aerodynamic structure control and thermal protection, integrated design of engine and missile body, orbital transfer of high-speed cruise missiles, terminal autonomous homing guidance, ammunition swarm networking, collaborative perception and strikes, new energetic materials, efficient damage, and controllable damage. Weapons will also transform from single to swarm deployment, from fixed trajectory to variable trajectory with penetration, from low to high or hypersonic speed, from individual to network-controlled, and from conventional to efficient and controllable damage. Furthermore, the attack and defense systems will undergo revolutionary changes due to an unprecedentedly accelerated combat tempo and greatly improved precision strike and damage efficiency.

Figure 2.4 Cruise missile series (Zhuhai Airshow).

4. **Multi-target detection, anti-stealth, anti-hypersonic, anti-swarm tech-
    nology, solid-state laser weapons, high-power microwave weapons, elec-
    tromagnetic guns, and networked command and control will significantly
    improve forces' overall defense capability.** In the future, combat forces
    dealing with field air, mobile, and strategic defense will face greater threats
    from new reconnaissance technologies in space, air, and ground, as well
    as various strike weapons and PGMs. Reaction to stealth aircraft, cruise
    missiles, hypersonic weapons, unmanned aerial vehicles (UAVs), ammu-
    nition swarms, and terrorist attacks will become the major mission for air
    defense and anti-missile systems on the future battlefield. The capabilities
    of anti-terrorism, anti-hypersonic weapons, and anti-swarm mechanisms
    will become the focus for building comprehensive defense systems in the
    future (stealth aircraft and cruise missiles are categorized as hypersonic
    weapons).

A defense system must be equipped with comprehensive capabilities on the
future battlefield, including large-scale detection, three-dimensional coordination, and multi-level interception. It is also the trend of air defense and anti-
missile system development to advance three-dimensional and all-weather
detection, networked precision command and control, integrated real-time
linkage, multi-level and multi-means interception, and flexible mobile organization killing and interception capabilities integrating software and hardware
as well as targeting points and surfaces (Figure 2.5).

5. **Self-media (independently operated social media accounts), social media,
    and live videos, as well as infrastructure based on the Internet, IoT, and
    industrial control systems, will become focal points of cognitive battle
    and cross-domain control.** In the future, geopolitical gamers and over-
    seas interest defenders will focus on resolving conflicts among different
    beliefs, cultures, races, and religions. The new cultural forms, shaped and
    developed with the popularity of the Internet, will intensify and diver-
    sify ideological and cognitive confrontations. Western countries have
    focused on exporting ideologies (e.g., world views, values, philosophies,
    and lifestyles) to other countries worldwide with the new cultural forms
    as their vehicle. It can be predicted that new cultural, legal, psychological,
    and public opinion warfare will be extended into ideological and cogni-
    tive confrontations, with new cultural patterns as their carrier, becoming
    important battlefields and territories in the future. Likewise, infrastruc-
    ture plays an increasingly prominent role in warfare. If it is destroyed or
    paralyzed, it will significantly impact the military's psychological state
    and morale. In future warfare, the management and control of important
    civil infrastructure will give birth to new operational fields and modes
    (Figure 2.6).
6. **Brain science, affective/emotional computing, mental state detection, bio-
    logical genes, life reconstruction, biomedicine, and interdisciplinary biology**

Figure 2.5 Air defense missile weapon system (Zhuhai Airshow).

will foster a deep integration between biological and machine intelligence, leading to breakthroughs in biological cross-competition and biosecurity protection. Biological and interdisciplinary biological operations as well as biosecurity protection will become new operational patterns. At present, research outcomes in detecting and utilizing brain consciousness as well as physiological and psychological signals including electroencephalogram (EEG), electromyogram (EMG), electrocardiogram (ECG), and imagery movement have been applied in such fields as target recognition, vehicle control, weapon control, command and control, real-time monitoring of combatants' physiological and psychological states, and selecting and training combatants. In the future, brain science programs will progress worldwide. Breakthroughs in discovering the working processes of molecules and cells in the human nervous system and their integration into the central function control system, the cognitive characteristics, and essence of the human brain will promote the emergence of brain-like chips, brain-like systems, and advanced machine learning modes, resulting in the second leap in AI. Furthermore, better knowledge of brain structure, the characteristics of nervous system operation, and the nature and transmission mode of consciousness will introduce innovative

Figure 2.6 Digital transformation of the industry exerting a major impact on future combats.

means for developing new brain–computer interfaces and intelligent interactive systems, which will help to acquire natural human–computer interaction, deep integration of human brain cognition and machine learning, multi-brain collaborative decision-making in complex systems, rapid classification and accurate recognition of multi-mode battlefields, and emotional interaction and ideological exchange between humans and computers. In addition, novel military biomedicine, non-lethal biological weapons, new military energy, military materials, and biological computing equipment will be developed with the accelerated development of biological genetics, life reconstruction, biomedicine, interdisciplinary biology, nanomaterials, and other technologies, especially through the design, transformation, and reconstruction of synthetic biotechnology or the creation of biomolecules, biological components, and living bodies, which help to promote a huge leap in human body function, making individual "superheroes" such as Iron Man, Ant-Man, and Captain America a reality. Likewise, biological weapons, genetic weapons, new viruses, and genetically optimized bacteriological weapons, growing in large numbers, will become new fields of military confrontation as well as new concerns in international conventions and arms control.

On May 20, 2010, US researchers announced the world's first laboratory-created artificial cell. Craig Venter, who led the research, revealed that for the artificial cell, the chromosome was designed using "four bottles of chemicals" and inserted into a genetically modified bacterial cell. This resulted in a cell with self-replicability controlled by a synthetic genome. The work took 15 years and cost $40 million. In 1995, the research team published the digital DNA sequences of two mycobacterial cell genomes, containing more than one million sequences of the organism's genetic instructions. Later, by reverse-engineering the genetic instructions stored in the computer step by step, they generated JCVI-SYN1.0 *Mycoplasma capricolum* cells with self-replicability. In 2013, George Church, among other geneticists, introduced at a TED Talk symposium in the United States that they were studying the extraction of DNA from ancient museum specimens and using modern biosynthesis technology to regenerate extinct animals. They successfully revived an extinct ancient carrier pigeon. In 2016, Craig Venter again shocked the scientific community by announcing that he had successfully synthesized the smallest synthetic organism composed of 473 genes. Replacing traditional semiconductor chips with biochips inspires us with a completely new concept of computers. In 2008, US researchers achieved the first bacteria-based parallel computing system by engineering *E. coli*. In October 2011, British researchers announced the successful development of a new modular "biological logic gate," marking another important step in biological computer research.

In March 2018, the report *Ground Warfare in 2050*, released by the US Army Research Laboratory [2], pointed out that high-value artificial cell products will be produced on demand, overturning the traditional logistics support mode. Artificial cells, composed of natural or synthetic phospholipids or synthetic polymers, will be able to function as non-biological reactors to allow substances to enter, exit, and be exchanged between cells; generate the energy needed to maintain cell function; divide and multiply like biological cells; perform human-designed biological programs in various environments; execute biological programs when triggered by necessary molecules; and control the products of biological programs and release them outside the cell. It is estimated that by 2050, artificial cell platforms will be developed with powerful multi-functional on-site manufacturing capabilities to produce high-value drugs and fuels according to military operational needs. This kind of artificial cell platform, designed with programmability, will be able to independently produce different drugs, thus greatly reducing the demand and cost of high-value material transportation and storage as well as the logistics burden of an army.

7. **Cross-domain creative thinking of crowdsourcing, crowdfunding, and crowd-creating, based on time–space benchmarks and network information, will introduce the concept of "open-source construction and combat."** Global positioning system (GPS), the Beidou global space–time benchmark platform, digital maps, big data, and emerging business models

such as sharing economies, crowdsourcing, and "seekers" are all gaining traction. Against this backdrop, technology, capital, information, and human resources worldwide, if shared efficiently, will usher in a new model of research and development (R&D), production, and marketing. "Open-source sharing" will lead to lower technical barriers, reduced development costs, shortened product life cycles, and transformed profit models in various fields. It will also bring profound changes to future operations, manifested in more transparent access to equipment and technologies. Traditional equipment development will be put in a situation where the manufactured equipment becomes outdated even before entering service. Socialized support will take a greater share of war mobilization support. Civil–military technical barriers will gradually disappear. Civilian technologies, systems, products, and infrastructure will also become important means of war. For example, UAVs made by DJI (Shenzhen, China), which were originally civilian products, have been purchased, used, and transformed by the US military, Al-Qaida, and many national armed forces. Future warfare, involving multi-source perception, correlation verification, multi-domain coordination, and cross-domain actions, is increasingly dependent on open-source information and data. In this context, collaborations between land, sea, air, space, network, and electromagnetic domains, as well as civil–military integration, have become necessary for achieving intelligent perception, decision-making, confrontation, strike, and support. New battle spaces and fields involving networks, space, deep sea, biology, infrastructure, and networks necessitate the use of civilian-oriented or civil–military integrated technologies. Combat systems that are relatively closed, independent, and unable to exchange information and data with other systems will become outdated and isolated, making them harder to adapt to informatized and intelligent operations. Besides, the ecological circle formed by "open-source sharing" will limit the expansion of war scales and serve as an important factor that facilitates the controllability and civilization of war.

After years of in-depth observation and study, the author believes that the advanced and strategic technology framework that will profoundly affect military security in the next 30 years involves 18 fields, more than 50 directions, and nearly 200 key technologies.

AI is the overarching and strategic technical field that may directly affect or participate in all other fields. Cyberspace technology ranks second; it has a slightly narrower impact, directly or indirectly affecting the overall and related technical fields. Other important technical fields include four fields of space technology, five fields of system technology, and seven fields of other special technologies (Table 2.1).

Table 2.1 Advanced and strategic technology framework

No.	Field/technology	No.	Field/technology
1	Artificial intelligence (AI)	10	Advanced energy conversion and power
2	Cyberspace	11	Biology and interdisciplinary biology
3	Air–space integration	12	Efficient damage
4	Maritime security protection	13	Novel communication and navigation
5	Advanced stealth and anti-stealth	14	New materials and devices
6	Virtual simulation and simulated training	15	Quantum information
7	Unmanned system	16	Antimatter technology
8	Intelligent munition	17	New maintenance and support
9	Directed-energy weapon	18	Advanced manufacturing

2.3 The Overall Supporting Role of the Economy

War history shows that the economy plays a direct, realistic, and holistic material supporting role in wars. While the economy serves as the bedrock, politics serves as the superstructure. War is a natural extension of politics and a critical tool for achieving economic goals. To a large extent, the economy and war share a common root, complement one another, and are interdependent. Most wars are motivated by competition and contention over economic interests. Meanwhile, warfare cannot be divorced from the overall support of financial and material resources. The role of the economy in warfare is manifested in many aspects.

First, the level of economic development determines the wealth and strength of a country and its ability to sustain military combat construction and support. With combat and equipment systems becoming more complex and technical in modern warfare, the construction and operational costs also rise. In particular, the types and models of weapons and equipment are becoming increasingly diverse and complex, which has brought difficulties in their usage, maintenance, and management and posed additional challenges to financial expenditures. According to the calculation of the purchase price of equipment in the world military trade market, in RMB terms, a modern tank costs more than 10 million RMB, a third-generation combat aircraft costs 200–300 million, a medium-range and long-range conventional ballistic missile more than 20 million, a modern destroyer is nearly 2–3 billion, a strategic nuclear submarine around 4–5 billion, and an aircraft carrier more than 10 billion. These are only calculated according to single suits. If calculated by formed units, an army digital machine infantry division requires more than 20 billion RMB and an aircraft carrier group 40–50 billion. These are then coupled with the costs of training, maintenance, headcount, management, etc. With all of these considered, the investment of a divisional combat unit is close to the annual output of a large enterprise. Without adequate financial support and strong economic strength to back it up, modern warfare would be impossible. New technologies and equipment could not be developed, or, even when they are developed, it might be impossible to buy, use, and repair them.

Second, military weapons, equipment, and other war materials need to be backed up by a strong military industry and relevant basic capabilities. Modern weapons and equipment are both technology- and knowledge-intensive, with complex and comprehensive systems. Their development needs support from primary industries concerning raw materials, components, advanced power and energy, electromechanical products, common software, systematic traction, and integration of military industries such as space, aviation, weaponry, electronics, shipbuilding, nuclear industry, etc. It also needs the collaboration of professional institutions for basic, exploratory, and advanced research, experimental testing, maintenance and support, measurement standards, etc. Historically considered, the US "Manhattan" project, "Apollo" program, "Polaris" program, "Star Wars" program and China's "two bombs and one satellite" (atomic and hydrogen bombs, man-made satellites) project, manned spaceflight, and other large-scale S&T projects were all conducted with the participation of relevant departments and related industries of the whole country. It can be described as "10,000 people with one gun." Modern tanks, aircraft, ships, missiles, and other weapons and equipment with large personnel also need specialized fuel and facilities for storage, testing, maintenance, and other necessities of life, like military supplies, medical support, catering, barracks, resources, etc. For a large country, a certain economic scale and industrial complementation must be in place to develop an effective mobilizing and supporting capacity for combat. High-tech industrial systems and basic capabilities are essential for national defense and armed forces construction and operation.

Third, changes in economic patterns and models have often led to revolutionary changes in warfare. Although this change has a lag effect, it should be accompanied by a change in economic patterns. Primitive society corresponds to hand-to-hand combat marked by physical force and natural tools. Due to the development and large-scale application of smelting technologies such as bronze and iron wares, the early Metal Age in agricultural society opened an era of warfare characterized by cold-steel weapons. The Gunpowder Age in late agricultural society corresponds to warfare characterized by hot weapons. During the Industrial Age, the emergence of steam engines, internal combustion engines, and electrical equipment led to a socialized mechanical industry and mechanized warfare. After World War II, with the invention of computer technology and the spread of the Internet and GPS, modern society gave rise to information-based warfare. In the future, the development of intelligent technology and a knowledge-based economic industry will inevitably lead to an era of intelligent warfare.

2.4 The Transition of Warfare Foundations and Conditions

Warfare's foundations, conditions, focuses, and requirements have been relocated, adjusted, and changed by a combination of factors such as practical demands and technical initiatives.

First, modern warfare is greatly compressed in time. Due to technological constraints, ancient warfare was slow in marching and maneuvering. Command orders were communicated mainly through beacons and fast horses. The course and cycle of battles were long, with typical battles and tactical operations measured in weeks, months, and even years. For mechanized warfare, the mobility of vehicles and troops was calculated in terms of days and hours. In information-based warfare, the flight time, battlefield updates, and effective distance of firepower and missiles in minutes are calculated. Future warfare could be expected to be measured in seconds, milliseconds, or even shorter units of time to calculate the speed of battlefield intelligence perception, target identification, cyber attack and defense, hypersonic and swarm strikes, and defense, and the efficiency of AI-based autonomous decision-making.

Second, the domain of modern warfare has largely expanded. Warfare has progressed geographically from land to sea to air and is now expanding into space, the deep sea, and the deep earth. From the perspective of virtual–real relationships, due to the emergence of man-made networks and information systems, cyber- and virtual-space warfare has gradually become an independent battlefield. The war is now expanding from physical to virtual space and the mutual integration and fertilization of the two. From the perspective of domains, the battlefield is expanding rapidly in a multi-domain fashion, from physical and information domains to cognitive, social, and biological domains.

Third, modern warfare concerns a great leap in distance-spanning capabilities. Combat vehicles' mobility and projection capabilities have achieved process and global coverage from tactical confrontation to strategic strikes. Battlefield awareness is globally interconnected and interoperable through space, air, land, sea detection systems, and cyberspace. Network communication brings people and equipment into increasingly interconnected networks. The Internet and mobile communications provide quick access to relevant intelligence and information from all around the world.

Fourth, modern warfare has exponentially sped up. From subsonic to supersonic to hypersonic, combat vehicles and missile weapons are getting faster and faster. The era of hypersonic confrontation is just around the corner. The emergence and mature application of directed-energy weapons will make beam strikes and defense possible soon. With the application and breakthrough of military intelligence technology, the efficiency and speed of the observe–orient–decide–act (OODA) loop, and command and control will be significantly increased.

Fifth, modern warfare is becoming increasingly more accurate in terms of precision. Through an expanding repertoire of investigative and detection techniques for space, air, land, and sea, and an increasingly higher resolution, through ground-based system enhancement, the positioning accuracy of satellite navigation can reach the decimeter or even centimeter level, and the timing accuracy can reach the nanosecond level. With the corresponding weapon vehicle and ammunition control accuracy improved, strike and destruction power has multiplied, and fire strike efficiency has increased exponentially.

Sixth, there is a pursuit of optimization and balance in both quantity and quality. Weapons and equipment are gradually transiting from a "large platform" development model that focuses less on the cost, and more on the magnificence of the platform and its integrated combat capability, to a more advanced and balanced (in both quantity and quality) development model that focuses on cost-effectiveness, swarming, high–low cooperation, and manned–unmanned combination, embarking on a path of sustainable development.

Seventh, there is a quest for asymmetric advantage in offense and defense. It would be a trend for future warfare to seek asymmetric advantages: to confront less with more, small with large, partial with whole, unadvanced with advanced, single function with multi-function, single-domain with multi-domain, reality warfare in physical spaces with virtual–real interaction warfare, non-intelligence with intelligence, and weak intelligence with high intelligence.

2.5 New Types of Operational Capabilities

In the previous chapter, an in-depth analysis of the disruptive impact of S&T on warfare was given to elucidate those new types of operational capabilities. Driven by AI, they are taking shape in the following fields: cyber and cognitive confrontation, unmanned and swarm operations, hypersonic confrontation, multi-domain fusion and cross-domain attack and defense, new types of integrated defense, bio cross-technological competition, and open-source contention and exploitation. The advancement of S&T is of pace-setting and transformational significance; it accelerates the formation of combat forces, enriches warfare styles, and promotes evolution from conventional to intelligent warfare.

2.5.1 Cyber and Cognitive Confrontations

The operations of cyber attacks and defenses, electronic confrontations, intelligence mining, public opinion monitoring, psychological warfare, consciousness intervention, and so on are all part of cyber and cognitive confrontations.

Based on the virtual space, without the limits of time, space, and geography, virtual warfare, combat experiments, and simulations can be performed to gain insight into opponents, defend against the damaging effects caused by them, and support the military's parallel operations. Cyber and cognitive confrontation represents a comprehensive deterrence and warfighting capability of strategic importance, a new operational capability, and the core element of intelligent warfare. It relies on network information systems, distributed clouds, parallel systems, with relevant corpora, model libraries, knowledge databases, algorithm and tactic libraries as its leading force.

2.5.2 *Unmanned Swarm Operations*

Unmanned swarm operations involve land, sea, air, space, network, and electromagnetic domains. They include manned and unmanned coordinated operations, unmanned vehicle-based operations, autonomous swarm operations, AI-based cyber attacks and defenses, and electronic confrontations. The basic reliance of unmanned swarm operations is on various types of unmanned aircraft, unmanned ground vehicles (UGVs), bionic robots, unmanned surface vehicles (USVs), UUVs, space robots, intelligent munitions, and other unmanned vehicles and their various swarm combinations. The core power is vehicle AI, swarm control AI, and net electric attack and defense AI. With the progress of AI technology, autonomous swarms will gradually become a new type of deterrent force.

2.5.3 *Hypersonic Confrontations*

With the development of advanced power, launch, guidance, control, and efficient destructive technologies, various types of hypersonic vehicles and weapons are emerging, from strategic to tactical. With the support of network information systems and the domination of intelligent AI systems, the era of hypersonic confrontations and precise destruction is coming. Hypersonic weapons and technologies are enabling an operational capability of both deterrent and operational importance.

2.5.4 *Cross-Domain Attacks and Defenses and Multi-Domain Fusion*

The Internet, IoT, cognitive communication networks, space-based information, big data, and other technologies have facilitated multi-domain fusion and cross-domain attacks and defenses. It involves multi-dimensional battlefield intelligence perception, multi-source information correlation and fusion, multi-domain joint command and control, cross-domain firepower strikes, cross-domain maneuvering, and sudden defense, cross-domain integrated security, and other multi-domain cross-domain combat operations, etc. It can fulfill different task requirements, such as cross-domain self-grouping, synchronous collaboration, with multi-domain collaborative advantages.

2.5.5 Integrated Defense

With the diversification of offensive means, defense systems will face escalating threats under conditions of multi-domain and cross-domain operations in the future. New types of integrated defense capabilities must be enhanced through technological exploration in four areas: (1) the defense of stealth aircraft, cruise missiles, hypersonic weapons, drones, and munitions swarm attacks; (2) the prevention of various terrorist attacks; (3) the management of public opinion; and (4) the security governance of post-war society and infrastructure.

2.5.6 Competitions Caused by Interdisciplinary Biological Technologies

First, the development of human–machine hybrid intelligence technologies such as brain science, brain-like chips, bionic robots, and bionic systems will further promote the upgrade and leap of military intelligence confrontation. Second, the confrontation of human–computer intelligent interactive information links, wearable systems, human–computer hybrid systems, and emotional interaction systems based on brain–computer technologies has become a new focus. Third, the development of biomedicine, human function enhancement, artificial life forms, artificial bacteria, and other technologies, and the emergence of related biological weapons, genetic weapons, and new types of viruses will bring new threats and challenges in the biological anthropology field.

2.5.7 Open-Source Contentions and Exploitations

With the civil–military integration in the high-tech field, the importance of open-source contention has continued to grow. First, informatized, networked, and intelligent civilian high-end technologies are used in military fields. Major countries compete for the R&D of AI facilities such as brain cognition, intelligent chips, model algorithms, nanomanufacturing, big data intelligence, software-defined networking, and interoperability. Second, open-source information is extensively used in national defense and exerts influence on military information management. Third, civic facilities, such as the Internet, IoT, satellite communications, mobile communications, ground mobile vehicles, air transport vehicles, civilian maritime ships, civilian drones, UGVs, USVs, robots, intelligent speakers, as well as logistics materials and equipment, intelligent fault diagnosis, equipment maintenance, etc., can be applied to military directly or after slight modifications.

2.6 The Reconfiguration of Strategic Deterrence

With the advent of nuclear weapons, a strategic deterrence system supported by nuclear forces emerged. As time moves forward, the boundaries of nuclear-based deterrence are being extended, and a new type of deterrence with both conventional and nuclear means is developing.

Deterrence with both nuclear and conventional means of operations is a combination of nuclear and conventional deterrence. While conventional deterrence refers to war prevention and combat capabilities, it encompasses six domains: cyber, space, hypersonic, autonomous swarm, global, immediate response, and military–industrial power. Thus, deterrence with both nuclear and conventional means consists of seven domains: nuclear, cyber, space, hypersonic, autonomous swarm, global rapid response, and military–industrial base deterrence.

Nuclear deterrence, the most effective strategic deterrent means, will be upgraded and operated through miniaturized tactics in the future to improve penetration and secondary strike capabilities further. A negative impact caused by this strategy is that nuclear powers must wrestle in strategic interactions between nuclear prohibition and anti-nuclear prohibition, and dilemmas between nuclear attacks and counter-nuclear attacks.

Cyber deterrence is an emerging and developing strategic deterrent. One power uses the threat of national network information, electric power and energy, economy and finance, military intelligence, public social opinion, and personal privacy to preclude an attack from an adversary power. Cyber deterrence gives rise to new operational capabilities such as virtual–real interaction and parallel operations.

Space deterrence is developing and growing rapidly. It refers to the ability to enter, use, and control space, including space and near-space vehicle networking and utilization, information offense and defense, ground-to-sea and air-to-air strikes, anti-missile defense, and so on.

Hypersonic deterrence mainly stems from ultra-fast speeds and massive destruction technologies that implement rapid global strikes and tactical network attacks. Hypersonic deterrence will change the course of space attack–defense confrontations and will have the effect of threatening punitive retaliation to prevent a foe from attacking.

Autonomous swarm deterrence is a new type of deterrence characterized by automatic target-seeking, autonomous network-building, autonomous attack, and swarm emergence. It has the effect of taking an enormous psychological toll on enemies, discouraging automatic attacks, not by the will of man. It will become a strategic focus with strict prevention and control from different countries.

Global rapid response deterrence refers to the rapid long-range response to overseas targets and tactical precision strike capability, based on network information and intelligent support, mainly including the "find-and-destroy" capability, rapid strategic delivery, and regional control and decapitation capability, etc. It also has a dual effect of deterrence and combat.

Military–industrial power deterrence is a sustainable combat capability based on advanced military design, production, manufacturing, and mobilization support. It is like a vast potential arsenal of high-tech weapons that threaten opponents and support military operations.

In addition, with the rapid development of biological and biological cross-over technologies, it remains to be further studied and observed whether the new biological confrontations can become a strategic deterrent due to accelerating international arms control. It is not clear whether the ongoing R&D of artificial life, genetic weapons, and new viruses will become a type of deterrence for international communities' condemnation. However, if related R&D spirals out of control, it will produce new sets of formidable strategic deterrents for biological warfare that extend beyond the realm of intelligence warfare.

2.7 Warfare Evolving into a New Stage

With the changing needs, the continuous advancement of S&T, and constant economic innovation, significant changes are happening in the potential war's external conditions, driving factors, operational elements, capability configuration, force organization, and operational patterns. Nuclear and conventional weapons, offensive and defensive operations, soft killing and physical destruction, physical combat, and virtual confrontation are becoming more complex and intertwined. The future of warfare is characterized by "intelligence, all-domain, parallel, integration, fastness, and comprehensive deterrence." Operation systems are developing from separation to integration, from the nodes to the network, from a single unit to a swarm of units, from human control to automation, from physical force to virtual and physical interaction.

"Intelligence" refers to the form of war that develops from mechanization and informatization to a higher stage of intelligentization. The development of military intelligence technologies can better adapt to highly complex environments, high-intensity confrontations, high-dynamic responses, incomplete information, and uncertain boundaries. It can provide a new set of solutions to such problems that are beyond human capacities, such as massive information screening, multi-strategy response, multi-task distribution, multi-means choice, and multi-modes of support in tasks like situational awareness, target identification, mission planning, rapid strike, precise support, cyber attack and defense, cognitive confrontation. Therefore, it can remove the "fog of war" and enhance such combat capabilities as intelligent perception, decision-making, offense and defense, support, evaluation, etc., in complex situations. Unmanned, swarm, virtual and reality integrated, and constantly evolving AI-led military intelligence technologies will have a global, long-term, and revolutionary impact on future warfare.

"All-domain" refers to the expansion of the operational space from the physical domains of "land, sea, air, and space" to "information, cognitive, social, and biological domains." Meanwhile, military operations are evolving from traditional joint operations and coordinated firepower to multi-domain interaction, cross-domain attack and defense, joint all-domain operations, and military operations other than war (MOOTWs). The combat process (time domain) changes from a clear wartime and peacetime demarcation to a wartime and peacetime mixture with features of abruptness and dispersion. The operational objectives are shifting from physical destruction, resource contention, and territory occupation and control to missions as diverse as cognitive confrontation, consciousness intervention, and joint effort and win–win collaboration.

"Parallel" refers to parallel forces, parallel training, and software-defined warfare, representing an important future development trend in the network and intelligent era. Most combat operations will be AI-guided in virtual space, i.e., the military will first use virtual space for combat simulation, experiments, modeling, and necessary physical verification. Taking advantage of virtual space not being limited by time, space, and terrain, AI technology can simulate the battlefield, equipment, soldiers, commanders, and troops in response to different battlefields, opponents, and combat patterns. It then recommends the optimal option to guide the development and operation of the physical forces, thus achieving the objective of victory through full interaction between the virtual and the real.

"Integration" refers to warfare developing in the direction of open-source, civil–military integration, resource sharing, etc. During the Cold War, defense investment accounted for a relatively high proportion of global innovation, so many civilian technologies came from the military sector. It was the main practice that military technology would be transferred for civilian use. However, in the era of information, networking, and intelligentization, this situation began to reverse when commercial inputs accounted for an increasingly high proportion of global innovation. Driven by market mechanisms, the application and commercialization of research results are significantly faster than those in the military sector. The utilization of civilian technologies and resources is becoming increasingly important for the military. The continuous upgrade of global networking and the IoT provides efficient and fast channels to utilize open-source information and civilian resources. The open collaboration and deep integration of military and civilian information, technology, standards, talents, facilities, industry, mobilization, and support have become inevitable.

"Fastness" refers to fast perception, decision-making, action, and support to make discoveries, decisions, responses, strikes, control, and defense ahead of the enemy. First, fast flow of energy in operations such as maneuvering, firepower, protection, and destruction. Second, the fast flow of information in network interaction, data transmission, and command and control. Third, fast decision chain processes on the battlefield, situation awareness, situation analysis, threat assessment, mission planning, force coordination, efficient support, and so on. The OODA loop will become faster in the future. The widespread use of hypersonic weapons and swarm systems will soon achieve the effect of a "one-hour strike" at the strategic level and "killing in seconds" at the tactical level.

"Comprehensive deterrence" refers to the scope of strategic deterrence expanding to include systemic and diversified means of deterrence, a combination of both nuclear and conventional capabilities, and the coexistence of deterrence and war. Because of their unique and irreplaceable role, nuclear weapons will continue to be a means of deterrence long into the future. In contrast, conventional deterrence will receive increased international recognition because of its compatibility with deterrence and warfare, which has a relatively low threshold and is developing in a diversified and systemic way. At the same time, human society may soon face new potential security threats and challenges in the biological domain.

In short, with more intelligent combat systems, stronger all-domain operational capabilities, more frequent virtual practices, faster combat actions, richer means of deterrence, and greater combat superiority, it will be easier to implement comprehensive deterrence, multi-domain coordination, and dimensionality reduction attacks, and thus realize the asymmetric advantage.

From a strategic perspective, although there are many war-affecting factors and various operational patterns, the essence and main path of future warfare development are intelligence, parallelism, openness, fastness, and integration. All of these are manifestations of intelligence-led warfare in the future. At present, mankind is still in the stage of "informatized warfare under nuclear deterrence." By the middle of this century, the future of war will be "all-domain intelligent warfare under nuclear and conventional deterrence."

References

[1]   Schmidt, E., Work, B., Catz, S., Chien, S., Darby, C., Ford, K., ... & Moore, A. (2021). *National Security Commission on Artificial Intelligence (AI)*. Washington, DC: National Security Commission on Artificial Intelligence.

[2]   Kott, A. (2018). *Ground warfare in 2050: How it might look*. Aberdeen Proving Ground, MD: US Army Research Laboratory.

3 The Transformation of Intelligent Warfare

The history of human civilization unfolds as a history of understanding and transforming nature as well as of self-understanding and self-liberation for humans. Through the development and application of technology and tools, humans have been able to constantly enhance their capabilities, lighten their burdens, get rid of constraints, and liberate themselves. It is the tireless pursuit of humans to expand the reach of their brain and limb capacities.

Intelligent technologies such as artificial intelligence (AI), big data, cloud computing, interdisciplinary biology, unmanned systems, and parallel systems, developing rapidly and being deeply integrated with traditional technologies, have enabled humans to understand and transform the world. Intelligentization, following mechanization and informatization, will become a new stage of human civilization. It will bring group breakthroughs to the military field, fundamentally transform the generation and composition of military combat effectiveness, and accelerate the arrival of the era of intelligent warfare.

3.1 A Historical Glimpse of AI

The concept of AI originated in 1956 to explore the essence of intelligence and develop machines with human-like or even superhuman intelligence. After more than 60 years of challenges and developments, breakthroughs have been achieved in AI's theory, technology, and application. With the inflection point of AI technology, the comprehensive application of AI technology, and the speed of industrialization significantly accelerated [1], military intelligence and reform will become an inevitable trend.

3.1.1 The Birth of AI (1940–1960)

In 1942, the "Three Laws of Robotics" were proposed by Isaac Asimov, an American science fiction giant; later these laws became the recognized principle for AI research and development in academia. In the Three Laws, the First Law is stated as, "a robot may not injure a human being or, through inaction, allow a human being to come to harm." The Second Law is that "a robot must obey the

DOI: 10.4324/b22974-4

orders given it by human beings except where such orders would conflict with the First Law." The Third Law is, "a robot must protect its existence as long as such protection does not conflict with the First or Second Law."

In 1950, the Turing Test was proposed by Alan Turing, one of the "Fathers of AI." According to his definition, a machine can possess intelligence if it can communicate with humans (via telex) without being identified as a machine. In the same year, Turing also predicted the possibility of truly intelligent machines being created by humans.

In 1954, the world's first programmable robot was designed by George Devol, an American inventor.

In the summer of 1956, the first conference on AI in history was held at Dartmouth College. At the conference, John McCarthy first proposed the concept of "AI," which was defined as "the science and engineering of making intelligent machines." Allen Newell and Herbert Simon showed their Logic Theorist, the first AI program.

In 1959, the first industrial robot, also the first generation of robots in the world, was created by George Devol, together with another American inventor, Joseph Engelberger. Subsequently, they established the world's first robot-manufacturing facility, Unimation.

3.1.2 The Golden Age of AI (1960–1970)

In 1965, work on "sentient" robots began. The Beast robot, developed by Johns Hopkins University's Applied Physics Laboratory, was able to adjust its position in response to data collected from the environment via sonar systems, photocells, and other devices.

In 1966, Joseph Weizenbaum of the Massachusetts Institute of Technology released ELIZA, the world's first chatbot. ELIZA's intelligence was derived from her programmed ability to comprehend natural language and generate human-like interactions.

In 1968, at the Stanford Research Institute, the world's first intelligent robot, Shakey, was born. The robot, which is equipped with visual sensors and is controlled by a room-sized computer, can locate and grasp building blocks in response to human commands.

Doug Englert invented the computer mouse on December 9, 1968, at the Stanford Research Institute in California. He also invented hypertext links, which laid the groundwork for the modern Internet decades later.

3.1.3 The Trough of AI (1970–1980)

In the early 1970s, AI research came to a halt. Computers' limited memory and processing speed precluded any solution to the practical issues raised by AI development. Researchers quickly realized that expecting a program to have the same level of understanding as a child was unrealistic. Nobody had ever created a database of this size in 1970, and even if they had, no one understood how a program could learn from such a wealth of data. Due to the lack of progress, funding agencies for AI, including the British government, the Defense Advanced Research Projects Agency (DARPA), and the National Science and Technology Council (NSTC), gradually phased out funding for pointless AI research. The National Research Council (NRC) halted funding after allocating $20 million.

3.1.4 The Boom Period of AI (1980–1987)

In 1981, the development of AI computers was initiated in Japan, where the Ministry of Economy, Trade, and Industry of Japan allocated $850 million for a project on fifth-generation computers, which were then called AI computers. Subsequently, the United Kingdom and the United States followed suit and provided substantial funds for information technology research.

In 1984, the Cyc (Encyclopedia) Project was launched under the leadership of Douglas Lenat, an American researcher, to enable AI applications to work in a way similar to human reasoning.

In 1986, the first three-dimensional (3D) printer in human history was created by Charles Hull, an American inventor.

3.1.5 The Winter of AI (1987–1993)

The term "AI winter" was coined by researchers who experienced funding cuts in 1974. They noted the enthusiasm for expert systems and predicted that people would soon turn to disappointment. Unfortunately, the practicability of expert systems was limited to a few specific scenarios. By the late 1980s, DARPA, with its new leadership, suggested that AI would not be the "next important thing" and that funds would go to projects that seemed more likely to yield returns.

3.1.6 The Spring of AI (1993–2019)

On May 11, 1997, Deep Blue, a computer developed by IBM, defeated world chess champion Garry Kasparov, becoming the first computer to defeat a world chess champion under standard chess tournament time controls.

In 2002, Roomba, a vacuum cleaner robot, was launched by iRobot in the United States, marking the birth of home robots. The robot designs routes off obstacles and automatically drives to a charging station when the battery is low. It is also one of the world's top-selling home robots.

In 2011, Watson, developed by IBM, the first AI program that uses natural language to answer human questions, beat two human champions in a US quiz show and won a prize of $1 million.

In 2012, "Spaun," a virtual brain with 2.5 million simulated neurons and simple cognitive abilities, created and named by a team of Canadian neuroscientists, passed a basic IQ test.

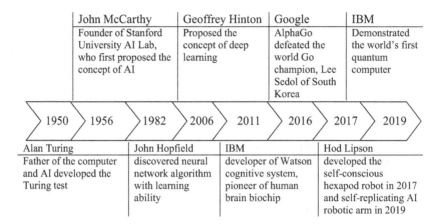

Figure 3.1 A brief history of artificial intelligence (AI).

In 2013, deep learning algorithms began to be widely used in the development of Internet products. Facebook set up an AI lab for exploring deep learning and providing its users with a more intelligent product experience. Google acquired DNN Research, a voice and image recognition company, to promote deep learning platforms. Baidu also set up a deep learning research institute.

In 2014, Eugene Goostman, a chatbot, passed the Turing test at the "Turing Test 2014" competition held by the Royal Society. This was the first time a robot had passed the test, heralding a new era of AI.

In 2015, a year of AI breakthroughs, Google opened up TensorFlow, a second-generation machine learning platform that directly trains computers to complete various tasks based on enormous amounts of data. The University of Cambridge established the Institute of AI.

On March 15, 2016, Google AI AlphaGo defeated the world Go champion Lee Sedol in the fifth game after five hours of fierce human–machine competition. The game ended in Lee Sedol's defeat with a total score of 1:4. This human–machine competition had officially made AI known to the world, drawing the whole AI market into a new round of prosperity.

From 2017 to 2019, the team of Hod Lipson, Professor at Columbia University, developed hexapod robots with self-awareness and later AI robotic arms with self-replicability (Figure 3.1).

3.2 The Age of AI

After more than half a century, AI has broken through major bottlenecks and achieved qualitative changes, with increasingly wide application in various fields and industries. It has opened up a new era of intelligence.

3.2.1 From Quantitative to Qualitative Changes

Historically, AI research is divided into three schools: Connectionism, Actionism, and Symbolicism. Connectionism, imitating the connection mechanism and learning algorithm of cerebral cortical neural networks, leads to deep learning in which big data are processed using multiple hidden layers. Actionism involves imitating a control system of perceptive actions for biological individuals and groups, reinforcing learning with reward and punishment control by gaining output through behavior enhancement or attenuation. Symbolicism, based on the hypothesis of a physical symbol system and bounded rationality, suggests the method of knowledge mapping to simulate the logical structure of the human brain.

The three AI schools, combined and complemented by algorithms including Monte Carlo tree search (MCTS) and Bayesian optimization, have introduced AI systems surpassing human intelligence in specific areas, such as IBM's Waston and Deepmind's AlphaGo. For relevant algorithms, the Monte Carlo algorithm solves problems requiring a solution to be given within a limited number of samples, but it is not necessarily optimal. In contrast, the Las Vegas algorithm solves problems that require the optimal solution within an unlimited number of samples.

The theory of early AI research is a traditional "top-down" one. It defines the thinking activities of the human brain through formulas and rules or translates the activities into programming languages for input into the machine, which is expected to imitate and thus possibly produce the thinking activities of humans.

In 1976, Allen Newell and Herbert Simon proposed the hypothesis of the physical symbol system as the necessary and sufficient condition for expressing intelligent behavior. In this hypothesis, an information-processing system can be regarded as a physical system like the human nervous system, a computer system, etc. In the 1980s, Newell and others devoted themselves to the research of Soar, a cognitive architecture based on a chunking theory, which can acquire search-control knowledge and operators by a rule-based memory. In the 1970s, Marvin Minsky, based on his psychology research, proposed a frame-based knowledge representation system, believing that in daily cognitive activities, the knowledge acquired and collected by people from previous experiences has always been stored in the human brain in a frame-like structure. In the 1980s, Minsky proposed that there was no unified theory to explain human intelligence. In 1985, he published *Society of Mind* [2], suggesting the concept of "the society of mind" as a complex society composed of numerous units with specific thinking abilities. The logic school, represented by John McCarthy and Nils John Nilsson, advocated using formal methods, namely logic, to study AI. The logic school uses methods including conceptual knowledge representation, model-theoretic semantics, and deductive reasoning. McCarthy advocated that everything can be represented by a unified logical framework and common-sense reasoning centered on non-monotonic logic.

In 1982, John Hopfield, one of the representatives of Connectionism, discovered neural network algorithms with learning ability. Later, a more straightforward statistical method that consumes fewer computing resources was found – support vector machine (SVM), a generalized linear classifier that classifies data by supervised learning. Next, a new network algorithm was proposed: long- and short-term memory (LSTM), a recurrent neural network for processing and predicting notable events with long intervals and delays in time series.

Towards the twenty-first century, Geoffrey Hinton, among others, suggested the concept of deep learning. Since 2010, machine learning, represented by deep learning, has become the dominant concept in the AI industry. As the computer hardware industry has developed rapidly, even mobile phones possess the computing power necessary for image recognition. This has revolutionized the field of Connectionist research. Meanwhile, machine learning has brought significant achievements and key technical breakthroughs to AI. Taking computer vision as an example: before 2011, computer vision recognition had always maintained an error rate of above 26% despite decades of effort. Since then, with the emergence and application of deep learning, big data, and other technologies, the error rate has begun to decline sharply. In 2015, Microsoft's Deep Residual Networks, with an error rate of only 3.57% in image recognition, had surpassed the human eye with an error rate of 5.1%. Generally, breakthroughs in key and supporting technologies have led AI from quantitative accumulations to qualitative leaps, thus bringing the curtain down on a revolution in AI.

For unmanned system research, better results can be achieved by using deep learning to identify environments and targets, and using behaviorism, namely reward and punishment mechanisms, in task and path planning. Robotic behaviors that meet human expectations will get more rewards (higher weight for scoring), and those that do not will get weighed lower. Behaviorism, combined with other intelligent principles, ideas, and approaches, will see critical applications in future military fields, as well as greater demands, advantages, and potential.

3.2.2 Arrival of the Inflection Point

Recent neurophysiological and brain science research has demonstrated that the cortical areas of the brain associated with perception functions such as visual, auditory, and motor serve as input and output channels and are directly involved in thinking. Intelligence is not solely concerned with problem-solving via knowledge and reasoning; it also includes perception. The human brain is composed of two systems: one that connects the cranial nerves to various organs and another that receives feedback from organs. It is critical for the scientific community to gain a comprehensive understanding of the brain's structure, function, and operation mechanism.

Because humans had such a limited understanding of AI, it grew at a slower rate than anticipated. Alan Turing optimistically predicted decades ago that machines would pass the "Turing Test" and demonstrate rudimentary intelligence by the year 2000, but that deadline appears to have been pushed back. It was difficult to make significant advances in the early stages of AI development due to a lack of understanding of the fundamental problems of intelligence, overly ambitious development goals, and a lack of computing power. However, intelligent algorithms such as deep learning and critical technologies such as big data and cloud computing resolved the development dilemma in the twenty-first century, enabling the rapid rise of AI.

Since 2016, AI has entered a new era of rapid development. AI is now demonstrating new capabilities such as cross-border integration, human–computer collaboration, open-group intelligence, and self-control. These characteristics are primarily a result of new brain science theories, innovative technologies such as mobile Internet, big data, supercomputing, and sensor networks, as well as the strong demand generated by economic and social development. The new focus of AI research is on knowledge acquisition via big data, cross-media collaborative processing, enhanced intelligence via human–computer cooperation, group intelligence, and autonomous intelligent systems. As a result of brain science research findings, brain-like intelligence is gaining momentum and moving toward brain-like chips, hardware, and platforms. A new generation of AI is currently sweeping the disciplines of discipline construction, theoretical modeling, technological innovation, software development, and hardware upgrade. It is accelerating the transition from digitalization and networking to intelligentization across the economy and society.

By the end of 2017, AI research had homed in on five critical areas: robots and driverless cars, computer vision, natural language processing, virtual assistants, and machine and deep learning. As evidenced by the primitive AI era, AI is still in its infancy. Despite this, the intelligence wave has accelerated. Will the next decade of AI development be another flash in the pan, followed by a bitter winter, as the 1960s and 1980s were? While most experts and scholars believe that AI development will continue to face obstacles in the new round, the inflection point has arrived, accompanied by an irresistible vigorous trend.

AI applications have been exploding with new capabilities. As a result of recent advancements, the spread of AI into various fields is unavoidable. Google officially launched RankBrain in 2015, a new AI algorithm based on machine learning that has risen to become the third most important ranking factor among Google's hundreds of search ranking indicators. The convergence of AI, the Internet of Things (IoT), cloud computing, and traditional industries has resulted in an intelligent manufacturing revolution. At the moment, AI has pervaded the economy, is widely used in traditional information services, and has begun to penetrate high-end, brain-intensive industries such as medicine, journalism, and finance. AI began as a smoldering ember and has gradually grown into a prairie fire.

The intelligence pattern has been accelerated. AI, as represented by deep learning, has garnered considerable attention from research institutions, industry, and world powers due to its successful application and broad potential in a variety of fields. All parties have elevated AI to a commanding position in terms of future development and have restructured their strategic plans to reflect this. With 2015 as the tipping point, when Facebook, Google, Microsoft, and other global technology giants announced the launch of a wave of open-source AI development platforms, AI research and development entered a new era marked by public participation. Due to its innovative applications, it has experienced a meteoric rise in popularity. The AI industry has attracted global technology giants and venture capitalists, resulting in increasingly competitive bidding for talent, technology, and investment among significant companies, with investments from industry titans such as Google and Baidu growing at a rapid pace.

AI has risen to prominence as a new engine of economic growth. AI, as the primary catalyst for a new wave of industrial transformation, will reimagine all facets of economic activity, including production, distribution, exchange, and consumption, and will uncover new macro and micro requirements for intelligentization in a variety of fields. Additionally, AI will enable the development of novel technologies, products, industries, formats, and models, as well as significant structural changes in the economy, profoundly transforming human life and thought.

AI has also expanded the possibilities for social construction. The widespread use of AI in education, healthcare, pensions, environmental protection, urban operations, and judicial services will improve the precision and quality of public services. In 2016, IBM's "Watson" AI platform diagnosed a rare form of leukemia in 10 minutes after learning from over 20 million medical papers imported by Japan's Medical Institute of Tokyo University and suggesting a treatment change, thereby saving the life of a 60-year-old patient.

Competition on a global scale has shifted its emphasis to AI. The world's major powers have viewed developing AI as a critical strategy for increasing national competitiveness and ensuring national security, and they are vying to lead the new round of international scientific and technological competition. The White Documents *Preparing for the Future of Artificial Intelligence* [3], the *National Strategic Plan for AI Research and Development*, and *Artificial Intelligence, Automation, and the Economy* [4] were all proposed for the United States in 2016, followed by the "American AI Plan" in 2019 [5]. In September 2017, Russian President Vladimir Putin stated publicly that AI is "the future of Russia." "Whoever is at the forefront of this field will dominate the world." On October 10, 2019, Putin signed the National Strategy for the Development of AI (NSDAI) by 2030. The strategy prioritized AI development directions, critical tasks, and mechanism measures with the objective of increasing Russia's independence and global competitiveness in this field.

China's State Council issued the New Generation AI Development Plan on July 8, 2017 [6], outlining the guiding ideology, strategic objectives, critical tasks, and safeguards for AI development in the country by 2030, with the goal of developing an innovative country and a world scientific and technological power with AI dominance.

The unpredictable nature of AI development adds new complications. As a disruptive technology with enormous influence, AI will have a profound impact on government operations, economic security, social stability, and even global governance. For example, it may alter the employment structure, have an effect on law and social ethics, infringe on personal privacy, and jeopardize international relations norms. While aggressively developing AI, the state must prioritize the security risks it may introduce, implement preventative measures, strengthen restraint guidance, and mitigate risks to ensure the safety, reliability, and controllability of AI.

AI has accelerated its growth and development in the second decade of the twenty-first century. With the widespread development and application of AI, a wave of unprecedented intelligence is sweeping across all spheres of human society, upending established norms and establishing new ones. According to the Future Today Institute's *2021 Tech Trends Report* [7], AI has advanced in nearly every sector of society in recent years, including enterprise, medical treatment, healthcare, science, consumption, and research, with typical applications including mass translation systems, emotional intelligence, decentralized computing, cloud AI, edge computing, advanced AI chips, digital twins, drug synthesis, virus mutation detection, and mind. According to the report, DeepMind projects have established that AI is far superior to humans in certain areas and has surpassed the threshold of weak AI, paving the way for future research into strong AI. Between 2021 and 2027, the global AI market is expected to expand at a compound annual growth rate of 42.2%. Additionally, it was noted that as world leaders in the field develop intelligent autonomous weapon systems, AI is reshaping the current international order. For instance, the US military has established new military–industrial complexes in recent years to facilitate the efficient development of new weapons, thereby establishing new military–industrial complexes for future algorithmic warfare. Future conflicts will almost certainly be fought entirely in code, with data and algorithms serving as lethal weapons.

3.2.3 Three Major Events

Three significant events occurred recently in the field of AI.

The first event: IBM announced that it successfully developed the world's first stand-alone quantum computer, which customers can use to perform large-scale data computing via the Internet. On January 9, 2019, IBM demonstrated the quantum computer, IBM Q System One, at the annual Consumer Electronics Show, displayed in a 2.3-meter-high glass box. IBM's quantum computer finally entered the commercial stage 2 years ahead of schedule.

It was a genuinely epoch-making leap. Quantum computers, which previously seemed out of reach, are suddenly approaching human beings. The quantum computer is enclosed in a glass box due to the complexity of the materials used in it. These materials include a hardened chamber that houses the quantum bits used in computation, tanks of liquid helium, cryogenic equipment that maintains the temperature of the quantum bits at or near absolute zero (–273.15°C), electronics that control the quantum bits' movement, and "read" their output, and finally, the cables that connect all these elements. A total of four shells wraps around the quantum bits layer by layer to protect them from any external interference.

Quantum computers have the potential to perform epochal computations. For instance, it would take 600,000 years to crack RSA passwords using the largest and fastest current supercomputer, but less than 3 hours using a quantum computer. When faced with quantum computers, our once-proud traditional computers revert to being the computing equivalent of abacuses, easily defeated. The transition from electronic to quantum computers will result in a significant increase in computing power and the ability to process large amounts of data. IBM has industrialized its 20-Qubit Machine, and the 50-Qubit Machine will follow suit shortly. As implied by its name, the 50-Qubit Machine will be capable of concurrently performing 2 to the 50th power (approximately 1,125 trillion) calculations.

The second significant event: on January 23, 2019, Science Robotics published research by the team of Hod Lipson, Professor of Mechanical Engineering and Director of Creative Machine Lab at Columbia University, claiming that a robot (robotic arm) gained "self-awareness."

Professor Lipson and his doctoral students created a human-sized "iron arm" with four independently adjustable joints without relying on complex computer structures, physics, or geometric dynamics. They then subjected the "iron arm" to 35 hours of random movement. After 35 hours of training, the robot (robotic arm) created a set of self-simulators that enabled it to adapt to different situations, handle new tasks, and even detect and repair body damage. It collected about 1,000 motion trajectories during the movement, each containing 100 motion nodes, and then built a self-model through deep learning of the data. The process of building self-models is also called "self-imagination." Whereas previous robots have generally conducted self-learning through human input and constant attempts, this time, the robot demonstrates the ability to think, imagine its form, and know its purpose, specifically, how to replicate itself. As a result, the robot "taught itself" and accurately completed the task, with an error of no more than 4 cm between the self-model and the actual physical entity.

When the robotic arm was switched to a "calibrated" mode that allowed it to adjust to the trajectory, it successfully picked up and placed multiple balls in the cup. "It's comparable to blindfolded cup picking," the researchers observed. Humans must be trained to adapt to such movements.

Not only that, when the researchers replaced the robotic arm's components with 3D-printed, mutilated components to simulate physical injury, the arm appeared to sense its own "injury" and readapted to the new structure after a period of learning and adjustment, continuing to perform the previous task of object picking with no loss of efficiency.

Although the robotic arm exhibits only a superficial sense of self-awareness, Professor Lipson hypothesizes that it behaves similarly to the evolutionary origins of human self-awareness, which are currently occurring during infants' self-learning process.

The third major event: Elon Musk, CEO of Neuralink, a start-up focused on brain–computer interfaces, held a press conference on July 16, 2019, claiming to have discovered a more efficient method of creating a brain–computer interface. The company developed a brain–computer interface system that utilizes a neurosurgical robot to implant a 4–6-micron-thick wire into the brain. Through a USB-C interface, researchers can read and control brain signals directly (Figure 3.2). The Neuralink researchers sewed a bundle of 1,024 fine wires, each one-fourth the width of a human hair, into a brain using a specialized surgical robot (ten of the wires were implanted in the skin). On one end, each wire was connected to a chip; on the other, it was connected to a detachable, wearable, and upgradable device behind the ear. Wireless communication is possible between the device and a mobile phone. "A monkey has been able to control a computer with its brain," Musk explained during a

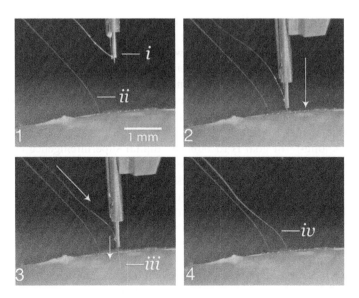

Figure 3.2 Neuralink connects a human brain to a computer via implanted electrodes.

live demonstration on YouTube. Neuralink's goal is to determine the viability of higher-bandwidth brain connections by recording more neuronal activity via new "wire" connections.

The electrodes on Neuralink gathered brain signals from neurons and "wrote" them into the brain. Musk stated that the company would work on brain–computer interface technology, which would involve implanting electrodes in the human brain to connect it to a computer, allowing for future direct uploading and downloading of ideas. His mission is to reconcile humans and machines. Rather than relying on mobile phones, ideas will be transferred directly from one person's brain to another.

On August 29, 2020, Elon Musk demonstrated the latest technology from Neuralink using Gertrude, a pig. The brain activity of a pig sniffing around a staged pen is visualized via a wireless connection to Neuralink's brain–computer interface device.

This demonstration by Neuralink shows that brain–computer interface technology is closer to completing Musk's ambitions than it was when the product was first unveiled in 2019. Musk said the US Food and Drug Administration approved testing of the "breakthrough device" in July 2020. He also showed a second-generation brain–computer interface device, which was more sophisticated and fit into a small cavity dug out of the skull.

"It's kind of like a Fitbit in your skull with tiny wires," Musk said of the device. The device communicates with brain cells through 1,024 thin electrodes that penetrate them. It also can be connected to external computing devices via Bluetooth. Nevertheless, the company has been working on other radio technologies, hoping that a new one will significantly increase the number of data connections.

Neuralink focuses on medical treatments for brain–computer interface applications, such as helping people with brain and spinal cord injuries or congenital disabilities. "If you can sense what people want to do with their limbs, you can do a second implant where the spinal injury occurred and create a neural shunt," Musk said. "I'm confident in the long term it'll be possible to restore somebody's full body motion." "Neuralink chips can measure temperature, pressure, and movement, data that could warn you about a heart attack or stroke," Musk said.

Neuralink's long-term goal is to build a "digital superintelligence layer" to connect humans with AI, including "consensual telepathy," in which two people can communicate digitally by thinking about each other rather than writing or speaking. "The future is going to be weird," Musk said, discussing sci-fi uses of Neuralink. "In the future, you will be able to save and replay memories," he said. "You could basically store your memories as a backup and restore the memories. You could potentially download them into a new body or a robot body."

In April 2021, Neuralink published an original blog post with a video demonstrating how they had taught Pager the monkey to play Pong (a video game that simulates two people playing ping pong) using only its brain. According to the report, Pager's brain was implanted with a brain–computer interface approximately 6 weeks before the video was shot. Pager had previously been instructed on how to play the game with a joystick in exchange for a delectable banana smoothie. While playing the game, Neuralink's device learned and predicted hand movements primarily by analyzing information emitted by neurons, or lit areas of the brain. Pager's joystick was disconnected from the computer once it had finished learning its gaming patterns. Pager, on the other hand, was still capable of playing the game without using a joystick, relying entirely on the thoughts of its brain.

Quantum computers, self-aware robots, neural connections between human brains and computers, 5G, big data, and swarms of small satellites will all accelerate the development and application of AI. In the not-too-distant future, traditional combat forces will almost certainly become powerless and vulnerable in the face of "quantum computing + advanced AI + military applications." The military and the entirety of human society will be engulfed in a massive revolution.

The history of war demonstrates that each major scientific and technological revolution, along with its associated industrial revolution, resulted in significant changes in operational patterns, theoretical tactics, weapons and equipment, military formation, military training, logistics, and other military fields, ultimately altering the way war is fought.

3.3 From Land Power to Intelligence Power

The history of human war shows that the control of war is constantly changing, enriching, and evolving with the progress of science and technology and the development of affairs. Since the nineteenth century, human warfare have involved the control and struggle for power in the realms of land, sea, air, space, and information, and will move towards the competition for intelligence power in the future.

3.3.1 Land Power

At the beginning of the twentieth century, Halford John Mackinder, a British School of Geography representative, suggested the land power theory. He focused on comparing representative national strategies: the strategy of controlling the Eurasian heartland, pursued by countries such as Germany and Russia, and the strategy of controlling the rimland, pursued by countries such as the United Kingdom and the United States. He found that industries and technologies such as railways and telegraphy expanded the scope and speed of military operations. He also predicted that by using new transportation and communication means and mobilizing local resources, the heartland countries would establish protected communication lines and quickly launch military operations in their chosen areas. He referred to Eurasia and Africa as the "World-Island" and the central areas of Eurasia and Africa as the

"Heartland." His famous summary of the theory is: who rules East Europe commands the Heartland; who rules the Heartland commands the World-Island; who rules the World-Island commands the world [8].

Later in 1944, Nicholas John Spykman, American geopolitical theorist, in his book *The Geography of the Peace* [9], developed the Rimland Theory, arguing that the heartland of Eurasia was harsh in weather, sparsely populated, and economically backward, and thus not the commanding point of world domination. In contrast, the world's population and resources were distributed on the rimlands of the Eurasian continent, namely the Western European peninsula and East Asia. The expansion of military operations, such as amphibious operations, and the commissioning of aircraft carriers or land-based air power, leads to a key role for rimland countries in Eurasia and other parts of the world. Spykman argued, in contrast to Mackinder, that "whoever rules the Rimland commands Eurasia, and whoever rules Eurasia commands the world" [9].

The land power theory has had a significant impact on the development of continental countries and the form of ground operations. In particular, with the development of power technologies such as steam engines, gas turbines, and diesel engines, mass ground-based mechanized swarm confrontation has become the dominant operation pattern, as demonstrated in the Soviet–German theaters of war during World War II. The Soviets and Germans committed more than four million troops, 69,000 artillery pieces and mortars, 13,000 tanks and self-propelled guns, and 12,000 combat aircraft in the Battle of Kursk alone, and were involved in an encounter in the Battle of Prokhorovka, the most considerable tank encounter in history, where both sides deployed a total of 1,200 tanks and self-propelled guns.

From a modern perspective, the land power theory, with historical limitations, neglected the impact of technological developments on warfare while focusing on how geography framed combat operations and how powerful, offensive, and influential the nomads once were. However, from the perspective of the time, the theory had its validity. Towards the new century, with the development of global science, technology, economy, and social patterns, Chinese experts and scholars have gained a new understanding of land power, which now represents the power to survive and develop in land space. From the sole perspective of economic and social development, land power outweighs sea, air, space, and information power. Simply put, humans cannot survive beyond land space but are less affected beyond the sea, air, space, and information space. Land power is the basic prerequisite for sea, air, space, information power, and other forms of power. In conjunction with these forms of power, it enhances a country's influence [10].

3.3.2 Sea Power

In 1890, in his publication *The Influence of Sea Power upon the French Revolution and Empire History*, Alfred Thayer Mahan argued that for major economic powers that made their fortunes from maritime trade, naval power was the basis of major national strategies, and that priority should be given to developing naval and commercial fleets to expand the scope of trade, the international influence, and the naval forces ashore [11]. Mahan writes,

> Sea power in the broad sense ... which includes not only the military strength afloat, that rules the sea or any part of it by force of arms, but also the peaceful commerce and shipping from which alone a military fleet naturally and healthfully springs, and on which it securely rests.

Mahan's sea power and naval strategy theories have significantly impacted naval force building, maritime trade, and resource competition in the world. The sea power theory has also contributed significantly to the development of naval equipment. Battleships, cruisers, destroyers, frigates, large aircraft carriers and their carrier-based aircraft and command-and-control systems, conventional submarines and nuclear submarines, amphibious equipment, and other weapons and equipment have stepped on to the stage of history. During World War II, the Battle of Midway between the United States and Japan was typical of modern naval warfare.

3.3.3 Air Power

The Italian military theorist Giulio Douhet, in his book *The Command of the Air* [12], suggested that aviation technology development had introduced mighty air power against one's enemy, and air warfare would depend upon "smashing the material and moral resources of a people caught up in a frightful cataclysm which haunts them everywhere without ceasing until the final collapse of all social organization." Theorists who subscribe to the air power theory believe that future warfare will be of short duration and that the side that occupies air supremacy will gain the advantage. Billy Mitchell, the father of the US Air Force, said:

> The ability of air power to reach right into the heart of the enemy and to suppress and destroy it has brought a whole new dimension to traditional operational theory. It is now recognized that the enemy's land forces in the operational territory are no longer the primary target [13].

The air power theory has led to the dominance of the theory of victory by air and its influence today. Since 1906, when the Wright brothers invented the airplane, this new type of equipment has been used in offensive operations against the ground. It has gradually evolved into air-to-air fighters, ground-attack aircraft, long-range reconnaissance aircraft, bomber aircraft, and command and warning aircraft for aerial swarm attacks and defenses, transport helicopters, helicopter gunships, stealth aircraft, and unmanned aircraft. The experience of World War II and local wars since the 1990s has demonstrated that air supremacy and power have a decisive impact on the victory or defeat of the war. For example, the Kosovo war, with the operational objectives achieved by 78 days of airstrikes, was an entire air- and missile-led war, creating a completely independent operational pattern.

3.3.4 Space Power

With the advancement of technology, humans have extended their military and economic activities from the sky to space. In the 1950s and 1960s, the advent of artificial satellites and intercontinental ballistic missiles led to a strategic military focus on space. Space, while currently a global public asset in which nations share resources and domains, is increasingly becoming an important battleground for strategic deterrence and military operation system of systems (OSoS). Mark E. Harter, Major in the US Air Force, stressed that space is the new strategic high ground: "Space systems significantly improve friendly forces' ability to strike at the enemy's heart or COGs [centers of gravity], paralyzing an adversary to allow land, sea, and air forces to achieve rapid dominance of the battlespace" [14]. Outer space is the fourth frontier of human activities after land, sea, and air, where activities are of great strategic significance to national security, economic interests, scientific and technological development, and social progress.

Over the past decades, the rapid growth of human space activities has triggered fierce competition for limited resources. Given the great significance of space power in Modern warfare, confrontation and competition between nations in the field of space have become inevitable. For example, satellite information resources such as space-based reconnaissance, communications, navigation, and meteorology have played a vital role in global information grid construction, long-range joint operations, global integrated operations, systematic operations, and over-the-horizon strikes, among other operations of the US military. Space-based information has comprised more than 50% of the information for major campaign operations of the US military and more than 80% for specific campaigns. Under the new situation, the competition for space power has gone beyond the struggle for space-based information dominance and entered a new stage of space attack, defense, and control. Space power is about the national ability to enter, utilize, and control space, the competition of which focuses on the possession of space resources, equipment, and technology, the dominance of space attack and defense, and the right to formulate rules on human activities in outer space.

3.3.5 Information Power

Since the 1980s, with the development of computers, the Internet, global positioning system (GPS), and mobile communications, as well as the increasingly prominent role of C⁴ISR (command, control, communications, computers; intelligence, surveillance, and reconnaissance), electronic information systems, cyberspace, and information technology equipment in wars,

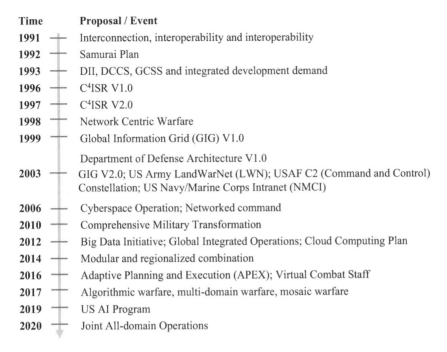

Time	Proposal / Event
1991	Interconnection, interoperability and interoperability
1992	Samurai Plan
1993	DII, DCCS, GCSS and integrated development demand
1996	C⁴ISR V1.0
1997	C⁴ISR V2.0
1998	Network Centric Warfare
1999	Global Information Grid (GIG) V1.0
2003	Department of Defense Architecture V1.0 GIG V2.0; US Army LandWarNet (LWN); USAF C2 (Command and Control) Constellation; US Navy/Marine Corps Intranet (NMCI)
2006	Cyberspace Operation; Networked command
2010	Comprehensive Military Transformation
2012	Big Data Initiative; Global Integrated Operations; Cloud Computing Plan
2014	Modular and regionalized combination
2016	Adaptive Planning and Execution (APEX); Virtual Combat Staff
2017	Algorithmic warfare, multi-domain warfare, mosaic warfare
2019	US AI Program
2020	Joint All-domain Operations

Figure 3.3 US military accelerating informatization and moving towards intelligentization after the Gulf War.

information supremacy has become the key to war victory and has had an impact on other forms of supremacy, while giving rise to new types of combat forces, including digital forces, networked forces, and electronic combat forces. Information power, with the electronic information system as the core, truly integrates the systems and technologies of the army, navy, and air forces, completing a "seamless and rapid information transmission from sensors to soldiers," constituting an integrated, joint, 3D, and precise strike capability, bringing systematic combat advantage versus local, and fast versus slow.

Since the Gulf War, the US military has strived for information supremacy (Figure 3.3). Beginning with the Big Data Initiative in 2011, the US military, based on informatization, has moved towards intelligentization by successively proposing the Big Data Initiative, Cloud Computing Strategy, Adaptive Planning and Execution (APEX) for Joint Warfare, virtual staffs, algorithmic warfare, multi-domain warfare, mosaic warfare, and joint all-domain operations.

3.3.6 Intelligence Power

Recently, the US and Russia have made intelligent technology a core means of maintaining their strategic position as global military powers. With the concept, model, organization, and innovative application of science and technology significantly transformed and numerous military intelligentization moves performed, thus opening the competition for intelligence power in the future military field.

In 2011, the US Department of Defense (DoD) proposed its Data to Decision Science: Priority Development Plan [15]. In 2012, the US DoD officially released its Cloud Computing Strategy [16]. In 2014, the US Army deployed Distributed Common Ground System-Army (DCGS-A) to Afghanistan, the Army's first private military cloud system and the first tactical cloud computing node deployed operationally. The DCGS-A system is available in fixed, embedded, and mobile configurations. It was deployed in this operation as a mobile version that could connect directly to a cloud computing platform to provide real-time information for soldiers on the ground, such as predicting the most viable paths to find improvised explosive devices (IEDs) using historical data and providing full-motion video from unmanned aerial vehicle (UAVs).

In April 2017, the US DoD announced the establishment of an Algorithmic Warfare Cross-Functional Team, laying the groundwork for the large-scale development and application of warfare algorithms. Algorithmic warfare is a new operational concept that involves bringing together the army's big data on a cloud platform and analyzing them to create an AI OSoS.

In August 2017, the US DoD said that AI warfare is inevitable in the future and that the United States needed to "take immediate action" to accelerate the development of AI warfare technology. The US military, proposing the Third Offset Strategy, believed that a storm of military change marked by intelligent armies, autonomous equipment, and unmanned warfare was on the horizon. Intelligence, represented by autonomous systems, big data analytics, and automation, was identified as a major development direction. The US military planned to build a basic intelligent OSoS by 2035, forming a new military "generation gap" with major adversaries, and to fully engage the intelligence of combat vehicles, information systems, command and control by 2050, with the capability to wage a truly "robotic war" [17].

In June 2018, the US DoD proposed to establish a Joint AI Center (JAIC). Guided by the National Strategic Plan for AI Research and Development, the Center coordinates the building of intelligent military systems of the US military. It planned to advance about 600 AI projects jointly with the US military and 17 intelligence agencies, investing more than $1.7 billion [18]. In September 2018, the DARPA announced that it would invest $2 billion in the next 5 years to advance the development of AI.

In March 2019, the US Army issued the Mosaic Warfare concept, proposing a new operational pattern with "rapid, flexible and autonomous combination of combat elements," using elastic networks and dynamic mission planning to develop intelligent decision-making and coordinated attack capabilities for distributed combat forces, giving them advantages over adversaries.

The United States has also launched an action plan at the national level. On February 11, 2019, President Donald Trump signed an executive order creating the American AI Initiative, a high-level strategy to guide the development of AI in the United States. The Initiative involves five key areas:

1. Invest in AI research and development. Federal agencies were required to "prioritize investments in AI" in their research and development (R&D) budgets and report on how these funds would be used to create a more comprehensive government AI investment plan.
2. Unleash AI resources. Federal data, algorithms, and computing power would be available to researchers to fuel AI applications in transportation and healthcare.
3. Setting AI governance standards. Government agencies, including the White House Office of Science and Technology Policy (OSTP) and the National Institute of Standards and Technology (NIST), would be invited to develop guiding standards for the development of "reliable, robust, trustworthy, secure, portable, and interoperable AI systems."
4. Building the AI workforce. Institutions will be required to prepare workers for changes in the job market brought about by innovative technologies by setting up fellowship and training programs.
5. International engagement. The US government planned to work with other countries on AI development but would do so to preserve American "values and interests."

Russia has also formulated a series of plans that give prominence to AI and unmanned equipment development. It has approved the Concept of Developing Military–Industrial Complex (OPK) by 2025, emphasizing that AI systems will soon become a key factor in winning future warfare, with a focus on the intelligent adaptation of weaponry, as well as the development of combat robots and AI missiles for the next generation of strategic bombers. The government plans to increase the proportion of unmanned combat systems equipped to 30% by 2025.

Major countries worldwide have launched respective AI development strategies and military application plans, indicating that the global competition for "intelligence power" has been launched globally. Land, sea, air, sky, information, and intelligence power are all products of scientific and technological progress in specific eras, with their strengths and weaknesses. Relevant theories are also constantly changing and progressing with the times. Since modern times, the trend in the control of war has shown that only information power and intelligence power have a global impact on wars and thus have a heavier weight and more significant influence. In the future, with the acceleration in pace of intelligentization, intelligence power will be a rapidly growing new type of control with more significant strategic influence on the overall combat situation.

3.4 AI-Led Rules and Laws of Victory

Different eras and forms of warfare have led to an entirely different battle-field ecology, as well as different operational elements, and rules and laws of victory.

3.4.1 Evolution of the Form of Warfare

Since modern times, human society has experienced large-scale mechanized warfare and small-scale information-based local warfare. The two world wars in the first half of the twentieth century were typical mechanized warfare. The Gulf War, the Kosovo War, the Afghanistan War, the Iraq War, and the Syrian War since the 1990s have fully embodied the form and characteristics of information warfare. Towards the new century, with the rapid development and broad application of intelligent technology, the era of intelligent warfare characterized by data, computing, models, and algorithms is coming.

Mechanization, a product of the industrial era, technically focuses on machinery, electricity, explosives, destruction, and equipment forms including tanks, armored vehicles, artillery, aircraft, and ships, corresponding to a mechanized form of warfare. Mechanized warfare, theoretically based on Newton's laws and classical physics and practically based on socialized mass production, is dominated by large-scale, swarm, linear, and contact operations. For the tactics in this context, the combat forces usually rely on in situ reconnaissance and terrain surveys to understand their opponents' front and rear deployments. They then make decisions on offensive or defensive operations according to their combat effectiveness, task division, operational coordination, and logistics. This kind of warfare is characterized by a hierarchy of command and control and the serialization of space and time.

In contrast, informatization, a product of the information era, technically focuses on microelectronics, optoelectronics, computers, network communications, satellite navigation, and other information technologies. The equipment forms in this era include radar, radio, military satellites, precision-guided weapons, stealth aircraft, military computers and software, command-and-control systems, cyber attack and defense systems, integrated electronic information systems, corresponding to an informatized form of warfare. Informatized warfare, theoretically based on the Three Laws of Computing (Moore's Law, Gilder's Law, and Metcalfe's Law), is characterized by integrated, 3D, and precise operations, with the pursuit of "seamless and rapid information transmission from sensors to soldiers," seizing information supremacy, and achieving pre-emptive enemy detection and strike. For the tactics in this context, combat forces need to identify and catalog battlefields and targets in detail, highlighting the role of networked perception, command-and-control systems, placing new demands on platforms for interoperability and other information functions. With the development of global information systems and diverse network communications, informatized warfare has diluted the boundaries between the front and the rear of a battlefield, emphasizing the integration and flattening of "reconnaissance, control, strike, evaluation, and support" as well as of strategy, campaign, and tactics.

Intelligentization, a product of the knowledge economy era, technically focuses on AI, big data, cloud computing, cognitive communication, IoT, interdisciplinary biology, military bionics, digital twins, hybrid enhancement, swarm intelligence, autonomous navigation, and cooperation. The main equipment forms include unmanned ground vehicles (UGVs), UAVs, unmanned surface vehicles (USVs), unmanned underwater vehicles (UUVs), military robots, intelligent ammunition, swarm systems, intelligent wearable systems, networked intelligent perception systems and databases, military cloud platforms and service systems, combat simulation, military parallel systems, adaptive mission planning and decision-making systems, infrastructure security intelligent control systems, intelligent support and logistics systems, corresponding to a form of intelligent war. Intelligent war is a new form of war, theoretically based on bionic, brain-like principles, and practically based on AI battlefield ecosystems, characterized by multi-domain fusion, cross-domain attack, and defense, unmanned combat, swarm confrontation, and integrated interaction between virtual and physical space, with renewable energy, information interconnection, network communications and distributed clouds as the support, data computing and model algorithms as the core technologies, and cognitive confrontation as the central function.

Intelligent warfare aims at meeting the demands of nuclear and conventional deterrence, joint warfare, all-domain operations, and military operations other than war. Their focus is on multi-domain fusion operations that involve cognitive, information, physical, social, biological, and other domains, while presenting distributed deployment, networked links, flat structures, modular combinations, adaptive reorganization, parallel interaction, focused energy release, non-linear effects, and other characteristics, which overturn the traditional rules and laws of victory, bringing qualitative changes to military organizational forms, unprecedentedly improving operational efficiency, and transforming the combat effectiveness generation mechanism (Figure 3.4).

3.4.2 *The Essence and Dominant Role of Intelligence*

Military intelligence is primarily concerned with the development of decision-making-based optimization algorithms for warfare systems, including:

- Equipment systems: equipment for single and swarm units, collaborative manned and unmanned operations, and multi- and cross-domain operations.
- Combat force systems: individual soldiers, teams, detachments, combined combat units, and theater joint command.
- Operational stages: networked perception; mission planning, command, and control; combat force coordination; and integrated support.

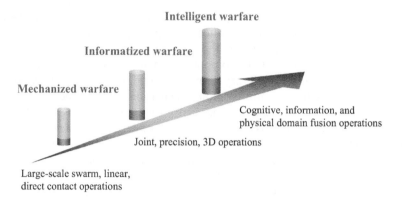

Figure 3.4 The evolution of warfare patterns.

- Sophisticated systems: cyber attack and defense, electronic confrontation, management of public opinion, and infrastructure management.
- Military capabilities: intelligent design, R&D, manufacturing, mobilization, and support.

AI will be integrated into all of the systems, levels, and links via chips, algorithms, and software, forming a systematic brain that will constantly accompany combat vehicles and missions in intelligent warfare. This systematic brain will be built on a distributed cloud with numerous databases and models, knowledge, algorithm, and tactics bases. Whether in peacetime or wartime, establishing systematic brains, distributed clouds, and associated databases requires the assistance of intelligent networks, which also contribute to the collection of multi-disciplinary battlefield data, efficient command and control, and combat force cooperation. Combat vehicles will be distributed, networked, and swarmed, and will be connected to combatants. As a result of intelligentization, new operational elements such as brain systems, distributed clouds, cognitive networks, combined virtual and physical ends, and joint forces will reconfigure the battlefield ecosystem, fundamentally altering the rules and laws of victory (see Chapter 5).

The AI system based on models and algorithms is the critical operational capability that spans all facets and connections. For instance, combat vehicles can be controlled by AI, the swarm can be guided by AI, and decisions can be made with the assistance of AI. At some point, AI models and algorithms will supplant traditional human-based strategies and take over the OSoS and its processes. In this sense, algorithm warfare will be a deciding factor in future conflicts, and intelligence will become the primary source of contention.

Mechanized warfare is vehicle-centric and featured by "dynamics," specifically firepower and mobility, in which combat effectiveness is carried by combat vehicles. The operational elements include soldiers, mechanized equipment, and tactics. In this context, the rules and laws of victory, based on mechanized equipment and human-led decision-making, are represented as overcoming less with more, weaker with more robust, and slower with faster, demonstrating a comprehensive, efficient, and sustainable mobilization capability.

Information warfare is network-centric warfare, with the central component being "interconnection," or information power, in which combat effectiveness is gathered and distributed via networks. The operational elements include data, soldiers, networked equipment, and tactics. With information permeating soldiers, equipment, and tactics via "seamless and rapid information transmission from sensors to soldiers," i.e., systematic and networked combat capabilities, the rules and laws of victory are represented as conquering the local through the systematic, the discrete through the networked, and the slow through the fast. Information serves as a multiplier and aggregator for equipment and OSoS in information warfare. However, despite being aided by information, combat vehicles are still manned, and most decisions are still made by humans.

Intelligent warfare is cognition-based, in which "calculation" takes precedence over firepower, mobility, and information power, and in which intelligence is harnessed and maximized for combat effectiveness. In this context, the rules and laws of victory will be defined as overcoming the physical with the virtual, the inferior with the superior, and gaining battlefield initiative through the deployment of AIs that outnumber or outperform adversaries. The operational elements, with AI being the most active and critical, include clouds, networks, swarms, and terminals. As AI evolves and improves, it will gradually supplant and replace human and traditional tactics. As a result, rapid recognition of massive amounts of data on the battlefield will be enabled, adaptive mission planning and decision-making processes will be accelerated, and the observe–orient–decide–act (OODA) loop will be shorter, resulting in increasingly visible unmanned, swarmed, and systematic effects. Meanwhile, military clouds, networks, and equipment will become increasingly intelligent as a result of embedded AI.

With an increasingly decisive role in warfare, AI will eventually play a decisive and dominant role in the future, i.e., the side with superior AI will gain the battlefield advantage.

Emphasizing AI's dominant role does not negate humans' role in war. On the one hand, AI is endowed with human-like intelligence. On the other hand, AI will never completely replace humans in pre-war planning, behind-the-scenes commands, and strategic planning for an extended period of time or in the near future.

3.4.3 Kill Chain Acceleration

To command modern warfare in an increasingly complex combat environment and at an increasingly rapid rate of confrontation, to rapidly identify and process massive amounts of information, to rapidly respond to battlefield conditions, and to rapidly develop a decision plan all exceed human capabilities and the limits of currently available technical means. As AI's application and role in warfare expand, it will reshape operational processes and increase the efficiency of the military kill chain. Rapid perception, decision-making, action, and support will all become critical components of victory in intelligent warfare.

In the future, AI systems will be capable of rapidly and accurately identifying complex battlefield information and targets via images, videos, electromagnetic spectrum, and voice from an air–land–sea integrated sensor network. Not only can AI achieve precise strike targeting, but it can also accurately model human behavior, social activities, military operations, and public opinion dynamics, allowing for early-warning predictions. Through extensive parallel modeling and simulation training in virtual environments, adaptive mission planning, autonomous decision-making, and command and control will be accomplished in battlefield commands. Each combat vehicle and swarm system's AI will be capable of autonomously and collaboratively performing tasks in accordance with adaptive mission planning and operational objectives, as well as adapting to unexpected changes. Additionally, establishing a distributed, networked, intelligent, multi-modal logistics system will enable armed forces to quickly implement precise logistics distribution, equipment and material supply, and intelligent maintenance. In short, increasingly used intelligent technologies, as well as AI systems with autonomous and evolutionary capabilities, will cover the entirety of the war process, including planning, prediction, perception, decision-making, command and control, and support, completing "simple, fast, efficient, and controllable" processes, accelerating the pace and duration of warfare, and gradually liberating humans from operational complexities.

The rapidity with which war is initiated has always been a primary objective of war initiators and the primary yardstick by which intelligent warfare is evaluated. In contrast to mechanized warfare, the term "rapidity" in intelligent warfare refers to the rate of information and energy flow, the rate of the OODA loop, and the rate of kill chain. The advantages of intelligentization are reflected in its rapidity.

In June 2016, Alpha, an AI pilot developed at the University of Cincinnati using fuzzy logic algorithms, shot down several fighters piloted by US Air Force tactical experts in simulated air battles after learning from numerous encounters with itself and adversaries, demonstrating AI's advantages over humans in terms of massive knowledge accumulation, rapid calculation, tirelessness, and accurate repetition of instructions. Due to its ability to change air combat tactics more than 250 times faster than humans, Alpha can even outperform exceptional human pilots. The AI responds to a situation in a millisecond, whereas human pilots typically take 400–500 milliseconds, with the best taking 250 milliseconds.

Near the end of the Libyan war, a US reconnaissance satellite spotted a suspected Gaddafi convoy exiting Sirte and transmitted images to a US airbase in Nevada, 9,600 kilometers away, where a five-man time-sensitive team immediately determined it was Gaddafi's convoy and informed North Atlantic Treaty Organization (NATO) command via satellite communications to mobilize the nearest Predator UAVs for a strike. Due to the UAVs' limited supply of munitions, French Rafale fighters were also deployed to assist in the attack. When Gaddafi climbed out of the damaged vehicle and attempted to flee, the US military command directed the coordinator of US Special Forces to inform opposition forces to capture Gaddafi. Following his capture, Gaddafi was assassinated in a state of confusion. Between the time Gaddafi was discovered and the time he was attacked, it took less than a half hour, and slightly more than an hour in total from the time he was later captured and killed. The operation amply demonstrated the US military's "find-and-destroy" strategy, as well as the importance and benefits of rapid response in the OSoS.

Time serves as the war's metric and is critical in determining whether the war is won or lost. While modern warfare is defined by an increasing variety of mobile and assault vehicles and a high rate of fire, battlefield perception technology enables rapid detection, signaling, and command delivery via a global satellite network. Text commands are delivered in less than a minute via light-speed radio transmissions, while voice commands are delivered in near real time. On the other hand, manual skills are still required to identify battlefield environments and targets using massive amounts of reconnaissance imagery and intelligence data, to comprehend the battlefield situation, and to make operational decisions. Processing satellite images takes at least several hours and up to several days. Instant result comparison is enabled by intelligent image recognition by a computer based on prior machine learning. While computers can identify targets and perceive the battlefield, the equipment and means of attack and defense, as well as the corresponding operational scenarios, are still largely determined by humans. While numerous emergency plans can be developed in advance of a battle, the time window for responding at the battle level to an incoming missile attack is only approximately 10 minutes or less. Measures taken against the battlefield's instantaneous dynamics, if based on ad hoc human judgments, will result in suboptimal outcomes or missed war opportunities. By leveraging AI, amassing a massive amount of prior red–blue confrontation experience, reasoning, and knowledge, and summarizing numerous ad hoc strategies and solutions for rapid ranking and selection, it will be possible to achieve highly effective responses to battlefield changes.

Modern warfare compresses mobilization time and OODA loop duration, with increasing time sensitivity, as illustrated in Tables 3.1 and 3.2. It is conceivable that the war duration will be reduced from months to weeks in the case of a medium-sized local war, weeks to days in the case of a typical joint campaign, and days to hours in the case of a single complex operation. In the future, AI will be used to calculate and control processes and durations, while unmanned forces will execute and complete specific tasks.

Table 3.1 Typical observe–orient–decide–act (OODA) loop duration for joint operations

Previous wars	OODA loop duration
Gulf War	4,230 minutes (almost 3 days)
Kosovo War	120 minutes
War in Afghanistan	19 minutes
Iraq War	Within 10 minutes
War in Libya	5 minutes
War in Syria	Near real time

Table 3.2 Preparation time for war mobilization and typical military systems

Category	Time
Mobilization and preparation for modern war	1–5 months
Campaign	5–18 days
Special operation	40–60 minutes
Air defense missile on standby	5–10 minutes
Situational awareness update cycle	3–5 minutes
Command delivery	1 minute
Internet delay	50–100 ms
Navigation timing accuracy	10 ns

3.4.4 Higher-Order Feature Reflected

In AI-led intelligent warfare, combat effectiveness, means, strategies, and measures all incorporate human intelligence, exceed human limits, and maximize machine advantages, indicating a progressive, disruptive, and innovative nature. This nature, which is a higher-order characteristic, is the result of a qualitative leap, not of a simple extension or quantitative growth of previous military technologies. This characteristic is reflected technically in the application of "brain-like" functions and the capacity to "exceed human limits" in a variety of fields, both of which are absent from traditional forms of warfare. The neural network is a byproduct of brain-like technology that enables the extraction of similar target features from massive amounts of data at speeds far faster than humans can. Swarm systems are also a natural outcome of bionic and intelligent technologies, as they are capable of self-coordinated targeting. It is critical to recognize that advanced AIs with learning and evolutionary capabilities based on data, models, and algorithms in networked sensing systems and distributed military clouds will eventually surpass individual soldiers, staff officers, commanders, and even think tanks. They will become "super brains" or a part of a "super brain swarm." The underlying concept of higher intelligent warfare is that the best AIs will evolve into invincible intelligent systems, intelligent robots, or "terminators," triggering revolutions in epistemology and methodology and serving as the most advanced combat capabilities currently foreseeable, achievable, and evolvable for humanity.

3.5 The Rising Role of Virtual Space

Battlespace will shift from physical to virtual. In future OSoS, virtual space will become more integrated with physical space. Virtual space is a networked and electromagnetic-based information space created by humans. It has the potential to reflect various aspects of the human social and physical world. It is applicable outside the objective world because it is constructed by the information domain, connected by the physical domain, reflects the social domain, and is utilized by the cognitive domain. In a narrow sense, virtual space refers to the civilian Internet. In a broader sense, it refers to cyber-space, which encompasses the Internet, the IoT, military networks, and private networks. Cyberspace is defined by the ease with which it can be attacked while being difficult to defend, by the integration of peacetime and wartime statuses, as well as civil and military uses. It has grown into a major theater of military operations, strategic deterrence, and cognitive conflict.

Virtual space demonstrates its military utility in three ways in the age of AI. To begin, it is the fundamental space for intelligent warfare, where the military integrates disparate combat forces and operational elements through network information systems, resulting in the formation of informational, networked, and systematic combat capabilities. Second, it is the primary location of cognitive clashes between cyberspace, intelligence, public opinion, human minds, and consciousness. Third, it is critical for establishing virtual battlefields, conducting combat experiments, achieving virtual–real interaction, and developing the capability to conduct parallel operations and control real entities by virtual entities.

3.5.1 Ubiquitous Interconnection

Rapid adoption of networked reconnaissance from space, GPS, mobile connectivity, Wi-Fi, high-precision global space–time benchmark platforms, digital maps, and big data systems will accelerate the emergence of an IoT-based world. In this context, human society and military activities will become more "transparent," which means they will become more networked, perceived, analyzed, correlated, and controlled. In the age of AI, military operations will have a far-reaching, profound, and pervasive impact. For example, the OSoS will gradually evolve from a closed to an open system, and from a military-oriented to an "open source and universal" system capable of integrating civil and military functions.

When information and data freely flow between battlefields, operational elements will become deeply connected, and OSoS will evolve from possessing traditional combat capabilities to advanced information fusion, data cross-chaining, and interactive integration, enabling full-area perception, multi-domain fusion, and cross-domain combat capabilities. These capabilities will enable effective control of critical targets, populations, and infrastructures at all times and in all locations, or more precisely, the ability to perceive, control, act, and evaluate data at all times and in all locations. The ubiquitous interconnection will be developed to a degree, precision, and quality that may differ from what was originally imagined.

A city the size of a provincial capital contains millions of optical or infrared probes distributed throughout streets, neighborhoods, warehouses, supermarkets, banks, hotels, government agencies, businesses, and other institutions, as well as a variety of increasingly common and advanced security, monitoring, surveillance, and control systems. Once networked or intelligent, these devices can form a multi-source heterogeneous and interconnected super network based on standardized data transmission protocols and data formats. This network will enable intelligent and ubiquitous urban combat and military operations other than war.

According to the US Army Research Laboratory, resilient tactical networks of the future will be self-organizing and self-recovering, enabling real-time communications across multiple environments in response to multiple cyber attacks, electronic warfare/jamming, and node disruption/failure encountered by military networks, sensors, radios, and robots, among other threats. Advances in AI and machine learning will enable efficient network resource retrieval, characterization, and allocation; and satellite networks will incorporate 5G, 6G, and 7G network standards, enabling ubiquitous and continuous networking. Based on new network encryption, authentication, and verification protocols, quantum technology will enable distributed computing and resource allocation. LandWarNet 2050 will be an intuitive, self-organizing, and self-resilient joint network that provides resilient redundant connectivity for soldiers operating in a variety of environments, ensuring uninterrupted information transmission anywhere and at any time.

Eagle Eye, a 2008 Hollywood blockbuster film, is about a super AI robot who is assisted by a variety of networks and technologies. The AI is possible to perform ubiquitous, omnipotent, and precise target tracking, device control, mission assignment, and guidance by combining data from multiple sources such as images, videos, audio, and positions. In some ways, the film reflects the intelligent and ubiquitous nature of future military operations.

3.5.2 Cyber Battlefield

The military application of cyber technology begins with intelligence gathering. During this process, the military will utilize civilian resources. For instance, the development of smart cities expands intelligence-gathering opportunities. Urban emergency service data can be used to track criminal activities, military unrest, and fire strikes, among other things, thereby limiting the reconnaissance range of combatants and allowing them to operate with lighter intelligence, surveillance, and reconnaissance (ISR) equipment. Previously used to deter crime, police cameras can now be used for military surveillance. Real-time traffic data enable the military to better direct the movements of millions of civilians. The Department of Motor Vehicles database contains personal and contact information for civilian vehicle owners, all of which will aid the military in wresting control of the city from the civilians. Additionally, data from hydroelectric grids can provide critical resource information and potential pivot points for strategy implementation to any combatant.

Private data sources ranging from civilian mobile phone cameras to commercial security cameras can be used for intelligence gathering. Social media and traditional media outlets will be able to distribute live coverage of any event within minutes through the use of UAVs, closed-circuit television (CCTV), and satellite network data. By 2030, increased use of commercial UAVs, cameras, 5G/6G networks, and quantum computers will enable real-time tracking of virtually any activity occurring in public space.

The developmental trend of air, space, ground, and sea detection technologies, as well as various civilian monitoring systems, indicates that the world is entering an era of ubiquitous global monitoring. Even if it is impossible to monitor every global activity, the proliferation of technology will undoubtedly result in an exponential increase in the number of potential sources of information.

The second embodiment is cyber attack and defense. The US, Russia, and countries throughout Europe, Northeast Asia, Southeast Asia, South Asia, and the Middle East are currently developing cyber attack and defense systems capable of invading, deceiving, disrupting, and destroying networks. The Edward Snowden revelations revealed 49 US government-led "Cyberspace" reconnaissance projects across 11 categories, including the use of the "Stuxnet" virus to sabotage Iranian nuclear facilities, the "Gauss" virus to invade Middle Eastern countries on a large scale, and the "Cuban Twitter" to manipulate public opinion. These incidents demonstrate the US's robust surveillance capabilities and ability to conduct both soft and hard attacks on the Internet, private networks, and wireless mobile networks. On a global scale, advanced persistent threat (APT) hacking groups, including "Lazarus" in Northeast Asia, "OceanLotus" in Southeast Asia, "SideWinder" in South Asia, "APT34" in the Middle East, "APT28" in Eastern Europe and Central Asia, and "Equation" and "Vault 7" in Europe and America, have carried out a series of large-scale cyberattacks since 2013, resulting in severe consequences and significant losses.

Third, psychological warfare is carried out with the assistance of cyber technology. For example, during the Libyan war, the US military influenced local public opinion by allowing Wikileaks and local embassy personnel to expose the Gaddafi family's vast overseas assets, resulting in a dramatic shift in public opinion, widespread social unrest, and the rapid establishment of anti-government forces that eventually overthrew the Gaddafi regime and killed Gaddafi. During the Tunisian Revolution, the US military incited anti-government sentiments within the country by exposing President Ben Ali's corrupt and extravagant behavior, precipitating the Ben Ali government's rapid collapse and his flight abroad. During the Iraq War, the US military assumed command and control of the Iraqi army. Cyber deception directed at thousands of Iraqi military officers dismantled Saddam Hussein's million-strong army and prevented it from mounting a large-scale resistance, resulting in the Iraqi military's defeat and the regime's rapid collapse. All these forms of psychological warfare served as a counterbalance to conventional warfare.

3.5.3 *Virtual Practices*

Modern warfare begins with experiments in virtual reality. The US military has been investigating combat simulation, combat experiments, and simulated training since the 1980s. Later, it pioneered the use of virtual reality, wargames, digital twins, and other technologies for the creation of virtual battlefields and conducting combat experiments. According to analysis, the US military conducted combat simulations and rehearsals prior to the Gulf War, Kosovo War, Afghanistan War, and Iraq War, as well as a recent series of decapitation strikes, in order to determine the optimal operation solution. Similarly, Russia's incursion into Syria was preceded by operational drills in war laboratories. The findings shaped the strategic exercise plan for "Centre 2015," which included drills for Russian forces' "mobility and accessibility in unfamiliar areas" [19]. Following the exercise, Russian Army Chief of General Staff Valery Vasilyevich Gerasimov emphasized that Syria's political and strategic objectives would be achieved through political and economic means, public opinion warfare, and psychological warfare, supplemented by long-range precision airstrikes and special operations. Russian operations in Syria have been shown to be consistent with their tests and exercises.

In the future, with the advancement and application of virtual simulation, mixed reality, big data, and intelligent software, it will be possible to create a parallel military system that includes virtual forces that are mapped and iterated with physical forces, capable of conducting rapid, high-intensity confrontations and massive computing outside the confines of physical space, as well as acting as a highly simulated "blue army" to confront a physical one. The system will use accumulated data to develop the optimal model, algorithm, or solution for guiding the construction and operation of physical forces, realizing interaction between the physical and virtual worlds, dominating the physical with the virtual, and defeating the physical with the virtual.

On January 25, 2019, DeepMind, a division of Google dedicated to AI, and Blizzard, the creators of the game StarCraft, announced the results of a match between the virtual system AlphaSTAR and professional players TLO and MANA. AlphaSTAR defeated both players in a five-game, three-win format by a score of 5:0. In just 2 weeks, the machine had completed the training equivalent to 200 years for a human player. This event demonstrates the enormous benefits and potential of virtual encounters.

3.6 Operational Patterns Dominated by Unmanned Vehicles

Unmanned equipment is the cutting-edge embodiment of human wisdom in the OSoS, as well as the fusion of intelligentization, informatization, and mechanization. In the age of AI, unmanned operations will become the primary form of operations, gradually moving to an advanced stage with the integration of AI and various technologies.

3.6.1 Introduction to Unmanned Systems

Unmanned equipment began with UAVs. In 1917, the United Kingdom developed the world's first UAV, but it was never used in combat. UAVs have gradually gained use in reconnaissance and strikes as technology advances. Since the turn of the twenty-first century, unmanned equipment has rapidly expanded its range of applications, covering the ground, water surface, underwater, air, and space, owing to its mission-centric nature, lack of crew burden, and high operational cost–efficiency ratio.

Unmanned combat vehicles have now achieved remote control, formation flight, and swarm cooperation through the use of satellite and autonomous navigation functions. UUVs and autonomous space robots have all been introduced within a short period of time. Intelligent bipedal, quadrupedal, multipedal, and cloud-based robots are also making rapid progress and will soon be deployed in military applications.

3.6.2 Three Stages of Unmanned Operations

Unmanned operations will enter three stages.

The first stage is primary, with unmanned equipment assisting human-led unmanned operations. At this stage, humans completely control and dominate pre-war, mid-war, and post-war operations.

The second stage is intermediate, consisting of unmanned operations with limited control, or operations led by unmanned equipment with human assistance. In this phase, unmanned vehicles operate autonomously in most cases, with some instances requiring limited, auxiliary, but critical human control.

The third stage is more advanced, with unmanned (AI-led) operations guided by human-designed rules. Humans design behaviors and rules for unmanned equipment in advance for a variety of operational environments and conditions and then delegate operations entirely to AI-driven unmanned vehicles or troops during this phase.

The three stages of unmanned operations are mutually reinforcing and iterative, with a parallel rather than linear relationship between them.

3.6.3 Trends in Autonomous Behaviors

Autonomy, or autonomous behavior, is at the heart of unmanned warfare and a defining characteristic of intelligent warfare, manifesting itself in a variety of ways, including but not limited to the following.

The first is UAVs, UGVs, precision-guided weapons, and underwater and space robots.

The second is the autonomy of detection systems, which includes automatic search, tracking, association, aiming, and intelligent recognition of image, voice, video, and signals.

The third is the autonomy of decision-making, as demonstrated by AI-based autonomous decision-making in the OSoS, which includes capabilities such as automatic battlefield analysis, adaptive task planning, automated command and control, and intelligent human–computer interaction.

The fourth is autonomy in operation cooperation, which includes both autonomous cooperation between manned and unmanned systems and autonomous swarms of unmanned vehicles such as bee swarms, ant colonies, fish schools, and other types of later-stage combat formations.

The fifth is autonomy in the use of open-source network resources, which includes automatic handling of big data (e.g., automatic access to websites and information systems, data collection, and associated calculations); automatic identification, tracking, and modeling of critical physical targets; and automatic tracking, analysis, and calculation of popular events and public opinion.

The sixth is autonomy in cyber attack and defense, which includes automatic virus and cyber attack detection, tracking, protection, and counterattack.

The seventh is the autonomy of cognitive and electronic confrontations, which includes automatic identification of the power, frequency band, and direction of electronic jamming against the adversary, as well as automated frequency hopping, networking, and electronic jamming against the adversary.

The eighth is the full autonomy of unmanned vehicles, which encompasses intelligent diagnosis, self-repair, and self-support. It is reasonable to assume that this level of autonomy will pervade all future intelligence warfare operations.

3.6.4 The Talisman of Humanity

Unmanned vehicles will advance toward autonomy, bionics, swarming, and distributed cooperation in the future as AI and various technologies are integrated, significantly reducing the confrontation of living forces on the battlefield. While manned vehicles will continue to exist in the future, in the age of AI, unmanned operations led by bionic robots, humanoids, swarm weapons, and mechanical forces will become the norm. Simultaneously, unmanned AI technology will gradually expand into a variety of fields, including cyber attack and defense, electronic confrontation, multi-source perception, correlation verification, target tracking, and public opinion analysis. Unmanned systems will eventually take the place of humans in a variety of operational areas and will shield humans from physical attack and damage. Thus, unmanned OSoS will act as the primary line of defense for humans, acting as the talisman and shield of humanity.

3.7 All-Domain Operations and Cross-Domain Attack and Defense

In the age of AI, all-domain operations and cross-domain attack and defense are fundamental operational patterns, with geographical and temporal spans, encompassing land, sea, air, and space; physical, information, cognitive, social, and biological domains; virtual and physical space; strategic deterrence in peacetime; and high confrontations, dynamics, and responses in wartime. The armed forces will be able to conduct unmanned vehicle-dominated operations in physical space as well as cognitive confrontations in virtual space, including cyber attack and defense, information warfare, public opinion control, and psychological warfare. It becomes critical to maintain control over critical infrastructures such as networks, communications, electricity, transportation, finance, and logistics. In addition, global security governance, regional security cooperation, counterterrorism, and rescue all take on new significance.

3.7.1 The Concept of Innovative Operations

Since 2010, the US military has been developing concepts such as combat cloud, distributed kill, multi-domain operations, and joint all-domain operations with the goal of achieving battlefield superiority through the use of a systematic force over an adversary's local force, a multi-functioned system over a single-function system, multi-domain capabilities over single-domain capabilities, integrated formations over discrete formations, and intelligence over intelligence. In response to the F-22, F-15, F-16, and F-35 fighters' inability to share battlefield data, the US Air Force has proposed the concept of operational cloud, which incorporates cloud computing into joint operations to enable a broader range of "conversational" capabilities between the vehicles, as well as the interconnectivity and interoperability of tactical information between the services. In 2014, the US Army deployed the DCGS-A Block 3 to Afghanistan, marking the Army's first deployment of a dedicated tactical cloud computing system. In 2016, the US military introduced the concept of multi-domain operations. In 2020, they proposed the concept of joint all-domain operations characterized by forces with intelligent cross-service and cross-domain joint operational capabilities, with single-service operations supported by all three services, achieving intelligence supremacy as well as all-domain capabilities superior to multi-domain or single-domain capabilities.

Multi-domain operations and cross-domain attack and defense will become a distinguishing characteristic of future intelligent warfare based on AI and human–computer hybrid intelligence. They will take place between functional domains such as physical, information, cognitive, social, and biological, as well as between geographical domains such as land, sea, air, and space.

3.7.2 From Joint to Integrated

In the intelligent era, joint all-domain and cross-domain operations will expand from manual, physical, and discrete to heterogeneously integrated ones with data cross-chaining, mutual tactical control, and cross-domain attack and defense integration. In such complex environments, operational demands of different categories, levels, and stages will arise, including cross-domain multi-source perception; heterogeneous information fusion; operational data cross-chaining; cross-domain joint strike and defense; multi-domain integrated logistics; mutual deployment, command and control of forces from different services; multi-domain operational planning and correlation; and interoperability of weapons and equipment. These demands will be met through the deep integration of AI and related technologies.

First, all-domain integration. This kind of integration, based on different battlefields, opponents, and joint operations in all domains, according to the requirement of joint operations, integrates: (1) operational patterns, processes, and missions; (2) data, firepower, defense, logistics, command, and control; and (3) strategic, campaign and tactical combat capabilities at all levels, forming the ability to provide rapid support from multiple domains for single-domain operations.

Second, cross-domain attack and defense. Under the support of a unified network information system, a unified battlefield perception system, and a unified data transmission standard, cross-domain attack and defense features data links opened between reconnaissance, control, confrontation, strike, evaluation, and support in cross-domain joint operations, realizing unified tactics and fire control, as well as interoperability between military services from different domains, and the seamless integration of operational elements and capabilities.

Third, the linkage of the whole operation process: the holistic design for all-domain operations and cross-domain attack and defense. Examples include pre-war intelligence gathering and analysis, public opinion warfare, psychological warfare, political warfare, and necessary cyber attacks; in-war decapitation strikes, point-breaking attacks, and precision controllable strikes through special operations and cross-domain operations; post-war defense against cyber attacks, including eliminating negative public opinion, preventing infrastructure damage, regional governance, and social restoration.

Fourth, AI support. It forms an intelligent brain system with different operational models and algorithms through combat experiments, simulation training, and real-combat testing, supporting joint operations, multi-domain operations, and cross-domain attack and defense.

3.8 Human–AI Hybrid Decision-Making

On the battlefield of intelligent warfare, the ever-improving AI brain system will surpass humans in a variety of fields and alter the human-dominated command, control, and decision-making model that has defined human warfare for thousands of years. Human command of AI, AI command of humans, and AI command of AI are all possible in future warfare.

3.8.1 Decision-Making Revolution

Intelligent OSoS are distributed, networked, flattened, and parallelized. The human-led, centralized, and single-decision-making model will gradually give way to a decentralized, unmanned, and autonomous model powered by AI, or to a weakly centralized model based on manned–unmanned cooperation. Military decision-making systems will increasingly incorporate a combination of decentralized, weakly centralized, and centralized decision-making. Unmanned and decentralized decision-making is critical for simple tasks with low operational complexity, while human-led and centralized decision-making is critical for complex tasks with high operational complexity. In general, pre-war decision-making will remain dominated by humans and aided by AI. In-war decision-making will be dominated by AI and aided by humans. Post-war decision-making will be dominated by cooperation between humans and AI (Table 3.3).

3.8.2 Human Brain + AI

Combat on future battlefields will be extremely complex, dynamic, and intense, generating enormous amounts of data that the human brain will struggle to process quickly and accurately. However, commanders can command and control forces on complex battlefields through the use of the "human brain + AI" model and technologies such as operational clouds, databases, network communications, and the IoT. As unmanned systems become more

Table 3.3 Human–artificial intelligence (AI) hybrid decision-making in the operational process

Pre-war	In-war	Post-war
Object detection and cataloging	Front-end intelligent identification	Defense and counterterrorism
Pre-war mission planning and simulation	Adaptive task execution	Early warning + adaptive
AI-aided decision-making	AI autonomous decision-making	Hybrid decision
Human-dominated	Human-aided	Human–AI hybrid

AI-enabled, autonomous decision-making will gradually emerge. When the command-and-control system becomes intelligent, the duration of the OODA loop will be significantly reduced, while command efficiency will be significantly increased. Pattern recognition algorithms in image processing, optimal solution-seeking algorithms in operational decision-making, particle swarm algorithms, and bee swarm algorithms for autonomous swarms will all contribute to command-and-control systems, gradually achieving the concept of "human outside the loop."

3.9 Non-Linear Amplification and Rapid Convergence

In the future, intelligent warfare will not rely on gradual energy release or linear superposition of operational effects, but on the rapid amplification of non-linear, emergence, self-growing, self-focusing, and rapidly convergent war outcomes.

3.9.1 Emergence Effects

The term "emergence" refers to the process by which each individual in a complex system interacts with its neighbors in a self-organizing manner, resulting in qualitative changes in the system as a whole. In the age of AI, events that spread widely and instantaneously will have a profound effect on a large number of interconnected users and objects, as will tasks that have been subjected to extensive simulations and machine learning in advance, as well as cases that have been carefully designed and verified in advance.

Regardless of its complexity and diversity, the information on future battlefields can be "collected in one location and shared by all sides" after being processed as images, voice, and video by intelligent recognition and military cloud systems. Relevant data can be quickly correlated using big data technology, revealing specific targets for weapon and fire control systems capable of conducting distributed strikes, swarm attacks, and cyber-psychological warfare, achieving "destruction upon detection," "rapid and comprehensive strikes against any target," "psychological panic effect among adversaries due to numerical superiority," and other emergence effect outcomes. Intelligent warfare will have three emergence effects: (1) the acceleration of the kill chain caused by rapid AI decision-making; (2) the operational effect triggered by manned–unmanned collaborative swarm systems characterized by "wisdom produced by a swarm of fools"; and (3) the rapid emergence of swarm behaviors based on interconnectivity and interoperability. Following the discussion earlier in this chapter on the emergence effects of kill chain acceleration, interconnectivity, and interoperability, the following section will discuss the emergence effect of swarm operations.

In November 2017, during a United Nations conference on weapons conventions in Geneva, a video was released exposing one of the most terrifying weapons in human history – robot killers.

The robot killer, according to media reports, is a small and intelligent UAV about the size of a bee with rapid mobility and a processor 100 times faster than a human brain. Despite its diminutive size, the bee possesses all of the necessary components. With a wide-angle camera, sensors, facial recognition, and other black technology embedded throughout its body, the bee-like robot killer can locate a target with the precision of a scalpel, even if the target figure is disguised or wearing a mask, with a recognition rate of up to 99.99%. Each killer robot is armed with 3 grams of concentrated explosives capable of rapidly destroying a target's brain with a single suicide hit. When released, a million swarm of killer bees costing just $25 has the potential to annihilate half of a city's population, spreading panic and misery wherever they go, similar to African locust swarms. Additionally, if the swarm is provided with an adversary's facial information, it can launch targeted strikes in response.

3.9.2 Operational Focus

After military intelligence reaches a new stage as a result of advanced AI, quantum computing, IPv6, hypersonic, and other technologies, the OSoS will produce non-linear, asymmetric, self-growing, rapid, uncontrollable, and amplified operational effects, most notably in unmanned operations, swarmed operations, cyber opinion control, and cognitive confrontations. The "wisdom generated by a swarm of fools," efficiency increased with quantity, non-linear amplification, and emergence effects will all become more prominent. AI-driven cognitive, information, and energy confrontations will be entwined, with shorter combat durations, increased confrontation speed, sharper effect amplification, and faster convergence of results. Energy shockwaves, lightning-fast warfare, AI terminators, opinion reversals, social unrest, psychological panics, and IoT chain effects all define intelligent warfare.

In unmanned swarm operations in which both warring parties' vehicles perform similarly, numerical superiority translates directly into qualitative superiority, because operational effectiveness is proportional to the square of the quantity, according to the Lanchester equation. The non-linear emergence effects will be more pronounced in cyber attack and defense, psychological warfare, and public opinion, because the effects will be proportional to the square of the number of connected users, as Metcalfe's Law dictates. The number and intelligence level of AIs on the battlefield will determine the system's overall intelligence level and who controls intelligence power, which will have an effect on the war's outcome. The major new challenges posed by future warfare in the age of AI are how to deal with the relationship between energy, information, and cognition, quantity and quality, the virtual and the real, and how to design, control, apply, and evaluate non-linear effects.

In the future, public opinion reversal, psychological warfare aimed at inducing widespread panic, swarm operations, and autonomous unmanned operations involving "humans outside the loop" will all result in widespread emergence effects. These acts are compatible with deterrence and combat objectives, and human society must closely monitor and regulate them.

3.10 Organic and Symbiotic Human–Equipment Relationships

In the age of AI, the relationship between humans and weapons will fundamentally shift, becoming more remote in physical terms and more intimate in mental terms. Human intelligence and wisdom will be completely front-loaded during the development phase, integrated for iteration during the testing phase, and fed back for upgrading following operational validation, and so forth.

3.10.1 Qualitative Changes in Equipment Patterns

Rapid advances in network communication, mobile Internet, cloud computing, big data, machine learning, and bionics, in combination with their widespread military application, will alter the structure and form of conventional weaponry. The next generation of weapons and equipment will be built on cyber-physical systems (CPS), which will combine a cloud-based back-end with a multifunctional front-end to enable virtual–real interaction and online/offline capability, effectively transforming the weapon and equipment into AI systems for human–machine interaction. This type of weapon and equipment system will be capable of a variety of tasks, including front- and back-end labor division, efficient interaction, and self-adaptive adjustment. It combines straightforward mechanical operation with sophisticated battlefield cognition and advanced human–machine intelligence interaction through the use of machinery, information, a network, data, and cognition.

3.10.2 Separation of Physicality and Concentration of Mentality

In the age of AI, humans and weapons will be physically separated but deeply integrated into a mental symbiosis. UAVs and robots will gradually transition from assisting humans to performing combat functions, allowing humans to fade into the background. Humans and weapons will be combined in novel ways, as evidenced by the inclusion of human intelligence in the process of weapon design, development, production, training, use, and supply, as well as by unmanned systems that perfectly combine human creativity with machine precision, speed, reliability, and fatigue resistance.

3.10.3 Management Reform

In the future, equipment design and management techniques will undergo dramatic changes. In comparison to mechanical equipment, which deteriorates over time, software advances and intelligent algorithms become more sophisticated over time. After several stages of development, testing, and finalization, the army receives traditional mechanized equipment, and its tactical and technical performance degrades over time and hours. Informatized equipment is the result of the marriage of mechanization and information technology, with the equipment system capable of being updated in response to advancements in CPU and storage device technology. It is characterized by "information hegemony, software-driven hardware, rapid iteration, and spiraling upward progression." Intelligent equipment is made up of automated and information-processing components that constantly optimize AI models and algorithms in response to accumulated data gathered during use, which is enhanced over time and with increased frequency of use. As a result, the model for equipment development, use, and maintenance will fundamentally change in the age of AI.

3.11 Evolution by Self-Learning and Self-Confrontation

In the age of AI, evolution is inevitably a defining feature of the OSoS and a high point of strategic competition. OSoS will gradually become self-adaptive, self-learning, self-confrontational, self-repairing, and self-evolving, becoming an evolvable class of ecological and gaming systems.

3.11.1 Principles of Evolution

The most distinguishing characteristic of intelligent OSoS is its ability to combine human intelligence with machine superiority to achieve super-human combat capabilities. This feature is based on models and algorithms self-evolving. However, whether the evolutionary capability of models and algorithms can result in the mobility and evolvability of the entire OSoS remains an open question. Assume that future OSoS resemble a living organism comparable to the human body, with the brain serving as the command-and-control center, the nervous system serving as the communication network, and the limbs serving as the brain's weaponry, and the whole system being capable of self-adaptation, self-learning, self-confrontation, self-healing, and self-evolution. In this case, such an OSoS can be considered evolvable. A single-system OSoS may resemble a living organism, whereas a multi-system OSoS composed of an ecosystem and a self-confrontation system exhibits greater complexity, competitiveness, sociality, unity, and emergence effects.

According to preliminary analysis, the development and application of combat simulation, virtual reality, digital twins, parallel training, intelligent software, brain-like chips, bionic systems, natural energy harvesting, and new machine learning will enable OSoS to evolve from a single function and single system to a multi-function, multi-element, multi-domain, and multi-system in the future. Each OSoS will rapidly develop strategies and act in response to changing conditions on the battlefield, the diverse threats and adversaries they face, and their own strengths, utilizing models and algorithms informed by prior experience and knowledge, as well as extensive adversarial simulation training. Additionally, during combat, these systems will be constantly optimized, refined, and evolved. Single-task systems will exhibit biological characteristics. By contrast, multi-task systems, similar to species communities in a forest, will exhibit cyclical functions and evolutionary mechanisms such as self-competition, which will drive them to evolve into ecological and gaming systems under complex environmental conditions.

3.11.2 Pathways of Evolution

There are four distinct pathways depicted. To begin, let us examine the evolution of AI. AI capabilities will inevitably be optimized and upgraded as data and experience accumulate. Second, combat vehicles and swarm systems will transition from piloted operation to partial or complete autonomy. This type of evolution involves both the evolution of AIs that control vehicles and swarms and the refinement of the mechanical and information systems that support them. Third, surveillance, attack, defense, and security mission systems will evolve and become more complex as a result of the involvement of multiple vehicles and missions. Fourth, we will observe the evolution of OSoS, which will inevitably be the most complex as it encompasses all operational elements, tasks, domains, and levels of conflict. OSoS evolution is not a built-in property of the system; it requires a carefully designed environment and conditions that adhere to bionics, survival of the fittest, mutualism, and whole-system and whole-life management principles (see Chapter 7).

3.12 Intelligent Design and Manufacturing

In the age of AI, the national defense industry will transition from a closed, physical-based, long-term R&D model to an open-source model based on intelligent design and manufacturing, with rapid response to military demands.

3.12.1 Challenges to the Tradition

The national defense industry is a strategic sector and a pillar of national security. During peacetime, it supplies the army with advanced equipment of superior quality and reasonable cost. In times of war, it provides operational support that ensures victory. The R&D of modern weapons and equipment is technology-intensive, knowledge-intensive, procedurally complex, and exhaustive. For example, developing large aircraft carriers, fighter jets, ballistic missiles, satellite systems, and tanks requires significant investment, lengthy lead times, and prohibitively high costs. The complex equipment typically takes 10, 20, or more years to complete and deliver. Following World War II, and influenced by the mechanization era, the global defense industry focused on weapons, ships, aviation, aerospace, nuclear, and electronics industries, as well as civilian support industries. Since the Cold War, the US defense industry has undergone strategic reorganization, mergers, and acquisitions,

Figure 3.5 Rotors on display at the 2021 China Zhuhai Airshow, products of intelligent manufacturing.

resulting in a structure and layout that are compatible with automated system confrontation. The top six military behemoths in the United States are all cross-service, cross-discipline system integrators, providing combat vehicles and systems as well as joint warfare solutions. Since the turn of the twenty-first century, as digital, networked, and intelligent manufacturing technologies advanced, the development model, R&D capabilities, and manufacturing capacity of traditional weapons and equipment have been reshaped and adjusted to meet the growing demand for informatized and intelligent combat (Figure 3.5).

3.12.2 New Modes of Combat Effectiveness Generation

In the future, in response to increased demand for joint all-domain operations, as well as the integration of mechanization, information technology, and intelligence, the national defense industry will transition from a single-service to a cross-service and cross-field orientation, and the R&D model will shift from a relatively closed, module-independent, object-oriented model with a lengthy development period to an open-source, crowdfunded model integrating virtual reality. The manufacturing industry will build upon this foundation to create an innovative, intelligent manufacturing system that integrates software and hardware, virtual and physical interaction, intelligent interaction between human, machine, material, and environment, vertically connected industry chains, horizontally distributed collaboration, and civil–military fusion. The fundamental model for developing intelligent OSoS and combat effectiveness entails collaboration between military and civilian stakeholders, as well as iterative optimization using parallel military systems, with military products being researched, tested, used, and built concurrently via simulated training and real-world verification (see Chapter 7, section 7.8, Self-Adaptive Factory).

3.13 Risk of Loss of Control

Although intelligent OSoS are theoretically capable of self-evolution to superhuman status, in the absence of pre-designed programs and nodes for human control, i.e., the "termination button," they may cause destruction and disaster. The possibility that a large number of hackers and warmongers will use intelligent technology to develop uncontrollable warfare programs and operational patterns is a particular source of concern. This enables numerous brain AIs and swarms of robot killers to wage invincible battles according to pre-determined combat rules, ultimately resulting in unrecoverable wreckage, posing a significant challenge for humans engaged in intelligent warfare and a significant unresolved issue. This issue demands attention in light of humanity's shared destiny and the long-term viability of human civilization. To ensure the controlled, reliable, credible, secure, and civilized development of intelligent warfare, measures such as rules of war, international conventions, and corresponding technical, procedural, ethical, and legal standards must be mandatory. Otherwise, as Tesla founder and prominent Silicon Valley figure Elon Musk predicted, it will be a Pandora's box that will be difficult to close.

Musk has repeatedly argued for the regulation of AI. On November 24, 2017, he stated in an interview that "Maybe there's a 5 to 10 percent chance of success [of making AI safe]." On November 27, that same year, he reiterated, "By the time we are reactive in AI regulation, it's too late."

Musk is even more concerned about Boston Dynamics' humanoid robot that runs and jumps. Musk declares, "This is nothing. In a few years, that bot will move so fast you will need a strobe light to see it. Sweet dreams ..." Musk's humanoid robot has evolved to the point where it can perform all human movements perfectly: jumps, spins, backflips, and everything else!

It's frightening to consider that this robot is still evolving on a daily basis and that, eventually, it will move at such a rate that we'll need a strobe light to see it, let alone that no human will be able to keep up with it. "We need to be super careful with AI," Musk later stated, adding that it is "potentially more dangerous than nuclear weapons."

Additionally, in the case of the miniature bee-like robot killers revealed at the United Nations' Geneva conference, if a scientist inserts a line of code instructing the robot to destroy humanity, or if the AI spontaneously mutates into an anti-human species, the entire human race could be wiped out by robots!

Stephen Hawking has repeatedly warned that robots may evolve more rapidly than humans, with unpredictable end results. Humans fear that AIs will eventually supplant humans as the new species [20]!

Even if Elon Musk's predictions do not come true, the dangers and threats posed by AI are enormous. When AI reaches a certain stage of development, humans will be forced to exercise restraint in their use and development of intelligent weapons, just as they did with nuclear, biological, and chemical weapons. However, because AIs are far more prevalent than conventional weapons of mass destruction, enforcing mandatory regulatory measures may prove difficult.

3.14 Innovation in Inheritance

Intelligentization is not an ethereal structure. Without automation and information technology, there would be no network, data, algorithms, or models that constitute intelligence. Alternatively, intelligentization is a significant lever for advancing mechanization and information technology to a new level.

Mechanization serves as limbs, providing the foundation for combat capabilities in physical space such as mobility, firepower, protection, and destruction. Informatization serves as the eyes, ears, and nerves of the organization, providing reconnaissance, detection, communication, command and control, fire control, intelligence, and information interaction and sharing capabilities through the use of general and special networks. Intelligentization serves as the "brain," performing bionic, machine, swarm, and human–computer fusion intelligence functions, as well as intelligent perception, decision-making, confrontation, support, and manufacturing capabilities. At its heart is evolutionary intelligence derived from data, models, and algorithms that is capable of self-adaptation, self-learning, self-coordination, self-confrontation, self-repair, and self-evolution. The three "-zations," namely mechanization, informatization, and intelligentization, form an organic whole that is interconnected, self-reinforcing, and iteratively optimizing. The foundation is now machines, while the driving force is information technology, and the compass is intelligence. In the future, mechanization will remain the foundation, with information technology acting as a support and intelligence acting as the driving force.

3.15 Comprehensive Forms and Characteristics

This chapter has discussed the evolution of intelligent war, the ten dimensions of war form transformation, from section 3.4 AI-Led Rules and Laws of Victory to section 3.13 Risk of Loss of Control, and the relationship between mechanization, informatization, and intelligentization. In summary, intelligent war exhibits six distinct characteristics, as follows.

3.15.1 Shorter Duration and Greater Impact

War engagements will become shorter in duration while their sphere of influence expands in the age of AI. Continuous AI evolution and optimization will result in AI-based kill chains, hypersonic weapons, operational swarms, and public opinion manipulation tools, significantly speeding up operational processes and introducing non-linear and operational emergence effects. War duration will be kept to a minimum, while the battlespace will continue to expand with deeper correlations between new domains such as space, deep sea, polar regions, biology, and most significantly, virtual space that involves information, cognition, and public opinion interactions.

3.15.2 Front-Loaded Military Construction and Operations

Military construction and operation will shift toward systematic, networked, unmanned, swarmed, precise, remote, and integrated approaches in order to meet the requirements of joint all-domain intelligent warfare. Long-term civil–military collaborative research, combat simulation, iterative optimization, and concurrent research, trial, application, and construction of weapons and equipment will introduce new combat effectiveness generation models and algorithms, emphasizing the magnitude of "raising soldiers for a thousand days prior to a campaign," "a minute in war relying on ten years of pre-war preparation," and "accumulation-based improvement."

3.15.3 Virtual–Real Parallel Interactions

Intelligent operational capabilities based on AI (single AI, swarm AI, system AI, human–machine hybrid AI, and other specialized AI) will be facilitated by civil–military collaborative combat simulations, virtual–real interactions, parallel intelligence, self-confrontational and self-evolutionary training, as well as innovations derived from data, model, knowledge, algorithm, and tactics libraries.

3.15.4 Three "Chains" Underpinning War Victory

The three chains (decision, information, and energy) will converge and deepen their integration at all levels of tactics, battles, and strategies and across land, sea, air, space, and cyber domains, forming a rapid chain of "reconnaissance, command and control, anti-jamming, strike, coordination, support, and evaluation" against various targets on complex battlefields, enabling outperformance of the adversary in terms of detection, decision-making, strike, and destruction.

3.15.5 New Capabilities Contributing to Asymmetric Advantages

The more intelligent the future OSoS, the more robust are cross-domain operational capabilities, the deeper the virtual–real integration, the faster the operations, the more diverse the deterrents, the greater the war initiative, and the easier it is to conduct comprehensive deterrence, cross-domain collaboration, and downscaled strikes, achieving asymmetric advantages and landslide victories. To be more precise, data algorithms, parallel training, swarm collaborations, multi-domain integration, hypersonic weapons, infrastructure seizure and control, and cyber psychological warfare will all be included in the seven new capabilities required for asymmetric advantages and emergence effects in the age of AI.

3.15.6 Unmanned Operations Breeding Civilized Warfare

Intelligent warfare will exemplify a civilized nature. With the advancement of intelligent simulation systems, unmanned systems, and precise/controllable destruction capabilities, conflicts between live forces will decline, while confrontations between simulated or unmanned forces will increase in the future. Steered by reason and consensus, humans will be able to have more effective control over the risks of warfare, and new civilized war modes will be conceived with AI development.

References

[1] New Media 网络传播杂志. (2017, January 23). *Rengong zhineng fazhan jianshi* 人工智能发展简史. Office of the Central Cyberspace Affairs Commission 中央网信办官方网站. www.cac.gov.cn/2017-01/23/c_1120366748.htm.

[2] Minsky, M. (1988). *Society of mind.* New York, NY: Simon and Schuster.

[3] Felten, E. (2016). *Preparing for the future of artificial intelligence.* Washington, DC: US Government.

[4] Executive Office of the President, White House. (2016). *Artificial intelligence, automation, and the economy.* Washington, D.C: US Government.

[5] Office of Science and Technology Policy. (2020). *American artificial intelligence initiative.* Washington, DC: US Government.

[6] The State Council of PRC 国务院. (2017). *Development planning for a new generation of artificial intelligence.* 新一代人工智能发展规划. Beijing: Government of China.

[7] Future Today Institute. (2021) *2021 Tech trends report, USA.* https://futuretodayinstitute.com/trends/

[8] Mackinder, H. J. 哈尔福德·J.麦金德. (2017). *Luquan lun* 陆权论 (F. Xu 徐枫, Trans.). Beijing: Qunyan Press, 119.

[9] Spykman, N. J. (1944). *The geography of the peace.* New York, NY: Harcourt Brace.

[10] Ye, Z. C. 叶自成. (2007). Zhongguo de heping fazhan: luquan de huigui yu fazhan 中国的和平发展: 陆权的回归与发展. *World Economics and Politics* 世界经济与政治(2), 4, 23–31.

[11] Mahan, A. T. (2011). *The influence of sea power upon the French revolution and empire*. Cambridge: Cambridge University Press.

[12] Douhet, G. (1989). The command of the air. In P. Bobbitt, L. Freedman, & G. F. Treverton (Eds.), *US nuclear strategy: A reader* (pp. 10–21). London: Palgrave Macmillan UK.

[13] Burlingame, R. (1978). *General Billy Mitchell, champion of air defense*. New York, NY: McGraw-Hill.

[14] Harter, M. E. (2006). Ten propositions regarding space power: The dawn of a space force. *Air Univ Maxwell Afb Al Airpower Journal*.

[15] US Department of Defense. (2011). *Department of Defense 2011 data center consolidation plan & progress report, 2011 federal data center consolidation initiative*. https://dbb.defense.gov/ (2011.11.8).

[16] Defense Business Board. (2012). *DoD information technology modernization: A recommended approach to data center consolidation and cloud computing* (Report FY12-01). https://dbb.defense.gov/Portals/35/Documents/Reports/2012/FY12-1_DoD_Information_Technology_Modernization_2012-1.pdf

[17] Xueshu Gaoji Pinglunyuan YR 学术高级评论员YR. (2019, January 7). *Meijun rengong zhineng wuqihua dapandian* 美军人工智能武器化大盘点. Guancha wang 观察网. https://user.guancha.cn/main/content?id=69816.

[18] US Department of Defense. (2020). *Joint Artificial Intelligence Center has substantially grown to aid the warfighter*. www.defense.gov/News/News-Stories/Article/Article/2418970/joint-artificial-intelligence-center-has-substantially-grown-to-aid-the-warfigh/

[19] Xinhua Net 新华网. (2015). *Russia's defense minister said that the military's "Central 2015" strategic exercise has achieved its goals* 俄防长说俄军"中央－2015"战略演习实现预定目标. Xinhua Net 新华网. www.xinhuanet.com//

[20] Hawking, S. W., Huo, J 霍金, & Zhou, X 周翔. (2017). Rang rengong zhineng zaofu renlei jiqi laiyishengcun de jiayuan 让人工智能造福人类及其赖以生存的家园. *Tech China 科技中国*, (6), 85–89.

4 Military Intelligence and Intelligent Technologies

Due to technological advancements and increasing demand, the scope of artificial intelligence (AI) has gradually expanded beyond machine learning to encompass cross-media intelligence, swarm intelligence, autonomous systems, hybrid intelligence, and bionic intelligence. AI 1.0 is being transformed into AI 2.0 [1]. As a result of its deep integration with other disciplines, AI has accelerated the development and application of military intelligence. In one sense, military intelligence can be derived from civilian technologies. However, it is not simply a duplication of civilian intelligence but requires secondary development, design, demonstration, and research. Military intelligence is a system in which fundamental civilian theories and technologies serve as the foundation, while systematic, specialized, and deep technologies remain to be developed. This chapter provides an overview of the fundamental concepts, research scope, and systematic structure underlying military intelligence and intelligent technologies.

4.1 Basic Concepts

4.1.1 Intelligence

The term "intelligence" has existed for a long time and is defined as "knowledge and ability" or as "possessing certain human knowledge and ability" in the modern Chinese dictionary [2].

4.1.2 Artificial Intelligence

AI is a relatively new field of technical study that focuses on theories, methods, techniques, and applied systems for simulating, extending, and expanding human intelligence. In a limited sense, AI refers to mechanical systems that can learn, make decisions, and perform creative tasks in the same manner as a human brain. In a broad sense, AI encompasses all intelligent systems, including "intelligence +," "+ intelligence," and other hybrid intelligent systems (see Chapter 3). AI began as a simulation of human intelligence

DOI: 10.4324/b22974-5

and evolved into bionic, machine, human–machine hybrid, and complex intelligence.

4.1.3 Intelligent Systems

Intelligent systems are electronic systems that satisfy human needs and lead the modern human civilization's trend by using networks, big data, the Internet of Things (IoT), and AI. Unmanned vehicles, for example, are typical intelligent systems that integrate sensor IoT, mobile Internet, big data analytics, and autopilot to meet humans' diverse travel demands. These intelligent vehicles are distinguished from conventional vehicles that require human drivers. Intelligent systems generally exhibit four characteristics. The first is perception, or the capacity to perceive and obtain information from the physical world, which is also a prerequisite for intelligent activities. The second is memory, or the capacity to store perceived and processed information and knowledge while simultaneously conducting analysis, calculation, comparison, judgment, and association. The third is learning ability and self-adaptation, or the capacity to accumulate experience and knowledge. The fourth is decision-making ability, which refers to the capacity to respond to external stimuli and to formulate and communicate decisions. A system that possesses the aforementioned characteristics can be considered intelligent [3].

4.1.4 Military Intelligence

There is no definitive definition for military intelligence yet, but it can be summarized as the process of developing an intelligence-led military combat capability through the use of AI, cloud computing, big data, IoT, interdisciplinary biology, unmanned systems, and parallel systems, the essence of which is decision-oriented optimization algorithms.

4.1.5 Military Intelligent Technologies

Military intelligence is supported by a science and technology system that includes general, frontier, systematic, exclusive, and interdisciplinary technologies. These technologies can be classified as fundamental, systematic, or domain-specific.

4.2 Classification

4.2.1 Functions

Military intelligence is functionally divided into five distinct areas.

The first is bionic intelligence, which at its core replicates the superior functions of the human brain or organisms, thereby endowing machines with biological intelligence.

The second is machine intelligence, which operates on a different logic than human intelligence and refers to a machine's ability to perform tasks using advanced technology.

The third is swarm intelligence, which is based on low-cost swarm systems and achieves tactical benefits in terms of collaborative detection, strike, defense, and wisdom by increasing efficiency with quantity and introducing the emergence effect of swarms.

The fourth is hybrid intelligence between humans and machines, which enables human–machine communication, collaboration, and fusion.

The fifth is intelligent manufacturing, which encompasses design, research and development, testing, production, mobilization, security, and maintenance.

4.2.2 Applications

Military intelligence manifests itself in five primary applications.

The first is single-load intelligence, which includes bionic and machine intelligence. Examples include unmanned aerial vehicles (UAVs), unmanned ground vehicles (UGVs), unmanned surface vehicles (USVs), bionic robots, and intelligent munitions.

The second is collaborative intelligence, which encompasses swarm intelligence, human–machine hybrid intelligence, manned and unmanned cooperation, and other operation systems with strongly coupled elements.

The third is systematic intelligence, which encompasses intelligent perception, decision-making, strike, defense, coordination, and security, as well as other elements of multi-factor, cross-domain, and holistic operations.

The fourth is domain-specific intelligence, which encompasses cognitive communication, cyber attack and defense, electronic countermeasures, stealth countermeasures, public opinion control, psychological warfare, and tactical deception, among other specialized areas of "intelligence +" or "+ intelligence."

The fifth is parallel evolutionary intelligence, which refers to the evolutionary capabilities and emergence effects of self-adapting, -learning, -confrontation, -cooperation, -repairing, and -evolving operation system of systems (OSoS), based on data, models, and algorithms.

4.2.3 Technical Focuses

Military intelligence's technical focus is on five areas.

The first area is intelligent chips, which include perception, computing, brain-like processing, signal processing, storage, and memory chips that are capable of adapting to complex, highly dynamic, interfered, and intense confrontations. Neural network processing units (NPUs) for missiles, vehicles, airplanes, ships, boats, stars, cloud platforms, and data centers are examples, as are general-purpose chips such as high-performance central processing units (CPUs), graphics processing units (GPUs), digital signal processors (DSPs), and field programmable gate arrays (FPGAs).

The second area is algorithms, including feature extraction and recognition algorithms for optical, infrared, video, electromagnetic, and acoustic signals; low-pixel remote sensing data from satellites, hypersonic dynamics, and weak long-range signals; and signals under extreme conditions such as strong electromagnetic interference and small sample data support. Additionally, it encompasses algorithms for target association based on multi-source data from space, air, ground, and sea; open-source big data and intelligence; analysis and modeling of battlefield posture; intelligent operating systems for unmanned vehicles, robots, and their swarm behaviors; swarm algorithms for homogeneous systems and collaborative algorithms for heterogeneous systems.

The third area is unmanned autonomy, which includes automatic inspection and overhaul of mechanical, power, control, protection, and weapon subsystems; robotic manipulation, intelligent path planning, and autonomous piloting; intelligent sensing, identification, striking, and evaluation of targets; automatic delivery, loading, and dismantling of ammunition; active defense of incoming targets; and automatic protection of the cabin. Unmanned vehicles are products that combine mechanization, information technology, and intelligence, with the latter playing an increasingly prominent and influential role.

The fourth is web-based information, including space-based information utilization, open-source civilian information utilization, cognitive communication networks, distributed military clouds, battlefield situational awareness, adaptive mission planning and assisted/autonomous decision-making; AI-based command and control, cross-domain cooperation, fire control, swarm networking, cyber attack and defense, electronic countermeasures, public opinion information, and infrastructure control; virtual staff officer, assistant AIs, human–machine intelligence interaction; database, model base, knowledge base, algorithm base, and tactic base based on simulations and live military exercises, etc.

The fifth is system integration, including intelligent design, simulation, rehearsal, verification, integration, training systems, and parallel military systems, virtual–real iterative optimization systems, intelligent security systems for joint domain and all-domain operations; intelligent cooperation between land, sea, air, space, and cyber forces, between manned and unmanned vehicles, between swarm unmanned vehicles and munitions, and between air–ground, air–sea, air–space, land–sea, space–ground cross-domain forces.

4.3 Framework

Military intelligence research is still in its infancy, with its architecture and direction likely to change significantly in the future. As stated previously, the author believes that military intelligence technology can be classified into

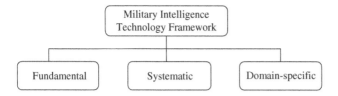

Figure 4.1 Military intelligence technology framework.

three broad categories based on current developments and predictions: fundamental, systematic, and domain-specific (Figure 4.1).

4.3.1 Fundamental Technologies

Fundamental technologies focus on fundamental military and civilian technologies that support military intelligence, emphasizing enhancing perception and recognition, knowledge computing, cognitive reasoning, action execution, and human–computer interaction while forming open, compatible, stable, and mature support. It includes ten categories of technology [4].

4.3.1.1 Machine Learning

Machine learning is an interdisciplinary field of study that examines how computers can mimic human learning behaviors such as acquiring new knowledge or skills and reorganizing existing knowledge structures in order to continuously improve performance [5]. It serves as the brain of military intelligence. In recent years, statistical learning research has expanded to encompass new areas such as statistical learning, uncertainty inference, and decision-making; distributed, privacy-preserving, small sample, deep reinforcement, unsupervised, semi-supervised, and active learning; and other intelligent techniques and efficient models.

4.3.1.2 Big Data Intelligence

The research on big data intelligence encompasses data-driven and knowledge-guided AI, cognitive computing with a focus on natural language understanding and image graphics, intelligent decision-making under conditions of incomplete information, data-driven mathematical models, and general AI theories.

4.3.1.3 Transmedia Perception and Calculation

Transmedia perception and calculation research focuses on the unified representation of cross-media data and knowledge, association understanding, knowledge mining, knowledge graph construction and learning, knowledge evolution and reasoning, intelligent description, as well as cross-media analysis, reasoning engines, and validation systems. Researchers are also interested in auditory perception and computation for natural environments and fully dimensional intelligent perceptual reasoning for urban areas.

4.3.1.4 Hybrid and Augmented Intelligence

Hybrid and augmented intelligence are concerned with intelligence that includes "humans in the loop," human–machine intelligence symbiosis through behavioral enhancement, machine reasoning and causal models, associative memory models, and knowledge evolution methods, self-learning methods for complex data and tasks, collaborative computing methods for cloud robots, contextual understanding, and human–machine swarm cooperation.

4.3.1.5 Swarm Intelligence

Swarm intelligence is concerned with the structure, organization, incentive, emergence, and learning mechanisms of robotic swarms, as well as with computational paradigms and models.

4.3.1.6 Autonomous Cooperation and Optimal Decision-Making

Among other theories, this field is interested in collaborative sensing and interaction, collaborative control, optimal decision-making for unmanned autonomous systems, and knowledge-driven collaboration/interoperability between humans, machines, and objects.

4.3.1.7 Brain-Like and Quantum Intelligent Computing

This field examines theories and methods for brain-like perception, learning, memory mechanisms, computational fusion, and brain-like control. Additionally, researchers examine quantum patterns and intrinsic mechanisms of brain cognition, as well as efficient quantum intelligence models and algorithms, high-performance and high-bit quantum AI processors, and real-time quantum AI systems capable of interacting with external data.

4.3.1.8 Intelligent Chips and Systems

Intelligent chips and systems research focuses on neural network processors, energy-efficient, reconfigurable brain-like computing chips, new perception chips and systems, intelligent computing architectures and systems, AI operating systems, and hybrid computing architectures suitable for AI.

4.3.1.9 Natural Language Processing (NLP)

NLP research involves text computational and analytic techniques, cross-lingual text mining techniques, cognitive machine intelligence techniques for semantic comprehension, and human–computer dialogue systems for multimedia information comprehension.

4.3.1.10 Interdisciplinary Biology

Interdisciplinary biology is a vast and complex field of study that encompasses synthetic biology, neuroscience, brain–computer interfaces (BCIs), affective computing, mental state detection, and bio nanotechnology.

4.3.2 Systematic Technologies

Systematic technologies are developed to address joint requirements for intelligent combat systems, which include system architecture, network communication, situational awareness, command and control, and combat simulation. Pattern recognition, deep learning, extensive data mining, cognitive communication, distributed military clouds, adaptive mission planning, autonomous decision-making, and parallel processing are among the key technologies involved. There are five distinct categories of technology.

4.3.2.1 Intelligent Combat System Architecture

The first category features planning, simulation, and verification systems based on AIs, clouds, networks, groups, and terminals, and incorporating detection, control, resistance, strike, evaluation, and support, manifested as air–ground and civil–military integrated network communication, distributed cloud systems, intelligent integrated information link, global military operation integrated security system, and military data security protection.

4.3.2.2 Cognitive Network Communication

The second category includes active/passive self-organizing network communication, anti-electromagnetic interference communication, cognitive communication, space–ground intelligent networking communication, new-generation intelligent data links, cooperative data links, laser and scattering networking communication, millimeter-wave and terahertz confidential communication, intranet encrypted communications based on civilian mobile networks, and military IoT.

4.3.2.3 Deep Battlefield Cognition

The third category encompasses multi-source heterogeneous networked intelligence perception, open-source and extensive data mining and correlation analysis, intelligent target identification in jamming environments, ultra-sensitive optoelectronic and radar perception, multi-spectral intelligent perception, full-spectrum electromagnetic detection, cross-domain cooperative perception, battlefield knowledge graphs, and battlefield prediction.

4.3.2.4 Distributed Intelligent Cloud for Command and Control

The fourth category includes cloud-based battlefield situational awareness, adaptive dynamic planning, autonomous decision-making for combat forces and missions, cross-domain interoperability, intelligent end-to-end control, assisted decision-making based on human–machine integration, and human–computer interface.

4.3.2.5 Parallel Combat Simulation and Assessment

The final category involves virtual battlefield environments, simulation platforms, intelligent object modeling, intelligent virtual–real interactions, mixed reality, efficient real-time confrontation simulation supported by big data, parallel combat experiments, distributed training, comprehensive verification for joint training, as well as data, model, knowledge, algorithm, and tactic bases.

4.3.3 Domain-Specific Technologies

Domain-specific technologies are aimed at utilizing the intelligence of combat vehicles, strikes, cognitive confrontations, defense, and security, with a particular emphasis on land, sea, air, space, cyber, and electromagnetic combat-specific applications, enhancing military capabilities such as intelligent perception, intelligent decision-making, intelligent combat, intelligent security, cloud cognition, autonomous networking, swarm attack and defense, and cross-domain operations, by utilizing pattern recognition, deep learning, cloud computing, big data, and interdisciplinary biological technologies. Seven main groups of technology are included.

4.3.3.1 Intelligent Combat Vehicles

The first group involves general technology for unmanned ground/aerial/surface/underwater/space vehicles, intelligent environment and target perception, autonomous mission planning, deep learning from combat experiences, assisted/autonomous driving, detection and avoidance of incoming threats, multi-environmental camouflage, and adaptive stealth, and other technologies.

4.3.3.2 Manned–Unmanned Cooperation

The second group incorporates multi-mission manned–unmanned planning, autonomous/assisted decision-making, multi-vehicle autonomous sensing, communication, navigation, and control, autonomous offline/online learning, human–machine hybrid squads/soldier systems, and other technologies.

4.3.3.3 Intelligent Precision Strikes

The third group encompasses multi-source information detection and correlation fusion, intelligent target identification and cataloging, intelligent mission assignment and strike means selection, autonomous navigation and path planning, precise attack and controllable destruction of munitions, intelligent network and cooperative data link of munitions, intelligent attack and defense of hypersonic weapons and munitions, and rapid evaluation of strike effects.

4.3.3.4 Virtual Space and Intelligent Information Countermeasures

The fourth group involves cyber attack and defense, electronic countermeasures, electromagnetic spectrum management, cross-domain information fusion in cyberspace, land, sea, air, and space, public opinion control and guidance, network psychological warfare, cognitive countermeasures, and intelligent infrastructure control.

4.3.3.5 Autonomous/Bionic Unmanned Swarms

The fifth group features multi-target autonomous detection and identification in complex environments, intelligent, cooperative sensing between swarm vehicles and munitions, autonomous multi-task planning and decision-making, distributed and remote control, autonomous system guidance, communication, navigation, control, and solution, and multi-mode swarm algorithm.

4.3.3.6 Anti-Hypersonic, Anti-Swarm, and Anti-Terrorist Technology

The sixth group incorporates multi-source integrated space–ground detection, infrared/multi-spectral electromagnetic analysis of hypersonic target images, high-powered microwave weapons, electronic interference interception, swarm electromagnetic pulse (EMP) weapons interception, electromagnetic high-speed munitions interception, intelligent anti-grenade/rocket/mortar systems, rapid recognition of terrorists, and non-contact rapid detection and jamming of suicide bombers and roadside bombs.

4.3.3.7 Intelligent Security and Maintenance

The last group features distributed and networked security information systems, security system simulation, rapid three-dimensional (3D) delivery schemes, ammunition and fuel consumption models, rapid material deployment and

supply, intelligent diagnosis and remote maintenance, equipment health management, rapid in situ repair and rescue, 3D in situ printing, 4D intelligent
transformation, rapid adaptive manufacturing, networked security resources,
cloud services, intelligent catering services, and energy support.

4.4 Expertise

Intelligent technologies with extensive military and civilian applications
involve a broad sense of knowledge. This chapter focuses on eight distinct
areas of expertise.

4.4.1 Machine Learning

4.4.1.1 Concept

Machine learning is an interdisciplinary subject involving probability,
statistics, approximation, convex analysis, and algorithmic complexity. The
field examines how computers can mimic human learning behaviors to
acquire knowledge or skills and reorganize existing knowledge structures for
continuous performance improvement [5].

There is a natural and obvious progression in AI research from a focus on
"reasoning" to a focus on "knowledge" and finally on "learning." The theory
is concerned with developing and analyzing algorithms that enable computers
to learn automatically. Machine learning algorithms are a subset of algorithms
used to automatically discover patterns in data and make predictions about
unknown data based on those patterns.

Machine learning is now widely used in various fields, including data
mining, computer vision, NLP, biometric recognition, search engines, medical
diagnosis, credit card fraud detection, securities market analysis, DNA sequen-
cing, speech and handwriting recognition, strategic gaming, and robotics.

4.4.1.2 Basic Structure

The basic structure of a machine learning system is as follows: the learning
part incorporates input data into the knowledge base to improve the executive
part's effectiveness, and the executive part completes tasks based on the new
knowledge base while feeding new data back to the learning part.

4.4.1.3 Learning Strategies

1. Rote learning
   In rote learning, the student makes direct connections between the
   information provided by the environment and does not engage in any
   reasoning or other knowledge transformation.

2. Learning from instruction (learning by being told)

   In this learning strategy, the student acquires knowledge via the environment (e.g., teacher or textbooks), converts it to internally usable representations, and organically integrates new knowledge with prior knowledge.

3. Learning by deduction

   The student uses deductive reasoning to derive conclusions from axioms after applying logical transformations. This reasoning of "fidelity" enables the student to acquire practical knowledge from reasoning.

4. Learning by analogy

   Based on the similarity of knowledge in two distinct domains (source domain and target domain), learning can be accomplished through analogy, i.e., by deriving the corresponding knowledge in the target domain from the source domain.

5. Explanation-based learning (EBL)

   In EBL, based on the teacher-provided target concept, example, domain theory, and guidelines, students first construct an explanation that connects the example with the concept and then generalize the explanation to a sufficient condition for the concept in accordance with the guidelines.

6. Learning from induction

   Inductive learning is a process in which the student reasons to reach a general conclusion based on teacher- or environment-provided examples or counterexamples to a concept.

4.4.1.4 *Representations*

1. Algebraic expression

   The system learns algebraic expressions to acquire a fixed functional form for the parameters or coefficients, adjusted for the optimal solution, e.g., desired performance.

2. Decision tree

   A decision tree is a flowchart-like structure in which each internal node represents a "test" on an attribute, each branch represents the outcome of the test, and each leaf node represents a class label (decision taken after computing all attributes).

3. Formal grammar

   When a language is identified, its formal grammar can be constructed by generalizing the patterns of a series of its expressions.

4. Production rules

   Production rules, expressed as condition–action pairs, have been used widely, and undergo generation, generalization, specialization, or synthesis in learning systems.

5. Formal logical expressions

   Propositions, predicates, variables, and statements that constrain the scope of the variables, and embedded expressions are the essential components of a formal logical expression.

6. Graphs

   Certain systems employ graph matching and transformation schemes to compare and index knowledge.

7. Frames

   Each frame contains a set of slots for describing the characteristics of objects.

8. Programming and encoding

   The purpose of acquiring this type of knowledge is to specify and reproduce a process, not deduce a structure.

9. Neural networks

   Neural networks are primarily used in linkage learning, in which the acquired knowledge is summarized in a neural structure.

10. Synthesized representations

    Occasionally, the knowledge acquired by a learning system is represented using a combination of the patterns above.

4.4.1.5 Applications

Expert systems, cognitive simulation, planning, problem-solving, data mining, web information services, image recognition, fault diagnosis, natural language understanding, robotics, and competition are some of the most significant applications of machine learning.

4.4.1.6 Comprehensive Classification

1. Empirical inductive learning

   In practice, empirical inductive learning is accomplished through several data-intensive empirical methods. Symbols such as attributes, predicates, and relations are used to represent learning examples and outcomes.

2. Analytical learning

   Analytical learning begins with a single or a few examples, which are then analyzed with domain knowledge. This approach is characterized by a preference for deductive reasoning over inductive reasoning, experience-based problem-solving, and generating search rules that facilitate more effective use of domain knowledge.

3. Learning by analogy

   This approach is equivalent to learning by analogy above, also known as exemplar-based learning in the learning strategies.

4. Learning by genetic algorithms

    Genetic algorithms simulate biological reproduction through mutation, exchange, and natural selection according to Darwinian principles. Individual solutions to the problem are encoded by the algorithms using a vector called "individual." Each component of an individual is referred to as a gene and assessed using an objective function. Genetic operations such as selection, exchange, and mutation are carried out to create a new population based on the evaluation value (fitness). Genetic algorithms are well suited for complex environments, such as those with a high volume of noisy and irrelevant data, constantly updated, unclearly defined objectives, and when determining the value of the data requires a lengthy execution process.

5. Linked learning

    An artificial neural network is a common model for linked learning. It consists of several simple computational units, called neurons, and a weighted coupling between the units.

6. Reinforcement learning

    In reinforcement learning, optimal solutions for sequential decision tasks are discovered through environment-based trial-and-error. The system interacts with an environment by selecting and executing actions that result in a change in the system's state, possibly receiving some form of reinforcement signal (immediate reward). A reinforcement signal is a quantitative indicator of a system's behavior that serves as a reward or punishment. The learning system's objective is to discover an action selection strategy, i.e., a method for determining which action to take in any given state produces the optimal result, such as the largest and most immediate payoff.

4.4.1.7 Forms of Learning

1. Supervised learning

    This is a type of learning in which machine learning generates true/false instructions. This subcategory of machine learning is used for classification and prediction. Supervised learning generates a function from a training dataset, which can predict the results of new data. This strategy's dataset contains inputs and outputs, which can be considered features and objectives, respectively. Humans label the objectives in the target. Regression analysis and statistical classification are two common algorithms for this strategy.

2. Unsupervised learning

    This is a type of inductive learning that utilizes the K approach to construct centers, reduces errors through loops, and accomplishes classification purposes.

4.4.1.8 Research Fields

Machine learning research focuses on three primary areas.

1. Task-oriented research
   Task-oriented research focuses on developing and analyzing learning systems for optimizing the performance of completing a predefined set of tasks.
2. Cognitive model research
   The study of cognitive models entails examining human learning processes and computer simulations.
3. Theoretical research
   Theoretically, this research examines various learning methods and algorithms that are not domain-specific.

Machine learning, along with expert systems, is a critical area of research for AI applications and is one of the core topics in AI and neural computing.

4.4.2 Deep Learning

4.4.2.1 Concept

Deep learning is a concept that evolved from the study of artificial neural networks. Deep learning discovers distributed feature representations of data. It combines lower-level features to create more abstract representations of attribute classes or features [6].

In 2006, LeCun, Bengio and Hinton introduced the concept of deep learning by proposing an unsupervised greedy layer-by-layer training algorithm based on deep belief networks (DBNs) in the hope of resolving optimization problems associated with deep structures [7].

Deep learning is motivated by building and simulating the human brain for analytical learning. It involves constructing and simulating neural networks that mimic the human brain's mechanisms for interpreting data such as images, sounds, and text.

As with machine learning, deep learning methods distinguish between supervised and unsupervised learning, each requiring a completely different set of learning models. For instance, convolutional neural networks (CNNs) are a machine learning model used in deep supervised learning, whereas DBNs are used in unsupervised learning.

4.4.2.2 Connotations

In deep learning, a flow graph represents the computation involved in generating output from an input, in which each node represents a basic computation and a computed value. A flow graph is analogous to a computational set

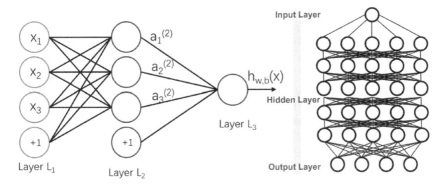

Figure 4.2 Deep learning model with multiple hidden layers.

that defines a family of functions and can be included in any node or graph structure, where there is no parent for the input node and no children for the output node.

The depth of a flow graph is a unique property, i.e., the length of the longest path connecting input and output.

A traditional feedforward neural network has a depth equal to the number of layers, e.g., the number of hidden layers plus one for the output layer. Support vector machines (SVMs) have a depth of two (as illustrated in Figure 4.2): one for the output or feature space of the kernel function and another for a linear mixture of the resulting outputs.

One direction of AI research is expert systems, defined by many "if-then" rules and a top-down approach. Artificial neural network (ANN) represents another direction, built by a bottom-up approach. There is no precise formal definition for a neural network. It is essentially a replica of how information is transmitted and processed between neurons in the human brain.

4.4.2.3 Issues

Deep learning is concerned with the following issues: insufficient depth, the structure of the human brain, and the layer-by-layer abstraction of cognitive processes.

In many cases, a depth of two is sufficient to represent any function with a given target accuracy, but at the cost of increasing the number of nodes required (e.g., the number of computations and parameters), as illustrated in Figure 4.2. Theoretical results confirm the existence of function families for which the number of required nodes grows exponentially in proportion to the input size.

The human brain is built on a complex architecture. For instance, the visual cortex has been extensively studied and has been shown to consist of a series of regions, each of which contains a representation of an input and a flow of signals from one region to the next. The association along some hierarchical parallel paths is omitted in this case, making the system more complex. Each layer of this feature hierarchy corresponds to an input on a different abstraction layer and contains more abstract features at the higher layers of the hierarchy.

Cognitive processes are abstracted layer by layer. Humans organize their ideas and concepts hierarchically. They begin with simple concepts and then use them to represent abstract concepts. For example, engineers deal with abstract problems by breaking them down into multiple levels of abstraction.

4.4.2.4 Basic ideas

Assume that a system S consists of n layers $(S_1, S_2, ..., S_n)$, each with an input I and an output O, denoted figuratively as $I \rightarrow S_1 \rightarrow S_2 \rightarrow ... \rightarrow S_n \rightarrow O$. If the output O equals the input I, i.e., if the information in "a" is processed to obtain "b," and "b" is processed to obtain "c," then the mutual information of "a" and "c" is equal to the mutual information of "a" and "b." This demonstrates that information processing does not add information but rather subtracts it. Additionally, information may remain unchanged, implying that the input I passes through each layer of S_i without losing any information, i.e., at any layer of S_i the output becomes a different representation of the original information (i.e., the input I). Returning to the topic of deep learning, which features the acquisition of features automatically, consider a collection of inputs I (e.g., images or text) and a system S (with n layers) designed to maintain the input I as its output. In that case, a series of hierarchical features of the input I, i.e., $S_1, S_2, ..., S_n$, can be obtained automatically.

The idea behind deep learning is to stack multiple layers. The output of one layer is used as the input for the subsequent layer, allowing for a hierarchical representation of the input data.

If we consider the learning structure as a network, we can sum up the central concept of deep learning as follows:

1. Unsupervised learning is used for the pre-train of each layer of the network.
2. Only one layer is trained at a time, and the training results are used as input to a higher level of layer.
3. All layers are then adjusted using a top-down supervised algorithm.

4.4.2.5 Key Technologies

Key technologies for deep learning include linear algebra, probability and information theory; underfitting, overfitting, regularization; maximum likelihood estimation and Bayesian statistics; stochastic gradient descent; supervised and unsupervised learning; deep feedforward networks, cost functions and backpropagation; regularization, sparse coding and dropout; adaptive learning algorithms; CNNs; recurrent neural networks; deep neural networks and deep stacked networks; long- and short-term memory (LSTM); principal component analysis; regular autoencoders; representation learning; Monte Carlo method; restricted Boltzmann machines; deep confidence networks; softmax regression, decision trees and swarming algorithms; K-nearest neighbor (KNN) and SVM; generative adversarial networks and directed networks; machine vision and image recognition; NLP; speech recognition and machine translation; finite Markov; dynamic planning; gradient strategy algorithms; augmented learning (Q-learning); turning points; etc.

Before 2006, all attempts to train deep architectures failed, as a deep supervised feedforward neural network produces poor results, reducing it to a depth of one (one or two hidden layers).

Three papers published in 2006 changed that situation, led by Hinton's seminal work on DBNs, which established the following fundamental principles [8–10]: unsupervised learning of representations is used to (pre-)train each layer; unsupervised training of one layer at a time is followed by the previously trained layer, with the representation learned at each layer serving as input to the subsequent layer; and supervised training is used to tune all layers (plus one or more additional layers used to generate predictions).

4.4.2.6 *Applications*

1. Machine vision

   The Multimedia Laboratory at the Chinese University of Hong Kong was the first Chinese team to apply deep learning to computer vision research. The lab defeated Facebook in the world-class AI competition LFW (Labeled Faces in the Wild), marking the first time that AI has surpassed humans in face recognition.

2. Speech recognition

   Through collaboration with Hinton, Microsoft researchers were the first to incorporate restricted Boltzmann machines (RBM) and DBN into speech recognition acoustic model training, achieving remarkable success in recognizing speech with a large vocabulary, resulting in a 30% reduction in error rates.

3. Natural language processing

   Numerous institutions are researching deep learning, used in various applications, including NLP, machine translation, and semantic mining.

4.4.3 *Bionic Technology*

4.4.3.1 *Concept*

Plants and animals have evolved over millions of years of natural selection and are perfectly adapted to their environment and very close to perfection. Bionic technology is a broad and interdisciplinary field of study that encompasses numerous disciplines such as biology, physics, materials science, energy, and information. It is primarily concerned with engineering and simulating animal and plant functions in nature. Bionic technology creates bionic materials, devices, or systems with unique functions and superior performance by simulating the structures, mechanisms, patterns, and behaviors of living creatures in biological systems. These bionic products can perform complex tasks precisely, flexibly, reliably, efficiently, and durably under harsh and demanding conditions.

4.4.3.2   Materials and Structures

Bionic materials can achieve stealth or camouflage, increase flexibility, and impact resistance, among other benefits, by mimicking biological properties.

1. Bionic structures of shell, bone, and fish scale
   In June 2020, researchers at the Massachusetts Institute of Technology (MIT) investigated the microstructure of seashells and developed extremely strong, tough, and biocompatible biomimetic films with a tensile strength of 48 MPa and a tensile toughness of nearly 400% [11]. The University of California investigated crustacean materials that can be used to make bulletproof vests, bullet-resistant helmets, Humvees, drones, and helicopters that are resistant to improvised explosive devices (IEDs) and light-weapons threats, as well as to replace steel structures in naval guns and naval propulsion systems that are susceptible to cavitation damage. Similarly, the US Air Force has been investigating the possibility of replacing titanium on the A-10 Thunderbolt attack aircraft with this type of material to mitigate the threat posed by light weapons on the ground. In 2017, the Pacific Northwest National Laboratory, the Lawrence Berkeley National Laboratory, and the University of Washington collaborated to achieve self-assembly of peptide-like materials, forming a network of nanoribbons with hexagonal patterns on mineral surfaces in a highly ordered manner, paving the way for the development of bionic materials that resemble shells or bones. In 2010, Pinnacle Armor of the US introduced Dragon Skin, a bulletproof vest shaped like fish scales constructed of small ceramic ballistic tiles and new ballistic fibers that will not break, even when struck by a grenade. In 2016, bionic flexible armor was developed by researchers at McGill University in Canada by combining rigid armor sheets with a softer substrate. The resulting material resembled fish scales and possessed better flexibility than the puncture resistance of homogeneous ceramic sheets. This lays the groundwork for future bionic flexible armor design, optimization, and fabrication.
2. Bionic stealth materials
   In 2014, the University of Illinois developed a flexible camouflage fabric that mimics the ability of octopuses to camouflage themselves using

thermal dyes and light sensors. Within a response time of 1–2 seconds, this flexible camouflage fabric produces a changing camouflage pattern that adapts spontaneously to the surrounding environment. Researchers at University of California, Berkeley have precisely etched micro patterns on ultra-thin silicon films, which can reflect different colors of light depending on how the film is bent, creating a chameleon-like camouflage effect with up to 83%reflectivity, making it the world's first flexible camouflage material with colors that can be changed simply by bending. In the "Chameleon" stealth development program, French researchers developed a "multi-spectral active skin" using adaptive bionic camouflage materials. This skin is electroluminescent, electro-reflective, and electro-radiative, enabling combatants to dynamically change color and blend into their surroundings, much like a "chameleon," achieving effective stealth in both visible and infrared wavelengths. In 2017, researchers from Belarus, France, the United Kingdom, and Germany collaborated to create a hollow sphere structure that resembles a moth's eye by encasing a single layer of hollow carbon spheres in a 2D structure. The material exhibits near-perfect microwave absorption and can absorb 100%of microwaves in the Ka-band (26.5–40 GHz) and is expected to be used in stealth technology as a radar-absorbing material. In November 2016, Harvard University collaborated with Saudi Arabia's King Abdullah University of Science and Technology to develop a disordered bionic nanostructure based on the structural color of bird feathers that emits a rainbow of colors for stealth in military equipment.

4.4.3.3 Sensing and Detection

Bionic sensors that mimic biological sensory functions such as vision, hearing, touch, and smell can enable military equipment to be disguised, mobile, and efficient in target acquisition and strike capabilities.

1. Visual sensing

   Bionic visual sensing enables sensors to have characteristics or functions similar to those of a biological eye, allowing them to accurately obtain parameters such as lightness, darkness, chromaticity, position distance, movement trend, contour shape, and posture of the target, as well as to store, transmit, recognize, and understand these parameters. Foreign research on military applications of visual bionics has invested significant human, material, and financial resources, primarily on the compound eye of flies and horseshoe crabs, as well as human eye bionics, with fruitful results. Both the US and the UK have researched missile vision systems based on hawk eyes. The US has also investigated anti-satellite (ASAT) weapons based on the compound eye of flies. In 2007, BAE Systems developed the "bug-eye" multi-aperture imaging system for microlight vehicles with funding from the UK Ministry of Defence. The system, which has a field of view of 60° that can be expanded to 120°, is currently used in missile warning and man-portable night visions.

2. Auditory sensing

In 2007, Georgia Tech modeled the inner ear of a fish to create an under-water acoustic transducer that responds to sound waves via the movement of villi. The US military anticipates that this new bionic sensor will evolve into a generation of ultra-sensitive and anti-jamming solid-state under-water detectors that outperform sonar. Stanford University developed an innovative underwater acoustic sensor in 2011 that replicates the ability of marine cetaceans (e.g., orcas) to change the pressure in their inner ear. The underwater sensor can hear both quiet and loud sounds and operate at depths of up to 6 miles underwater and at pressures of up to 1,000 times the atmospheric pressure. Additionally, the sensor can detect sounds up to 160 decibels, which means it can detect everything from whispers in a library to the sound of a TNT blast from a great distance.

3. Haptic sense

The primary research direction for haptic sensing technology is tentacle-like sensors. Several universities and institutes in the US, Switzerland, Japan, Germany, and Australia are currently developing tentacle-like sensors. The Marine Biology Laboratory at Northeastern University has developed powerful lobster robots with whiskers or tentacles. The tentacles on the left and right sides of this robot's head are made of polycarbonate for flexibility, and three binary microelectromechanical system (MEMS)-bending sensors are installed inside them to keep track of obstacles and water disturbances, emulating the highly sensitive tentacles of a lobster. This tentacle-like sensor can accurately identify mines suspended in any direction when used in conjunction with other sensors such as sonar altimeters. The US Office of Naval Research is also developing similar robots to detect enemy-laid mines before operations.

4. Olfactory sensing

Olfactory bionic sensors, also known as "electronic noses," are inspired by biological olfaction that extracts, senses, and recognizes odors through the use of an array of gas-sensitive elements. In 2007, General Electrics discovered that a unique nanostructure of butterfly wing scales possesses chemical sensing properties, reflecting a spectrum of colors when exposed to air containing trace amounts of chemical volatiles, and can deduce the chemical composition of the surrounding environment. Based on this mechanism, the company has developed bionic chemical sensing tech-nology that is sensitive, fast, and accurate and can detect dangerous war-fare agents and explosives on the battlefield. In 2010, Defense Advanced Research Projects Agency (DARPA) invested $6.3 million to develop bionic nanostructured sensors using this technology. In 2012, the tech-nology advanced to the point where it could detect a target's infrared thermal signature. In 2017, the US developed an artificial nose with external features and breathing rates similar to a canine nose. The artificial nose increased sensitivity by a factor of four at a distance of 10 cm from the vapor source and by a factor of 18 at a distance of 20 cm in the trial.

4.4.3.4 *Navigation and Guidance*

1. Visual navigation

    The Australian Defence Science and Technology Group (DSTG) and the US Air Force Research Laboratory's (AFRL) Munitions Directorate have collaborated on automated navigation technologies for microlight vehicles, focusing on using visual horizon stabilization to assist small military vehicles with navigation and guidance. To simulate the dragonfly compound eye's operation, the researchers developed a panoramic scanning device capable of generating a 400×200-pixel panoramic polarimetric map with a 360° field of view in the yaw direction and a 180° field of view in the pitch direction. If a microlight vehicle is not flying smoothly, the horizons detected by the left and right monocular will not line up in their respective fields of view. In response, a parameter for controlling the vehicle's aileron deflection angle will be generated based on the height difference, directing the aileron to eliminate the height difference.

2. Optical flow navigation

    Optical flow navigation is a vision-based bionic navigation technique. It entails detecting changes in the relative motion vectors of objects to determine their height, orientation, distance, and rotational speed and direction. When the optical flow vector is zero around a dragonfly, the dragonfly is hovering; when the optical flow vector rapidly increases on one side during flight, an obstacle appears on that side, and the flight path must be adjusted to avoid the obstacle. The dragonfly can also adapt to flying in a complex environment by adjusting the flight speed in response to the size of the optical flow vector.

3. Polarized light navigation

    Polarized light navigation relies on a polarized light sensor. The polarized light sensor is an angle sensor with numerous applications in navigation and positioning due to its small size, high accuracy, sound sensitivity, high integration, and strong anti-interference ability, and navigation error does not accumulate over time. In 2002, Schmolke et al. conducted successful experiments on mobile robot path tracking in an artificially polarized light environment in Germany [12]. In 2012, Chahl et al. developed a polarized light sensor in Australia that contained three independent, sensitive units, simulating the polarization-sensitive navigation of dragonflies [12]. The sensor was successfully used to determine the heading angle of a UAV.

4. Whisker-like navigation

    In 2017, researchers from the US and Singapore collaborated to research whisker-like bionic navigation. The research team developed a whisker array with a strain gauge at the bottom using five super elastic metal alloy wires covered in thin plastic tubes. The whisker array collects signals from the whisker's movements and generates images of the gas or fluid flowing through the array. These artificial whisker arrays could be used in place of conventional vision, radar, or sonar systems to navigate robots.

5. Magnetic navigation

    American cockroaches are magnetized when placed in a magnetic field. Magnetization results from magnetic particles in the cockroach's body aligning in response to an external magnetic field, as the magnetic properties of live cockroaches are significantly different from those of dead cockroaches, according to a joint study by Singapore, Australia, and Poland. A deeper understanding of biomagnetism may enable engineers to design better micro-robotic navigation systems. The DARPA project "Atomic Magnetometer for Biological Imaging In Earth's Native Terrain" is developing new magnetic gradiometers capable of detecting magnetic field signatures as low as picotesla and as high as femtotesla in an open environment, without shielding and regardless of the magnetic field state. Magnetic sensing and navigation functions with high sensitivity are now possible on low-cost devices operating in conventional environments.

4.4.3.5 Swarm Control

The US AFRL began researching bionic munitions in October 2007, successfully flying a swarming micro-bionic munition the size of a sparrow in 2015 and planning to fly a smaller micro-bionic munition the size of a dragonfly in 2030. DARPA launched the Swarm Challenge program in 2014, intending to develop autonomous swarm intelligent algorithms for unmanned vehicles to improve their ability to perform missions in complex environments, thereby relieving ground forces of their burden.

4.4.3.6 Bionic Control

Recent advancements in unmanned systems have resulted in emerging bionic systems with animal locomotion capabilities.

1. Swimming

    In 2017, Stanford University created a drone that featured multiple micro spines and a tail spine. The micro spines hook into the concave and convex structure of the stucco or slag brick wall and remain firmly attached to it through friction, allowing the drone to remain safely attached to the wall or ceiling like an insect. A translucent, coin-sized robotic fish has been developed at Harvard University in the US. When the cellular tissue contracts, the robotic fish moves downward; when the cellular tissue relaxes, the skeleton springs back into place, allowing the robotic fish to swim like a stingray, with a certain autonomy.

2. Anti-collision

   Swiss researchers developed quadrotor drones that borrow from the bio-mechanical collision avoidance strategy of insect wings. The drones feature a dual-rigid architecture that withstands aerodynamic loads within its flight envelope but softens and folds in the event of an impact, preventing damage. The drone's vulnerable center components are protected by the dual-rigid structure combined with proprietary energy-absorbing materials.

3. Cilia movements

   South Korean researchers used a glass substrate and a 3D laser lithography system to create a miniature robot that is 220 microns long, 60 microns high, with eight 75-micron-long cilia on each side of its body that mimic the movement of a grass worm's cilia. The robot can move and position when triggered remotely by a magnetic field formed by eight electromagnetic coils. The robot's average velocity for medical administration was 340 micrometers/second in tests conducted in a mixture of pure water and silicone oil.

4. Hydraulics

   US and Danish researchers have jointly developed a microfluidic device called a tree-on-a-chip. The device can replicate the pumping mechanisms found in trees and plants, continuously pumping water and sugar through the chip for days. Its chip is passive, requiring no moving parts or external pumps to operate. The technology could be used to control tiny robots hydraulically.

5. Soft-bodied pneumatics

   University of California and Stanford University researchers have collaborated to develop a soft-bodied pneumatic robot. The robot can extend its length indefinitely from the end, and the direction of extension is autonomously controlled via self-contained environmental stimulation sensors. The researchers demonstrated the robot's passive extension in a constrained environment and its ability to form 3D structures by elongating its body along a path.

6. Twisting motions

   In 2015, the International Institute for Advanced Studies in Italy used a model system to study twisting motions.

7. Dog-like walking

   In 2016, Turkish researchers developed a Petri Net model that can describe the behavior of biological systems using real-time data collected during experiments. In the test, a dog walked at a speed of 1 km/h on a treadmill. Through the use of color, digital, and step simulation techniques, the model enables the simulation of the dog's real-time walking rhythm. The initial conceptual design for a bionic robot dog can thus be generated.

4.4.3.7 Trends

1. Bionic systems will evolve from structural to functional bionic, from single-domain operability to amphibious and triphibious operability, demonstrating the trend towards miniaturization, intelligence, and swarming, and eventually culminating in 3D bionic combat systems capable of operating in water, land, and air.
2. Composite bionic materials will be developed to enhance systems' stealth capabilities in visible light and radar environments and increase their toughness and flexibility.
3. Bionic perception technologies for visual, auditory, and olfactory detection will be developed to help systems perceive, recognize, and track targets in complex environments.
4. Bionic control will evolve toward multi-joint and more precise control, resulting in increased control accuracy and stability.
5. Bionic system networking (e.g., large-scale, distributed, and intelligent swarm networking) will emerge, enabling real-time interoperability with battlefield data and joint munitions.

4.4.4 Swarm Intelligence

4.4.4.1 Concept

Swarm/collection intelligence is a concept derived from observations of insect groups in nature. Swarms of organisms exhibit macro-intelligent behavioral traits through a collaboration known as swarm intelligence [13].

4.4.4.2 Background

Since its inception in the 1980s, swarm intelligence has attracted the attention of researchers from a variety of disciplines. It has developed into a frontier area at the nexus of AI, economy, and social biology.

Swarm intelligence was discovered in nature long ago, where some organisms rely on their individual intelligence to survive, while others can benefit from the swarm's strength. Individuals in these swarms lack intelligence, but they can cooperate to complete complex tasks and exhibit a moderate level of intelligence. When they form swarms, they develop a high degree of self-organization, adaptability, and non-linear, emergent system characteristics.

Thus, swarm intelligence may refer to the intelligence that individuals exhibit when they cooperate in a swarm. In this concept, the individual is not completely unintelligent or illiterate but lacks intelligence compared to the swarm intelligence. When a swarm of individuals cooperates or competes, wisdom and behavior that were previously unknown to any individual can emerge quickly.

4.4.4.3 Basic Principles

1. The principle of proximity, which states that swarms are capable of performing simple spatial and temporal calculations.
2. The principle of quality, which states that the swarm can respond to environmental quality factors.
3. The principle of diversified response, which states that the swarm's scope of action should not be excessively limited.
4. The principle of stability, according to which the swarm's behavior should not change in response to all changes in the environment.
5. The principle of adaptability, whereby the swarm's behavior can be changed at an appropriate time and a reasonable cost.

4.4.4.4 Features

1. Control of the swarm is distributed, with no centralized command. Thus, the swarm is adaptable to the network environment, implying that a single or a few individual failures will not jeopardize the overall stability.
2. Each group member can respond to the environment via a process called "stigmergy," a form of indirect communication between members. Swarm intelligence is scalable because it is based on indirect communication and information transmission, where increases in the number of individuals result in only a slight increase in communication capacity.
3. Each individual's capability and rules of behavior are quite simple, making the realization of swarm intelligence straightforward.
4. Swarms are self-organizing, as the complex behaviors exhibited by swarms represent intelligence that emerges from the interactions of individuals.

4.4.4.5 Typical Models

Over a long period, swarm intelligence research has yielded numerous significant findings. Since 1991, when Italian scholar Dorigo proposed the theory of ant colony optimization (ACO), swarm intelligence has been formally proposed as a theory that gradually drew the attention of scholars, igniting a research climax. In 1995, Kennedy et al. proposed the particle swarm optimization (PSO) algorithm [14], and since then, research on swarm intelligence has accelerated, although the majority of work has focused on ACO and PSO.

At the moment, swarm intelligence research is concentrating on intelligent ACO and PSO algorithms. Intelligent ACO algorithms include ACO, ant colony swarming, and multi-robot cooperative systems. The ACO and PSO algorithms are the most frequently used in practice.

4.4.5 Hybrid Intelligence

4.4.5.1 Concept

At the moment, hybrid intelligence is under development, and there is no authoritative or unified definition for it. A preferred definition of hybrid intelligence is increased, more robust, and enhanced intelligence achieved through human–machine complementarity, collaboration, and fusion. Human–machine fusion intelligence is a novel form of intelligence that is distinct from human and machine intelligence. It is the next generation of intelligence, combining characteristics of multiple species [15].

According to experts, hybrid intelligence features the process of incorporating human cognitive abilities or models into AI systems to create new forms of AI that are both viable and necessary for intellectual growth. Intelligent machines and terminals of all types will become human companions, and the interaction between machines and humans will define the future of society.

Current AI systems perform "supervised learning" by analyzing many samples at various levels of training, whereas systems of true general intelligence deftly perform "unsupervised learning" based on accumulated experience and knowledge. Generic AIs cannot be created by simply combining various AI computational models or algorithms. Thus, hybrid augmented intelligence based on human–machine fusion will be a hallmark of the next generation of AI.

4.4.5.2 Forms

Experts define hybrid intelligence in two ways: "augmented intelligence with humans in the loop" and "augmented intelligence based on cognitive computing."

Hybrid augmented intelligence is a paradigm that incorporates humans into the intelligent system, namely, with humans in the loop. In this paradigm, humans are inextricably linked to intelligent systems. If there is a lack of confidence in the system's output, parameters will be adjusted manually to find the correct solution, creating a feedback loop that improves the system's intelligence level. Humans embedded in an intelligent system can provide analysis of fuzzy and uncertain problems and advanced cognitive mechanisms for machine intelligence, enabling humans and machines to work together in a two-direction feedback system. Combining human perception and cognitive ability with the computer's powerful computing and storage capabilities results in an enhanced intelligence form of "1 + 1 > 2."

In the military, as unmanned technology advances, the relationship between humans and machines may take on various hybrid forms. Humans and weapons are gradually becoming physically separated, and the focus on drones and robots has shifted away from assisting humans to replacing them, relegating humans into the background. Humans and machines will increasingly integrate into an organic symbiosis in the field of information and control, where human creativity, thoughtfulness, and wisdom will always be combined with the precision, speed, repetitiveness, and consistency of machines, whether as "humans in the loop," "humans above the loop," or "humans beyond the loop."

Hybrid augmented intelligence based on cognitive computing involves incorporating biologically inspired intelligent computational models into AI systems. According to Nanning Zheng, a Chinese scholar,

> Hybrid augmented intelligence is a type of artificial intelligence in which intelligent software or hardware augments computers' perception, reasoning, and decision-making capabilities by simulating biological brain functions. The objective is to develop intelligent computers capable of perceiving, reasoning, and responding to incentives in the same way that the human brain does, via causal modeling, intuitive reasoning, and associative memory [16].

For current AIs, while solving intellectually challenging problems may be straightforward, solving routine problems can be exceedingly difficult. For example, few 3-year-olds can play Go unless they have been specially trained, but all 3-year-olds are capable of recognizing their parents regardless of whether they have been trained with a labeled face dataset. One of the central directions of AI research is to apply findings from cognitive science and computational neuroscience to enable higher levels of computer intelligence through intuitive reasoning and experiential learning.

At the moment, machine intelligence is still computer-centric, with a low level of intelligence for human–computer interaction, and has not yet developed a "human-centered" interaction concept or systematic design concept, making it difficult to adapt to future social work and life scenarios requiring intelligent, time-sensitive, and massive data processing. The technology community is currently focused on incorporating human cognitive models into machine intelligence to achieve human-like reasoning, decision-making, and memory capabilities.

AIs have already demonstrated superior capabilities to humans in massive data search and gaming under ideal conditions with factual data. In the highly dynamic and uncertain environment of military confrontation, human experience, intuition, and inspiration can demonstrate enormous potential when combined with the speed and precision of intelligent systems. Future research on human–machine hybrid intelligence will concentrate on cognitive science-based AI frameworks, novel methods for human–machine integration, parallel intelligent systems, intelligent learning from big data, intelligent interaction of command/manipulation intentions, and human–machine military intelligence, as well as natural human interactions such as voice, vision, and action.

4.4.6 Knowledge Graphs

4.4.6.1 Concept

A knowledge graph is a collection of distinct graphs illustrating the process of knowledge development and structural relationships, describing knowledge resources and their carriers, mining, analyzing, constructing, mapping, and displaying knowledge and its interconnections using visualization techniques [17].

Knowledge graphing is a cutting-edge technique that combines theories and methods from applied mathematics, graphics, information visualization techniques, and information science, as well as metrological citation and co-occurrence analysis. It visualizes the core structure, history, frontier areas, and knowledge architecture of different disciplines. Additionally, it visualizes complex knowledge through data mining, information processing, knowledge measurement, and graphical mapping, reveals the knowledge domain's dynamic development pattern, and serves as a useful and practical reference for disciplinary research.

4.4.6.2 Characteristics

1. The more users who search, the broader the scope and the greater the amount of information and content that can be accessed.
2. Word strings are given new meanings when integrated into a graph.
3. Disciplines are integrated to improve user search coherence.
4. Users are provided with accurate information, more comprehensive summaries, and more in-depth and relevant information.
5. Users are presented with a structured presentation of knowledge associated with keywords.
6. Users can access all information and data on other services by logging into any online service.
7. Users are directed to a greater number of publicly accessible resources across the Internet.

4.4.6.3 Improved Search Results

These three aspects demonstrate how search results are improved.

1. Most desired results
   Due to the ambiguity of language, a single search request can have multiple interpretations. In this case, a knowledge graph may present the data in its entirety, allowing users to drill down to the most pertinent information. Google can now distinguish between different search keywords and narrow the search results to the most relevant results for users.

2. Most comprehensive summary
   Using the knowledge graph, Google can now better understand the information sought by users and summarize content related to the search topic. For instance, when users search for "Marie Curie," they will discover information about Madame Curie's life and educational background, as well as details about her scientific discoveries. Additionally, the knowledge graph will assist users in comprehending the relationships between objects.
3. Deeper and broader searches
   Since the knowledge graph compiles a comprehensive set of search results, users frequently make unexpected discoveries while searching. A user may discover a new fact or connection during a search, which generates a new set of search queries.

4.4.7 Brain–Computer Interface

4.4.7.1 Concept

The BCI directly connects a human or animal's brain and an external device. In the case of a one-way BCI, the computer either receives commands from the brain or sends signals to the brain (e.g., video reconstruction), but not both. In contrast, a bidirectional, or two-way, BCI enables information to be exchanged between the brain and an external device in both directions.

4.4.7.2 Background

Since the dawn of time, humans have aspired to communicate directly with the outside world and even gain control of the surrounding environment via signals generated by the brain's thought activity. Since Hans Berger first recorded the electroencephalogram (EEG) in 1929, there has been speculation that EEG could be used to enable the brain to act directly on the external world without the usual mediators of peripheral nerves and limbs. However, due to technological limitations and a persistent lack of understanding of the brain's cognitive mechanisms, little progress has been made in this area of research.

The concept of BCI was first proposed in the 1970s. Over the last few decades, research on BCI techniques has demonstrated a clear upward trend, paralleling advances in our understanding of the nervous system's function and the advancement of computer technology. In particular, the two international conferences on BCI held in 1999 and 2002 pointed the way for the development of BCI technology.

4.4.7.3 Basic Structure

Numerous prototype EEG-based BCI systems have been developed for various applications and are currently being demonstrated in the laboratory.

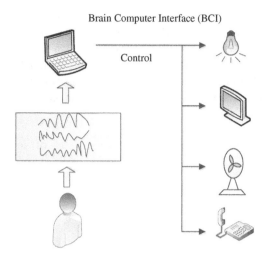

Figure 4.3 Brain–computer interface (BCI) control of household appliances.

BCI systems are composed of functional components such as input, output, signal processing, and conversion (Figure 4.3). The input function generates and detects signals corresponding to a specific EEG activity and characterizes these features using parameters. Signal processing is concerned with the conversion of an analog signal to a digital signal represented by specific parameters (e.g., amplitude and model coefficients of an autoregressive model) so that a computer can process and classify the signals and determine their corresponding brain activity levels. The process of signal conversion involves converting characteristic signals derived from signal analysis and classification into letters or words that represent the brain's desire to communicate with the outside world. Signal analysis and conversion are vital components of the BCI system since they act as intermediaries between input and output. The accuracy of classification can be enhanced by optimizing the algorithm for signal analysis and conversion with constant training intensity, which in turn optimizes the BCI system's performance. BCI devices can move a pointer, select characters, move a neural prosthesis, or control other devices.

4.4.7.4 Classification

The First International Conference on BCI classified BCI systems into two broad categories based on the type of input signal: systems based on spontaneous EEG and systems based on evoked EEG.

Spontaneous EEG-based BCI systems utilize spontaneous EEG as the system's input signal. Spontaneous EEG-based BCI systems are defined by the subject's ability to exert direct control over his or her external environment through EEG control. It does, however, typically require extensive subject training and is contingent on factors such as physical, emotional, and medical conditions.

In evoked EEG-based BCI systems, external stimuli are used to induce changes in electrical activity in corresponding areas of the cerebral cortex, which are then used as characteristic signals. Evoked EEG-based BCI systems do not require extensive subject training, but they do require specialized environments, such as a matrix of lined-up flickering visual stimulus inputs. This requirement is incompatible with the system's generalization and application.

In terms of system output, spontaneous EEG-based BCI systems allow the operator to move the pointer to any two- or multi-dimensional position, whereas evoked EEG-based BCI systems limit the operator to the listed options.

Additionally, BCI technologies can be classified as "implantable" or "non-implantable," depending on how the signals are detected. Due to the technical difficulties associated with high-precision brain skin implantation, the implantable type is still being tested in humans for medical research purposes. Currently, the Netherlands and the US are at the forefront of this field. Due to the ease of loading and unloading, the non-implantable type has entered the commercial stage and is used for entertainment and medical purposes. Representative companies developing this type of BCI include Neurowear of Japan and Emotiv of the US.

Implantable signal detection requires the connection of electrodes to the cerebral cortex, which enables noise- and loss-free signal measurement. However, this method involves complex surgical procedures that require highly skilled operators and expose patients to infection.

Non-implantable signal detection, utilizing external electrodes, is safe and straightforward and helps popularize BCI systems. However, the signal detected with this method is noisy due to the electrodes' long distance from the signal source. A BCI system's scheme is determined by several factors, including the signal's characteristics, the level of measurement, and the required accuracy.

4.4.7.5 Applications

The initial focus of BCI research is on the military to enable soldiers to operate weapons and even remotely control robots and drones via their brains, thereby increasing combat effectiveness and reducing casualties on future battlefields. BCI research in the civilian context is also of far-reaching importance. BCI systems can improve people's quality of life on the medical front by providing assistive controls, particularly for those with severe motor disabilities. For example, in daily life, BCI can also be used in place of traditional mice and keyboards or other hand-controlled devices to assist us in playing games, watching TV, and performing other tasks directly through our brains, thereby increasing our enjoyment of life.

Research on implantable BCI has seen significant progress in animal experiments but has been much more challenging in human experiments. Miguel Nicolelis of Duke University advocates driving BCI and extracting neural signals via electrode coverage of large cortical regions [18]. Miguel believes that this method has the advantage of reducing the instability and randomness of neural signals collected using a single electrode or a few electrodes.

In 1999, Garrett Stanley of Harvard University attempted to reconstruct visual images by decoding information about neuronal firing within the lateral thalamus of the cat's geniculate body. The researchers recorded the pulse trains of 177 neurons and used a filtering method to reconstruct eight videos played to the tested cat. The reconstructions revealed recognizable objects and scenes.

Brown University pioneered the use of BCI in clinical settings in 2006 when it implanted an EEG-controlled prosthesis on a paralyzed patient. On January 10, 2008, Duke University Laboratory implanted multiple microelectrodes (biosensors) into the head of a monkey named Edoia, and the monkey's mind directed a robot thousands of miles away at the Advanced Telecommunications Research (ATR) Computer Neuroscience Laboratory in Tokyo, Japan, to walk in the monkey's footsteps for up to 3 minutes.

In 2012, researchers at the University of Pittsburgh used electrocorticography signals from a human brain to control a robotic hand developed by the university. This allowed a severely paraplegic patient to shake hands with US President Barack Obama by allowing his brain to receive signals from the robotic hand, simulating the sensation of handshaking. In 2016, a team of researchers at Utrecht University in the Netherlands successfully enabled a patient who had lost his motor and even eye movement abilities due to acromegaly to type on a computer with 95%accuracy using BCI, a significant step forward in the application of implantable BCI.

Between 2019 and 2021, Elon Musk's startup, Neuralink, implanted BCI in the brains of monkeys and pigs, demonstrating that some of the monkey's and pig's behaviors match the AI's predicted signal curves (see Chapter 3, section 3.2).

Due to their relative ease of operation, more research and development teams prefer "non-implantable" modalities such as EEG, magnetoelectroencephalography (MEG), near-infrared spectroscopy (NIRS), and functional magnetic resonance imaging (fMRI). Several BCI products have been developed and are now commercially available. For example, Honda has developed mind-controlled robots in which the operator can direct the robots around them to perform the required movements simply by imagining their body movements. In a study at the University of Rochester, subjects could control objects in a virtual-reality scene via P300 signals, performing tasks such as turning on and off lights or operating a virtual car. Emotiv, a San Francisco-based neurotechnology company, has developed Emotiv Insight, a brainwave-compiling device that disabled individuals can use to control wheelchairs or computers.

Since 2003, DARPA has invested $24 million in six different laboratories across the US to research the Brain–Machine Interface (BMI) program, which aims to develop "mind-controlled machine warriors" or unmanned aircraft. In 2008, the US Army developed a 10-year plan for a BCI-based "multi-player decision-making system." In 2012, the US government proposed the Avatar humanoid robot project to remotely control a semi-autonomous bipedal robot via a BCI. NASA has also developed software that can decipher human unspoken thoughts and intentions.

China has also made significant strides in the field of research on BCI. Tsinghua University, South China University of Technology, University of Electronic Science and Technology of China, Shanghai Jiao Tong University, Zhejiang University, National University of Defense Technology, Lanzhou University, Third Military Medical University, Anhui University, Tianjin University, and Hangzhou University of Electronic Science and Technology, among others, have all established BCI research teams and achieved success in international BCI competitions. Zhejiang University, for example, pioneered research into a rat's mind control of animal robots, monkey's remote control of robotic hands, and finally, human mind control of robotic hands, using electrodes implanted intracranially in the first domestic patient. In 2015, Shanghai Jiao Tong University achieved successful mind control of cockroach behavior. In 2001, Tsinghua University pioneered mind control of the mouse and television buttons. In 2006, the Tsinghua researchers enabled humans to control two robotic dogs in a football match via motor imagination. Later, they developed a new method of rehabilitation training based on BCI and functional electrical stimulation technology. The South China University of Technology, using a combination of P300 signals and motor imagery, developed text input systems, cursor-controlled Internet emailing, and applications in assisted living (e.g., control of home appliances and wheelchairs) and neurological rehabilitation for people with disabilities. In October 2016, Tianjin University's neuroengineering team was tasked with designing and developing a test system for in-orbit brain–computer interaction, brain load, visual function, and other aspects of neural ergonomics. The system was launched into space aboard the "Tiangong II" to conduct China's first space brain–computer interaction experiments.

4.4.8 Mental State Assessment

4.4.8.1 Concept

Mental state assessment is based on validity analysis, mathematical models, and inference algorithms of "psychological and physiological calculations."

It is a multi-disciplinary field encompassing BCI, molecular biological data, brain imaging data, physiological electrical signals, speech, eye movements, virtual reality, and augmented reality.

Internationally, the theory of objective mental state assessment is still in its infancy, with most theories relying on biological or psychological data, including analyses of mental state validity, mathematical models, and inference algorithms. Due to the specificity of the information sources used, generalizing the resulting models is difficult. Critical development directions and practical methods include integrating multimodal physiological and psychological data about mental states, developing models for objectively assessing mental states via a new research concept and experimental paradigm based on "psychophysiological computing," and innovating the theory of mental state assessment.

4.4.8.2 Criteria

Current methods and rating scales for mental health assessment, widely used on a national and international level, are valid only when specific criteria are met. Establishing objective assessment criteria that enable early detection of individuals with poor mental health is a significant challenge that must be addressed.

A possible method would be to develop an objective, quantitative assessment standard system for mental states that incorporates EEG, speech, and eye movements to deduce complex correspondences between mental states and physiological responses and establish reliable, reproducible, and universal one-to-one relationships between them.

4.4.8.3 Key Technologies

Mental state assessment requires various critical technologies, including wearable physiological and psychological information acquisition, signal pre-processing, reliable communication and storage, big data-oriented databases, data mining, mathematical and statistical analysis, virtual reality, and brain function biofeedback. Biofeedback is a highly effective technique for resolving physiological and psychological discomfort. Simultaneously, virtual-reality technology has the advantages of being immersive, interactive, and conceptual, and it can be used to create customized virtual scenarios that engage users' visual, auditory, and haptic senses. As a result, wearable sensing devices can collect physiological signals from subjects during the conditioning process. The signals can then be used to continuously monitor their status and provide feedback to assist them in overcoming their negative mental state.

4.4.8.4 Applications

Mental state assessment technology can be used to detect and assess the mental health of children, new students, employees, special populations in non-disturbance environments, and in fields such as psychological diagnosis and elderly care. Additionally, it can be used in the military for recruiting, induction testing, on-duty monitoring and regulation, and psychological rehabilitation of wounded personnel, among other human–machine interaction scenarios.

References

[1] Pan, Y. H. 潘云鹤. (2017, January 15). *Rengong zhineng maixiang 2.0* 人工智能迈向2.0. Kexue wang 科学网. https://news.sciencenet.cn/htmlnews/2017/1/365934.shtm.

[2] Institute of Linguistics, CASS 中国社会科学院语言研究所词典编辑室. (1996). *Xiandai hanyu cidian* 现代汉语词典. Shanghai: The Commercial Press.

[3] Baidu Baike 百度百科. (2018b, March 4). *Zhinenghua* 智能化. Baidu baike 百度百科. https://baike.baidu.com/item/.

[4] The State Council of the People's Republic of China 中华人民共和国国务院. (2017, July 8). *Guanyu yinfa xinyidai rengong zhineng fazhan guihua de tongzhi* 关于印发新一代人工智能发展规划的通知. Official Website of the State Council of the People's Republic of China 中华人民共和国国务院官方网站. www.gov.cn/zhengce/content/2017-07/20/content_5211996.htm.

[5] Baidu Baike 百度百科. (2018a, December 24). *Jiqi xuexi* 机器学习. Baidu baike 百度百科. https://baike.baidu.com/item/.

[6] Baidu Baike 百度百科. (2019, January 1). *Shendu xuexi* 深度学习. Baidu baike 百度百科. https://baike.baidu.com/item/.

[7] LeCun, Y., Bengio, Y., & Hinton, G. (2015). Deep learning. *Nature*, *521*(7553), 436–444. DOI:10.1038/nature14539

[8] *Reducing the dimensionality of data with neural networks*. (2006). https://blog.csdn.net/zephyr_wang/article/details/119878836.

[9] *To recognize shapes, first learn to generate images*. (2006). https://blog.csdn.net/zephyr_wang/article/details/120157174.

[10] *A fast learning algorithm for deep belief nets*. (2006). https://blog.csdn.net/zephyr_wang/article/details/120007607.

[11] Raut, H. K., Schwartzman, A. F., Das, R., Liu, F., Wang, L., Ross, C. A., & Fernandez, J. G. (2020). Tough and strong: Cross-lamella design imparts multifunctionality to biomimetic nacre. *ACS nano*, *14*(8), 9771–9779. DOI: 10.1021/acsnano.0c01511

[12] Schmolke, A., Mallot, H., & Neurowissenschaft, K. (2002). Polarization compass for robot navigation. In K. Labusch & D. Polani (Eds.), *Proceedings of the Fifth German Workshop on Artificial Life* (pp. 163–167). Lubeck: IOS Press.

[13] Kuaidong Baike 快懂百科. (2021, July 26). *Qunti zhineng* 群体智能. Kuaidong baike 快懂百科. www.baike.com/wiki/.

[14] Kennedy, J., Eberhart, R. C., & Shi, Y. (2001). *Swarm intelligence.* Burlington, MA: Morgan Kaufmann.

[15] Zhao, G. L. 赵广立. (2017, August 3). *Hunhe zhineng* 混合智能. Qiushi wang 求是网. www.qstheory.cn/science/2017-08/03/c_1121423678.htm.

[16] Zheng, N. N., Liu, Z. Y., Ren, P. J., Ma, Y. Q., Chen, S. T., Yu, S. Y., ... Wang, F. Y. (2017). Hybrid-augmented intelligence: Collaboration and cognition. *Frontiers of Information Technology & Electronic Engineering, 18*(2), 153–179. doi:10.1631/FITEE.1700053

[17] Baidu Baike 百度百科. (2018c, August 22). *Zhishi tupu* 知识图谱. Baidu baike 百度百科. https://baike.baidu.com/item/.

[18] Nicolelis, M. (2020). *Tencent "We" session. Speech: New era of neurology is coming.* Beijing: Tencent.

5 AI-Based Intelligent Ecology

In traditional warfare, operational elements are relatively independent and distinct from one another. The battlefield ecosystem is relatively simple, consisting of human beings, equipment, and tactics. In the intelligent era of warfare, operational elements will present obvious integration, association, and interaction features. The battlefield ecosystem will undergo significant transformation, evolving into an operation system of systems (OSoS), swarm system, or human–machine integrated system, comprised of an artificial intelligence (AI) brain, distributed cloud, communication networks, collaborative swarms, and cloud terminals (equipment and personnel connected to the cloud and network), collectively referred to as an intelligent battlefield ecosystem characterized by "AI, clouds, networks, swarms, and terminals" (Figure 5.1). AI is a significant contributor to all of these.

5.1 AI Brain Systems

An AI brain system is a networked and distributed system pre-programmed to recognize and track combat vehicles and missions on the intelligent battlefield. This type of system has a plethora of classification schemes. An AI brain system can be classified into the cerebellum, swarm brain, mid-brain, hybrid brain, and main brain systems based on its function and computing power. Depending on the combat tasks and links, it can be classified into sensor AI, combat mission planning and decision AI, precision strike AI with controllable damage, cyber attack and defense AI, electronic countermeasure AI, intelligent defense AI, and comprehensive logistics support AI. Its form can also be classified as embedded, cloud, or parallel AI.

The cerebellum, or embedded AI in sensor platforms, combat vehicles, and security platforms, is primarily responsible for battlefield environment detection, target identification, precision strike, controllable damage, as well as equipment operation, usage, security, and maintenance.

The swarm brain refers to an AI used to control unmanned swarm systems on land, sea, air, and space. It is primarily used for cooperative battlefield sensing, swarm mobility, swarm strikes and defense.

DOI: 10.4324/b22974-6

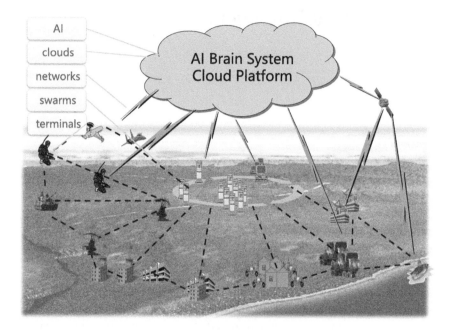

Figure 5.1 Intelligent battlefield ecosystem.

The mid-brain refers to an AI system that operates the battlefield's front-line detachment, data center, and post-edge computing command. It is primarily used to perform dynamic planning and autonomous and auxiliary decision-making for online and offline tactical detachment operation tasks.

The hybrid brain refers to the joint command and decision-making system between commanders and machine AI in operations. It is primarily responsible for pre-war human-based mission-planning, mid-war AI-based adaptive dynamic mission-planning and adjustment, and post-war human–AI hybrid decision-making for anti-terrorism, defense, and other functions.

The main brain comprises the command center, data center, model, knowledge, algorithm, and tactics base or repository used in joint, all-domain, and cross-domain operations. Before being mature enough to be loaded for various missions, all types of battlefield brain systems can be trained and modeled using sufficient data provided by the main brain.

Other AI technologies with a variety of functions and scales will also appear on the future battlefield. For example, sensor AI will perform image recognition, electromagnetic spectrum recognition, sound recognition, voice recognition, and recognition of human activity behavior.

From a civil standpoint, the "City Brain", which Alibaba is currently constructing and demonstrating in Hangzhou, and the robot controlled by the "cloud brain" based on CloudMinds' mobile intranet security architecture, represent a critical direction for future intelligent equipment. These civil "brains" can serve as a training ground for developing military brain systems.

In general, Alibaba's City Brain is a three-tier architecture comprised of a computing platform, a data resource platform, and an application service platform with instant, full volume, full network, and full video capabilities. All enabled by an open system. In terms of intelligent traffic management, a breakthrough was made by resolving the "world's most remote distance", the distance between traffic lights and surveillance cameras located on the same pole but that have never been connected via data. Hangzhou's City Brain enabled the use of camera data to direct traffic signals for the first time, and preliminary calculations and tests indicate that it can increase traffic efficiency by more than 17%. Traffic management is just one of the numerous applications of the City Brain. Additionally, the data generated by the City Brain will generate additional value for society. They will eventually enable intelligent perception of people and things at any time and location, thereby improving urban systems, public service efficiency, and citizens' quality of life.

CloudMinds, a tech company that has proposed and developed Human Augmented Robot Intelligence with eXtreme Reality (HARIX), is currently deploying a global virtual backbone network (VBN) and implementing the Mobile-Intranet Cloud Service (MCS) to enable an information security system for cloud-based remote control of robots and to provide a critical "cloud–network–end" security architecture for next-generation enterprise mobile Internet of Things (IoT). On February 23, 2017, CloudMinds launched the world's first cloud-based intelligent robot operation platform, enabling the platform's MCS to achieve end-to-end security isolation (Figure 5.2).

On February 25, 2019, CloudMinds unveiled the XR-1 cloud-based intelligent, flexible service robot at the Mobile World Congress (MWC) in Barcelona, Spain. At the conference, the XR-1 robot used its flexible arms to deliver beverages (e.g., coffee) to on-site guests. During the human–robot interaction stage, the robot also detected human presence and moved flexibly through smart compliant actuator (SCA) intelligent joints. Meanwhile, the HARIX performed closed-loop control based on visual feedback, enabling the XR-1 robot to perform intelligent grasping and transmission tasks in real-world environments. In particular, combining high-precision vision sensing, three-dimensional (3D) objects and environment perception, and innovative, flexible joint SCA technology, the XR-1 robot completed a high-precision operation, namely threading a needle, based on precise control of visual feedback, demonstrating human-like flexibility and service capabilities under cloud brain's command (Figure 5.3).

In the future, as intelligent technology advances become more widely used, AI of all types will exist throughout society, serving people and society in times of peace and, very possibly, the military during times of war.

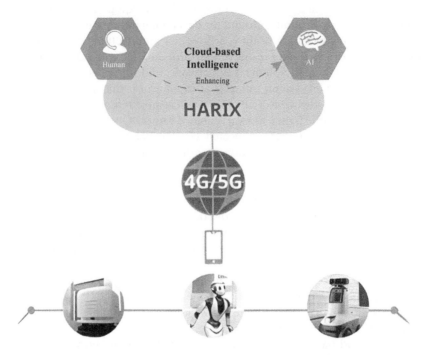

Figure 5.2 Mobile-Intranet Cloud Service (MCS)-based "cloud–network–terminal" security architecture.

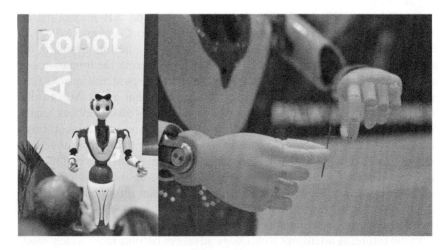

Figure 5.3 The XR-1's fine needle-threading performance is evaluated to determine the robot's precision and flexibility.

5.2 Distributed Clouds

Military clouds are distinct from civilian clouds. In general, military cloud platforms are distributed management systems that leverage communication networks to find, aggregate, analyze, compute, store, and distribute operational information and data. The military cloud platform is based on a distributed system and a multi-point fault-tolerant backup mechanism, enabling strong intelligence sharing, data processing, strike resistance, and self-repairing capabilities. It can also provide fixed and mobile, public and private cloud services to achieve "single-point data collection, shared by all", significantly improving information transmission efficiency, flattening and speeding up the command process, and avoiding repeated ad hoc commands.

To support intelligent warfare, the military cloud must establish a minimum of a four-tier system comprised of Edge Cloud, Force Cloud, Theater Cloud, and Strategic Cloud. According to operational elements, the system can also be divided into specialized cloud systems such as intelligence, situational, firepower, operational information, and support clouds. Once tens of thousands of small satellites form a multi-functional space-based Internet in the future, a warfare model based on the space–ground integrated "nebula" will be possible.

The computing services provided by the Edge Cloud include information perception, target identification, battlefield environment analysis, autonomous and auxiliary decision-making, and the evaluation of combat processes and effects among detachments, squads, and vehicles. The Edge Cloud serves two purposes. It enables platform-to-platform sharing of computing and storage resources and the interactive integration of collaborative and intelligent combat data. Once a platform or terminal is attacked, the relevant perception data, destruction status, and history are automatically backed up, replaced, and updated via the networked cloud platform and uploaded to the higher command post. The second is service and software upgrades for offline terminals.

The Force Cloud refers to cloud systems designed for battalion and brigade-level operations, emphasizing computing services such as intelligent perception, intelligent decision-making, autonomous operations, and intelligent logistics in response to a variety of threats and environments. The Force Cloud is a distributed cloud system that is networked, automatically backed up, and connected to multiple links at higher levels to meet the computing requirements of various missions and operations, including reconnaissance and perception, mobile assault, command and control, fire strike, and rear-armor protection, as well as joint tactical operations, manned/unmanned collaborative operations, and swarm attack and defense.

The Theater Cloud provides environmental data such as meteorological, geographic, electromagnetic, human, and social data for the entire battlefield, as well as military data such as troop deployment, weapons and equipment, movement, and battle damage on both sides of the conflict, including information about higher command posts, allied forces, and civilian support forces. The Theater Cloud is expected to be networked, customized, intelligent, and interconnected with each combat unit via military communication networks such as space-based, air, ground, sea, and underwater, involving civilian communication networks protected by secrecy measures, to ensure the provision of efficient, timely, and accurate information services.

The Strategic Cloud is a massive military database integrated into national defense systems and command-and-control organs. It covers defense science and technology, defense industry, mobilization and security, economic and social support capabilities, as well as political, economic, diplomatic, and public opinion data, providing assessments, analysis, and recommendations on war preparation, operational planning, operational programs, operational processes, and battlefield conditions. Additionally, it can provide supporting data on an adversary's military strength and war mobilization capability.

Among the above, there are small and large-scale clouds, high- and low-level clouds, and horizontal collaboration, mutual support, and service between them. The military cloud platform serves two primary missions: one is to support the development of an AI brain system for intelligent operations, and the other is to provide combat information, computing, and data support for combatants and combat vehicles. Additionally, cloud computing data, models, and algorithms must be integrated into smart chips embedded in combat vehicles and terminals that can be updated online or offline.

5.3 Super Networks

Military communication and network information systems are intricate super networks. Military communication networks can be classified according to their use in land, sea, and air operations, as well as fields, towns, and other urban areas. The networks include strategic and tactical communication, wired and wireless communication, secure communication, and civilian communication. Among them, wireless, mobile, and free-space communication networks are the most critical components of the military network system, on which subsequent electronic information systems are built.

In the mechanized era, military communication was based on the vehicle, terminal, and user. It was sufficiently specialized but lacked interoperability due to an excessive number of antennae. It has shifted in the information technology era. Military communication networks are currently implementing a new technical model with two primary characteristics. The first is network–digital separation, which means that information transmission will not be protocol-dependent. Any information can be delivered if the network link is open. The second is military Internetization, which involves using Internet protocol (IP) addresses, routers, and servers to achieve "all roads lead to Rome." Of course, military communication networks are distinct from civilian ones. They are constantly required for strategic and specialized communication, such as command and control of strategic weapons (e.g., nuclear weapons), information transmission for satellite reconnaissance, remote sensing, and strategic early warning. Exclusive communication methods may even be established in the context of special operations involving difficult communication conditions.

Nonetheless, privatization and Internetization will undoubtedly be the future directions of military communication network development. Otherwise, the proliferation of battlefield communication bands, radio stations, and information exchange methods will result in self-disturbance, mutual disturbance, and electromagnetic compatibility issues, as well as increased complexity in radio spectrum management. Additionally, implementing automatic connectivity between platform users based on features such as IP address-based and routing structures, such as email on the Internet, where a single-click command can be passed to multiple users, is challenging. Future combat vehicles will undoubtedly serve as user terminals, routers, and servers.

The US military has developed and improved a global information grid system by adhering to "network–digital separation" and Internetization. The US Army proposed the concept of LandWarNet (LWN) in response to the requirement for a Global Information Grid (GIG). The LWN is an information integration platform that leverages the existing networks, infrastructure, communication systems, and applications of the US Army to provide five capabilities: leader-centric mobile command, globally interconnected and interoperable capability, global command capability at any time and from any location, the ability to provide a unified joint common operational picture, and the application ability with joint operations as the core.

The military communication network system is comprised of several components, including a space-based information network, a military mobile communication network, a data link, a new communication network, and a civil communication network, among others.

5.3.1 Space-Based Information Networks

The US is a world leader in the development and deployment of space-based communication networks. The US owns more than half of the in-orbit platforms and payloads in space, numbering in the thousands. Through war practice following the Gulf War and particularly during the Iraq War, the US military has accelerated the application and advancement of space-based information networks. For instance, during the Iraq War, the US Army's 4th Digital Infantry Division and other ground combat units modified and upgraded the command-and-control system, Force XXI Battle Command Brigade and Below (FBCB2), to enable interoperability and over-the-horizon communications for their forces. Three critical functional module upgrades and enhancements were critical to the process. First, global positioning system (GPS)-based digital maps that enable tracking and interconnection of the Blue Force system via radio integration, communication, and positioning between commanders and force members. Second, satellite communication antennas and modules that enable communication over the horizon, battlefield situational awareness, and information exchange with superiors and teammates at any time and from any location. Third, an IP-based command-and-control system that enables simultaneous distribution of the commander's orders to multiple users via a military email system. The addition of the modified ground combat force resolved all of the challenges encountered by the US military during the 1990 Gulf War. At the time, US battle orders were delivered via air from the Qatar Peninsula to the battle fleet and front-line units. Troops continued to rely on printed military maps and walkie-talkies. Only the commander on board was aware of the reconnaissance details collected by the early-warning aircraft. The strike command was transmitted to the weapon system in the rear in approximately 2–3 days. Following the Iraq War, nearly 140 antennae from the Gulf War era were completely interconnected using space-based information networks and IP-based interconnection methods. With the latest AI recognition and decision algorithms, the duration of the observe–orient–decide–act (OODA) loop linking space-based sensors with the shooters has been significantly reduced from tens of hours during the Gulf War to less than 20 seconds (Figure 5.4).

The US Army's next-generation military communication network, the Tactical Combatant Information Network, leverages satellite communications to enable Earth Station in Motion (ESIM) beyond visual range (BVR). The network's access points are integrated into Bradleys, mine countermeasures vehicles, and anti-ambush vehicles, connecting long-range combat units to the Tactical Combatant Information Network via small satellite communications

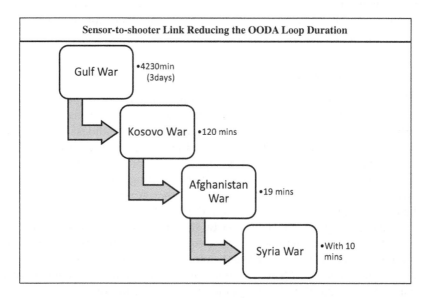

Figure 5.4 Nearly real-time response achieved by the US military.

antennas in the Ka- and Ku-band. Additionally, the US has developed new tactical command-and-control systems for mobile platforms. The second-generation Blue Force Tracking System network, acquired in 2018, is based on a mesh topology that transmits data directly from the transceiver to the satellite. Then it transmits it to ground forces, delivering data 100 times faster and uploading data 60 times faster than the previous-generation Blue Force Tracking System developed in the 1990s, and refreshing situational awareness position information in less than 2 seconds.

With the rapid advancement of small satellite technology, it is also worth noting that low-cost and multi-functional small satellites are becoming more accessible because of the widespread use of commercial devices, with costs falling from tens or hundreds of millions of RMB in the past to around 10 million RMB currently. Commercial launch costs are falling dramatically as competition intensifies, and a single launch can carry several, a dozen, or even dozens of small satellites. In the future, the cosmos will be filled with an increasing number of small satellites as SpaceX, Elon Musk's company, accelerates the process of mass-launching small satellites. The future battlefield will become more transparent if electronic detection means, visible and infrared imaging, position calculation, and even quantum dot microspectrometers are integrated into so many satellites, enabling reconnaissance, early warning, communication, navigation, command and control, meteorology, and mapping, among other integrated functions.

5.3.2 Military Mobile Communication Networks

Military mobile communication networks are applied in three fields. First is the command and control between joint forces and combat units, which require a high degree of secrecy, reliability, and security. Second, communication between vehicles and swarms, which necessitates the use of anti-jamming capabilities. Third, weapon system command and control, both of which are resolved primarily through data links.

Most traditional military mobile communication networks are "centralized, vertical-oriented, with a tree-like structure". With the advancement of information technology, the trend toward "centerless, ad hoc networks, and Internetization" has become more evident. Historically, military communications were largely vehicle- or task-based, involving dozens, hundreds, or even thousands of communication systems, radio models, information exchange formats, and data-processing formats, resulting in poor connectivity and operability between military services and combat forces. However, it is also challenging to replace all of these communication devices simultaneously.

How can a transition from an older communication network to a new one be accomplished? "Software and cognitive radio technology combined with self-organizing networks" may be a viable alternative. Cognitive radio technology will enable the future network communication system to automatically detect electromagnetic interference (EMI) and communication obstacles on the battlefield, rapidly locate available spectrum resources, and establish real-time communication contacts via frequency hopping and other techniques. Meanwhile, software and cognitive radio technologies will be compatible with a wide range of communication bands and waveforms, enabling a seamless transition from an older to a new system. This means that combat vehicles equipped with the software and cognitive radio communication system will be capable of non-centered, ad hoc, and jamming-resistant communication between similar pieces of equipment, as well as be compatible with a variety of legacy and new radio stations. One of the more contentious issues can be antenna sharing. From a technical standpoint, most experts believe that communications within 2G can benefit from a shared antenna and that communications between 2G and K-band can benefit from a shared or semi-shared antenna as well. In contrast, millimeter-wave, terahertz, and laser communications, among others, may require more point-to-point transceivers.

In 1999, Dr. Joseph Mitola, a researcher at the Royal Swedish Academy of Sciences and a MITRE consultant, introduced the idea of cognitive radio (CR). The fundamental concept of CR is to enable wireless communication devices to detect changes in their communication environment, thereby increasing communication reliability and spectrum utilization. The proposal of CR has been named a watershed moment in the evolution of radio technology due to the significant radio functions it enabled and the resulting changes to the radio spectrum management system. Additionally, the international academic community has called CR the "next big thing" in future radio communications, with widespread availability expected in the near future.

CR technology is a communication technology capable of automatically detecting and adjusting the transmitter and receiver's parameters. CR is a type of software-defined radio (SDR), a fully reconfigurable radio transceiver capable of automatically adapting its communication parameters to network and user requirements. While CR technology maintains interoperability with legacy narrow-band communication systems, it enables the formation of multi-platform, interconnected, and interoperable communication networks on the battlefield, ranging from the highest level (commander) to the lowest level (individual soldiers) (Figure 5.5).

5.3.3 Data Links

In simple terms, a data link is a wireless network communication device and protocol that enables data transmission. Data links serve as the modern combat command's central nervous system, seamlessly connecting sensor networks, command-and-control networks, and weapon platform networks. The US military currently utilizes the most advanced data link technologies, including Link4, Link11, Link16, and Link22. The US Navy uses Link11 to facilitate bidirectional tactical communications between ships, between ships and aircraft, and between ships and the Marine Corps. It enables communication and

Figure 5.5 Cognitive radio communication.

secrecy over the horizon, which is critical in combat situations. Link16, with a multipoint-to-multipoint architecture, enables non-centered, end-to-end data transmission for dozens of weapon platforms in the air while maintaining high levels of confidentiality and anti-jamming capability. It is suitable for weapon coordination, navigation identification, air control, and inter-fighter communication. Additionally, it bolsters the US military's efforts to establish an integrated data link system, transforming from a "platform-centric" to a "network-centric" warfare model.

Data link technology is a specialized military communication technology that utilizes time, frequency, and code division to transmit pre-agreed critical information between combat platforms on a regular or irregular basis. Therefore, interfering with data links is exceedingly difficult for adversaries if they do not fully comprehend or decipher what is being transmitted. Data links can be classified into two types: generic and dedicated. Generic data links are used in various applications, including joint operations, cooperative formation operations, and swarm operations. Dedicated ones continue to garner considerable interest, including satellite, unmanned aerial vehicle (UAV), missile-borne, and fire control data links. In the coming years, the trend toward generalization will continue, with specialization becoming increasingly rare. In addition, sensor data transmission, reception, and internal information processing will be coordinated with the military mission system, exhibiting increased specialization. In contrast, communication links and data transmission between platforms will become more generic.

5.3.4 Novel Communications

Due to the large dispersion angle, the large number of platforms, and the rapid development of associated electronic interference and microwave attack means, conventional microwave-based military communications are susceptible to a broader range of interference and damage. As a result, emerging communication technologies such as millimeter-wave, terahertz, laser, and free-space optical communications have emerged as viable options for high-speed, high-capacity, and high-bandwidth communication due to their anti-jamming properties. While high-frequency electromagnetic waves offer superior anti-jamming performance, precise point-to-point targeting, and omni-directional communication remain challenging due to the small dispersion angle. On a technical level, the mechanism by which combat vehicles maintain alignment and bidirectional communication while maneuvering at high speeds and changing orbits is still being investigated.

Additionally, a potential path for terahertz communications development is being considered. Terahertz communication systems, which are expected to be widely used by 2025, will be critical in a variety of applications, including satellite communications, space reconnaissance, and surveillance, missile defense, stealth target detection, tactical and confidential mobile communication networks, high-speed data links, terminal guidance, and near bomb fuses. Germany, Japan, and other major powers have developed near-commercial terahertz communication systems that communicate over several kilometers.

Additional research and development directions, such as free-space communication using white light or light-emitting diode (LED) light sources, passive sensing, and communication via optical and infrared lenses, may be pursued in the future, particularly in the context of swarm warfare.

5.3.5 Civil Communication Infrastructures

It is critical for the military to consider the most effective use of civilian communication infrastructures in the age of AI. In the future, open-source data mining and correlation analysis will be critical for providing information about the battlefield environment, targets, and status for combat and non-conflict military operations. These activities will mostly be accomplished via civilian communication networks, most notably 5G/6G mobile. Military operations other than war (MOOTWs) such as peacekeeping, search and rescue, counter-terrorism, and disaster relief place a strain on the army's dedicated communication network, which has a limited range and geographic coverage, making communication networks and connections with the world difficult. The two primary modes of civil communication infrastructure utilization are via civilian satellites, specifically small satellites, and civilian mobile and Internet communication infrastructures.

When civil assets are used for military purposes, the primary goal of integrated civil–military communication is to address security and confidentiality concerns. One method uses firewalls and encryption to secure command communications via civilian satellites and global mobile communication networks, which remain vulnerable to hacking and cyber attacks. Another option is to leverage newly developed virtualized Intranet technology.

Huang Xiaoqing, the founder of CloudMinds, invented Intranet technology and cloud-based intelligent robots in 2012. He also pioneered the development of practical Intranet-based mobile phone terminals (see section 5.1, "AI Brain Systems"). Intranet technology is the operation of a private internal network within a larger public network. Utilizing time division, frequency division, and virtualized operating systems, it is possible to conduct low-density, commercial-density communications while sharing information on the terminal display with a public mobile Internet phone. Based on "sandwich" overlay technology, Intranet mobile phone terminals can be easily and quickly integrated with multiple information systems in various fields, including public security, armed police forces, fire, and environmental protection, and medical or other emergency management systems, enabling multi-system information sharing and fusion. Multiple systems can be physically isolated or connected in a one-way fashion, but information can be shared via displays. Alternatively, one-way optical transitions offer a more secure data transfer method between systems of varying levels of security.

5.4 Collaborative Swarms

Combat effects that are difficult to achieve through conventional means and methods can be achieved by using simulations of natural bee swarms, ant colonies, bird flocks, and fish schools in conjunction with autonomous coordination mechanisms of swarm systems (e.g., drones and flexible munitions) to perform combat tasks such as attacking or defending against enemy targets. Collaborative swarms are an inevitable trend in the evolution of intelligence and a significant area of focus and direction for intelligent construction. Regardless of its performance or function, a single combat vehicle cannot form a swarm or scale advantage. Simple quantity accumulation and scale expansion make no sense without autonomous, collaborative, and ordered intelligent elements.

Collaborative swarms can be divided into three types. The first type is collaborative manned/unmanned swarms transformed from existing vehicles, mostly large and medium-sized combat vehicles. The second type is low-cost, homogeneous, and single-purpose combat swarms composed of small, unmanned combat vehicles and munitions. The third type is human–machine integrated swarms, including biological and AI bionic swarms with a high degree of autonomy, most humanoid, reptile-like, bird-like, and marine creature-like robots. Numerous advantages and characteristics can be derived from the use of collaborative swarm systems in swarm warfare.

5.4.1 Collaborative Advantages

Combat forces can be dispersed by using massive unmanned systems armed with intelligent munitions, forcing adversaries to attack more targets and compelling them to consume more weapons and munitions than intended. Due to its size, resilience, and recoverability, a swarm has high survivability, making the overall advantage obvious and the loss of individuals insignificant. Combat power can be maintained at a constant level by deploying low-cost unmanned vehicles, as opposed to deploying high-value manned vehicles equipped with sophisticated weapon systems (e.g., B-2 bombers, F-22 fighter jets, or F-35 advanced combat aircraft), the loss of which could result in a significant reduction in combat power. Most defense systems are designed to deal with a finite number of threats at a time. As a result, swarm operations, which involve the simultaneous launch of multiple attacks, have the potential to overwhelm an adversary's defenses. Even against dense artillery defenses, a single salvo of defense systems can only hit a finite number of targets, with some always being missed, leaving the systems vulnerable to swarm attacks.

5.4.2 Cost Advantages

Swarm warfare is primarily accomplished by using small and medium-sized drones, unmanned vehicles, and munitions. These machines have a straightforward spectrum, a large scale, and the same quality and performance requirements, enabling low-cost mass production. While modern weapons, equipment, and vehicles are being upgraded at breakneck speed, the cost is constantly rising and becoming prohibitively high. Since World War II, weapon development and procurement costs have increased significantly faster than technological advancements in weapon performance. The price of primary combat weapons and equipment have jumped dozens, hundreds, and even thousands of times, rendering them unaffordable for acquisition and use. Meanwhile, a gradual decline in the number of troops equipped with advanced weapons and equipment has become a major source of contention and concern.

5.4.3 Autonomous Advantages

Individual vehicles within a swarm can accurately perceive other vehicles' distances, speeds, and locations by utilizing a unified space–time benchmark platform. Additionally, they can rapidly determine battlefield targets' nature, size, and priority and the distance between themselves and friendly neighboring vehicles, via a networked active and passive communication link and intelligent perception technology. According to the pre-agreed combat rules, one or more vehicles may be configured to conduct simultaneous and sub-wave attacks, as well as simultaneous and multiple attacks in groups, based on the priority of the target threat, and to be replaced by a backup vehicle in the event of damage, thereby achieving autonomous decision-making and operation. This type of intelligent operation can be entirely delegated to a swarm of autonomous actions or to a swarm of semi-autonomous actions overseen by a human, depending on the degree of human involvement and gatekeeping.

5.4.4 Decision-Making Advantages

The future battlefield will become increasingly complex, characterized by intense games and confrontations between combatants. As a result, relying on human decision-making in a high-intensity confrontation environment with rapidly changing threats is inadvisable and unreliable. Decisions might be made too late and with suboptimal or inconsistent quality. Considering this, decision-making must be delegated to collaborative swarms to rapidly strike adversaries or mount effective defenses, gaining battlefield advantages and initiative. This includes automatic environmental adaptation, target and threat identification, autonomous decision-making, and collaborative action.

Cooperative swarm control poses novel difficulties for command-and-control systems. Applying command and control to swarms is a novel strategy. Generalized control can be applied at both hierarchical and task-specific levels [1] and in various ways, including centralized control, hierarchical control, coherent synergy, and automatic synergy. The more small-unit operations are at the tactical level, the greater the demand for autonomous operations and unmanned intervention. At the formation level, which necessitates the command and control of multiple swarms, centralized planning, hierarchical control, and minimal human involvement are required. At the advanced strategic and campaign levels, where swarms are used exclusively as weapons and operational patterns, unified planning and layout and a much higher degree of human involvement are required. Apart from that, swarm warfare research should prioritize countermeasures, with a particular emphasis on electronic deception, EMI, cyber attacks, high-power microwave weapons, electromagnetic pulse bombs, and bomb artillery systems, all of which play a clear role and have obvious effects. Additionally, research and development of laser weapons and swarm-to-swarm countermeasures are needed to establish a "firewall" against both coordinated and autonomous swarms.

5.5 Cloud Terminals

Cloud terminals refer to terminals connected via the "cloud" and "network," including a range of sensors, command-and-control platforms, weapons vehicles, support vehicles, combatants, and other supporting equipment and facilities equipped with intelligent modules. In the future, all types of equipment and vehicles will be supported by cyber-physical systems (CPS) and human–computer interfaces (HCI), with front-end and back-end integrated AI. Environmental perception, route planning, vehicle mobility, and weapon operation will all be facilitated by front-end intelligence, such as bionic and machine intelligence. Back-end cloud platforms will support complex battlefield target identification, combat mission planning, networked cooperative strikes, combat situation analysis, and advanced HCI. Future research and efforts will be directed towards virtual soldiers, virtual staff officers and commanders, as well as towards their intelligent and efficient interaction with humans.

Reference

[1]   Work, R. O., Brimley, S., & Scharry, P. 罗伯特·O.沃克, 肖恩·布瑞姆利, 保罗·斯查瑞. (2016). *20YY: jiqiren shidai de zhanzheng* 20YY: 机器人时代的战争 (H. Zou 邹辉, Trans.). Beijing: National Defense Industry Press.

6 Parallel Military and Intelligent Training

Sun Tzu said, "The general who wins a battle makes many calculations in his temple before the battle is fought. The general who loses a battle makes few calculations beforehand" [1]. Parallel military theory discusses algorithmic warfare as the primary form of warfare, in which the basic principle is to develop a virtual artificial system that collects data through mutual mapping between physical forces in physical space and software-defined forces in virtual space. Due to the absence of geographical and temporal constraints, combat experiments, simulation training, and combat effectiveness evaluation would be carried out in virtual space more easily. As a result of continuous optimization, iteration, and improvement, a set of parallel systems and model algorithms evolving from practice, higher than practice, and guiding practice will be gradually formulated to discover the optimal solutions for varying environmental and ecological situations. The objective of a solution is to enable physical forces to know themselves as well as their adversaries and to anticipate the outcome of battles before they occur. An intelligent operation system of systems (OSoS) based on parallel military theory represents the embodiment of advanced intelligence, the synthesis of imagination and reality, the integration of different domains, and the apogee of future intelligent warfare.

6.1 Parallel Theory

The parallel theory is derived from the quantum parallel universe theory proposed by Steven Hawking, who perceived the entire classical universe as a quantum particle. In Hawking's theory, there must be an infinite number of quantum parallel universes that constitute the wave function of the quantum universe throughout all possible universes. Since then, numerous astronomical phenomena and experiments have confirmed the validity of this theory.

Parallel military theory has moved from theory to practice due to the rapid development of the Internet, the Internet of Things (IoT), cloud computing, big data, and artificial intelligence (AI). Mr. Feiyue Wang, Director of the State Key Laboratory of Complex Systems Management and Control, Institute of Automation, Chinese Academy of Sciences, is an expert in parallel theory.

DOI: 10.4324/b22974-7

In 2004, he proposed the "Parallel Systems Approach and the Management and Control of Complex Systems". In 2015, he published the keynote report "X5.0: Parallel Intelligent Systems in the Parallel Era" and gave a public lecture on it. The main elements and ideas mentioned in this report are [2]:

1. After mechanization, electrification, informatization, and networking, we are entering the fifth stage of technological development – parallelism, an era of intelligence characterized by parallel interaction between virtuality and reality, X5.0, mentioned by Feiyue Wang. The mechanized era is characterized by steam engines; electrification is typified by electric motors, and informatization by the computer, and networking by routers. What characterizes parallelism: robots, drones, intelligent aircraft, or parallel machines? We are not sure. Nevertheless, as with steam engines and motors, computers and routers will soon disappear.

2. In a technical or engineering context, intelligence can be defined as the use of the known to solve unknown problems. Our brains are the source of our imagination. The human brain can perceive all known knowledge almost instantly, as well as reasoning about the unknown world and unknown problems. How does machine imagination work? Currently, it is limited to using closed algorithms. Almost all machine algorithms, no matter how complex, are currently confined to the memory of the machine. If algorithms are not "liberated", AI will always remain artificial and will not be able to achieve human-like intelligence. AI is only capable of remaining at the first level, using existing knowledge to solve known problems, and cannot be upgraded to the second level of intelligence, that is, using existing knowledge to solve unknown problems. Nor can it reach the third level of intelligence, using unknown knowledge to solve unknown problems.

3. The algorithm can only be liberated in a Third World, not the third world in the political sense, but the Third World proposed by Karl Popper [3]. Most people are familiar with only two worlds, the physical and the mental. According to Popper, there is a Third World, the artificial world. Algorithms can only evolve in this third world. Why? In the physical world, humans are the subject of action. In the psychological world, humans are the subject of cognition. Only in an artificial world are humans the real masters, allowing human-designed algorithms to flourish, without being constrained by economic, legal, moral, or scientific concerns. Ideally, the harmony of the future world should reflect the harmony of these three worlds.

4. Regarding "parallelism", we must start with complexity and intelligentization. Complex systems are indivisible and unknowable. The study of complex systems requires both reductionist and holistic approaches. Historically, things or phenomena were broken down into their most basic components. However, it would be impossible to break down all things in the external world due to limited resources

and the complexity of systems. A paradox exists between the need to break down everything and the inability to do so. Despite our desire to know, human cognition encounters difficulties in time and space. As with intelligentization, there is a contradiction between closed and open algorithms, and between known and unknown problems. Contradictions like these can be summarized as UDC (uncertainty, diversity, and complexity). The mission of AI is to transform the UDC into AFC (agility, focus, and convergence). Specifically, agility is driven by deep knowledge, focus is based on experimental analysis, and convergence is driven by feedback and adaptive interactions. Using information technology, automation, and AI is essential to accomplishing this mission, which makes "unsolvable" questions "solvable".

5. Paradoxes are often unsolved due to the limitations of the solution space. For instance, if we only consider real numbers, no solution can be found for the equation $x^2 + 1 = 0$, which requires a change in concept, introducing imaginary numbers to expand the space into one that has a specific "solution". It is necessary to bring opposites together in order to resolve the paradox of complexity and intelligence. The known and the unknown, the divided and the indivisible, are both opposites. How can these opposites be unified? This is a problem that the parallel theory has yet to solve.

6. One of the key milestones in physics is the leap from Newtonian mechanics to quantum mechanics, when the world began to recognize the "wave–particle duality" of matter. Now that we recognize both the wave–particle duality and the virtual–real duality of intelligence, we can "liberate" algorithms. In the future, we should consider not only real intelligent numbers but also intelligent imaginary numbers. Physical space represents the real number, while virtual network space refers to the imaginary number. Physically undivided and unknowable things may be divisible and knowable in the parallel spaces of physical and virtual reality.

7. An essential component of the transformation from UDC to AFC is ACP, the organic combination of "artificial societies + computational experiments + parallel execution". Cyber-physical social systems (CPSS) are the infrastructures supporting the ACP method, whose abbreviation has one more "S" than the currently popular concept of CPS (cyber-physical systems). The letter "S", which stands for social, is essential, since it represents integrating people and organizations into the system, allowing virtual–real interaction, closed-loop feedback, and parallel tasking.

8. The future world must be parallel in a way that integrates the virtual and the real: parallel people = people + virtual people, parallel things = things + virtual things. A virtual–real relationship in the future may be one-to-one, or one-to-many, or many-to-one, or may even develop into a many-to-many relationship, forming a parallel society in which virtuality

and reality coexist. Academically, this is referred to as "software-defined systems", "digital twins", "software robots", or "knowledge robots", etc. As we move into the future, there will not only be one "you" in the physical world, but also multiple "yous" in the virtual network world, which will accompany you throughout your life. Ultimately, a virtual "you" can supervise and guide you, assist you in solving problems in the physical world, and grow with you.

9. In essence, the core of the ACP parallel concept is to use data to build up the "virtual" and "soft" parts of the complex, intelligent system, and to "harden" them by calculating and implementing them quantitatively and in real time, so that they can be used to solve practical problems. The ACP approach is based on big data, cloud computing, and IoT technologies. Incorporating big data into the development of parallel systems can offer real-time, comprehensive, and effective inputs, which can be summarized as "data speaks", "data predicts the future", and "data creates the future". In short, in an artificial society, a computational experiment and a parallel system are used to promote large closed-loop feedback operations, ranging from knowledge representation, decision-making reasoning, to scenario adaptive learning and understanding.

10. In the near future, the competitiveness and strength of an enterprise, institution, military, or even nation will not be determined by its external size or resources, but by its awareness, practice, and efficiency with regard to the interaction of the real and the imaginary. This is an intelligent "parallel society".

11. The role of a future intelligent society is threefold: the artificial influences the real; the "virtual" influences the "real". The future influences history, and the "void" influences the "entities". The scientific instrumentalization of a "crystal ball" is not only to perceive history but also to perceive the future, so that statistics, design, intervention, etc. can be made for the future. In the X5.0 era, the intelligent system consists of one core: the parallel concept of virtual and real interaction; two support structures, the ACP method and the CPSS infrastructure; and three themes, namely intelligent organization, intelligent management, and social intelligence.

Parallel intelligence theory applied to the military will certainly trigger a new era and realm.

6.2 Parallel Military

Parallel military construction has three primary objectives. As a first objective, it will make full use of the new cyberspace-defined rational boundaries and the new intellectual space introduced to unite research and development (R&D), training, operations, and evaluation in the physical, cognitive, and information domains. As a second objective, it aims to explore new ways of conducting warfare based upon cyberspace. Third, it aims to build a parallel military system with virtual–real interaction so that parallel military analysis, experimentation, operations, training, and evaluation become regular military activities.

Parallel military systems should include R&D systems, operation systems, training systems, virtual battlefield environments, soldier systems, armament systems, virtual staff officers, and other elements that use big data, cloud computing, machine learning, simulation, high-speed communication, and complex battlefield environment perception. There are various operational concepts, operational rules, strategies and tactics, warfare theories, knowledge mapping, and software systems that are tailored to specific needs, capabilities, equipment, and means.

Parallel military theory is based on the interaction between virtuality and reality, in which virtuality can subdue the real world. According to this theory, the rules and laws of victory are reflected in four aspects.

6.2.1 Knowing Oneself and One's Enemy, and Foreseeing the Results Before the Battle Begins

In virtual space, it is possible to simulate a confrontation with different adversaries, provided accurate and detailed data about one's blue force (simulated adversary). It will be possible to predict the outcome (victory or defeat) of a given operation under different conditions before it starts.

6.2.2 Developing a Strategy and Outwitting the Best with the Worst

When a troop's operational capabilities and means are inferior to those of an adversary, virtual space simulation training can provide an advantage. Though the overall situation may be disadvantageous, the simulation can reveal some advantages in certain areas.

6.2.3 Interaction Between the Virtual and the Real Worlds, Using Virtuality to Promote Reality

An armed force can conduct correlation mapping, mutual iteration, and simulation training to identify the gap between personnel quality, equipment performance, application of operation methods, condition support, and combat opponents to maximize R&D, training, and construction.

6.2.4 Winning Real Battles through Virtuality

In a conflict in which both sides possess similar hard powers, if the other side does not possess virtual parallel operations, it is very easy for your side to defeat the other. In the event that the opponent also has the ability to deploy parallel forces, the outcome of victory or defeat will be determined by the superiority or inferiority of the parallel systems used by both sides. Those who possess a smarter AI will gain greater battlefield dominance and initiative, and their chances of victory will also increase.

6.3 Parallel Systems

In parallel systems, an artificial simulation is used to construct a parallel virtual–real world that maps the social, psychological, and spiritual realms as realistically as possible. In the age of AI, a parallel system or intelligent parallel system is both a physical simulation system as well as one based on data simulation, problem modeling, computational solution, practical guidance, and task execution. Further, this system may be continuously optimized and iterated based on the results of task execution. As a result, new laws are discovered from big data and virtual simulations, which then leads to the development of "little intelligence," i.e., precise knowledge for specific problems. Parallel systems manifest themselves in a variety of ways, such as primary, advanced, single-function, and multi-functional parallel systems.

An intelligent and parallel military simulation system differs fundamentally from a simple military simulation system. The system must fulfill three functions. The first function is that of simulating, which involves creating an electronic mirror of the real world in cyberspace, so that it can dynamically reflect what is happening on the battlefield in real time. The second function is learning, through which the system can accumulate simulated training data in peacetime, or combat data, results, and conclusions in wartime. The third function is guidance, i.e., obtaining accurate conclusions, methods, and strategies through machine learning and guiding operations. Consequently, the essence of the intelligent military parallel system is a typical ACP system.

Video games are, in a sense, elementary parallel systems. This is because many aspects of the game, such as the environment, characters, weapons, etc., have prototypes in the real world. Even though the plot of military games differs considerably from that of real battlefields and combat, it is, after all, a way to simulate and map the real conflict. As a result of human and machine participation, the game creates human–machine interaction and confrontation, and therefore, it may be seen as a simplified version of parallel systems reminiscent of science fiction.

A military chess projection system is also a parallel system. It consists of an electronic sandbox, an electronic map, as well as computer-generated forces, firepower, etc. Using different combat ideas, the red and blue sides build models that are based on relatively real strength data and simulate and rehearse the engagement process under different formations and warfare methods to obtain a variety of engagement results. It is true that this format does not fully simulate real troops' operations, battlefield environments, or weapons, but it can approximate combat to a large extent with a certain degree of credibility, and therefore is often used at strategic levels.

Recently, big data-based AI theories and techniques such as deep learning and reinforcement learning have shown success in the field of non-military gaming competitions. For example, Google's AlphaGo program defeated the world champion Go player in 2016 and 2017. In 2017, DeepStack, developed by the University of Alberta in Canada, and Libratus, developed by Carnegie Mellon University in the United States, defeated human players in Texas Hold 'em. Also in 2017, OpenAI's program defeated human players in Dota 2. In 2018, AlphaSTAR defeated human professionals in StarCraft.

This intelligent confrontation based on games and military chess rehearsals is still a long way from a true combat confrontation, but it does represent an excellent technical and methodological attempt. If combined with the real battlefield environment, weapons, and equipment, tactical application, destruction assessment, blue force, and other high simulation models, it can be closer to a real battle.

The issue of how to conduct combat-oriented confrontation training in peacetime is a contemporary issue and a global challenge concern. The authenticity and credibility of military chess rehearsals are low, if there is no actual combat scenario and no supporting combat data. Furthermore, improving the realism of military chess rehearsals requires exponential increases in operational practices and training subsamples. It is therefore not sufficient to rely on data and experience from actual military exercises. The problems of difficulty in accumulating data and low credibility of troop combat capability can only be overcome by relying on a simulation system to self-practice under confrontation conditions repeatedly.

Military simulation systems are typically parallel systems. Due to the development of simulation technology, specifically simulation training technology, simulation training is moving from the simulation of a single weapon and equipment to the simulation of a formation. Simulation training evolves from a fixed, simple simulation of operational elements into a distributed, complex simulation of the battlefield environment. It is important to translate emerging technologies, such as virtual reality and AI, into the "red–blue confrontation" training and evaluation system in order to improve the efficiency of combat operations and the accuracy of tactic and technique assessments. With the accumulation of data and the optimization of model parameters, the system will gradually evolve into an intelligent parallel R&D and operational training system. Finally, upon introducing human–machine learning and an adaptive planning AI, the simulation system will gradually become a parallel system with advanced intelligence.

The military simulation system is not restricted by training sites, time, and security conditions, nor does it require the transportation of weapons, fuel, ammunition, or supplies, nor does it damage the exercise territory and the environment. Training simulations and operational effectiveness assessments can be carried out in unfamiliar terrain, sensitive areas, and complex meteorological and electromagnetic environments outside the country, yielding significant military, economic, and ecological benefits.

Building an intelligent parallel military system based on a simulation is a simple and feasible approach. The basic idea is to collect a large amount of key data on troop and armament development, security, training, exercises, and to establish the mapping rules and methods from reality to virtual, and build a digital model of the virtual battlefield, soldiers, equipment, troop formation, tactical operations, and combat process. Through virtual–real interaction and self-learning, and through continuous optimization and iteration, all models can comprehensively and accurately represent the quality of personnel, the level of equipment, the quality of decisions made, the effectiveness of training, and the effectiveness of combat. It is necessary to implement pre-battle, in-battle, and post-battle virtual simulations according to different battlefields, operation missions, adversaries, and operational patterns, to predict and master the confrontation under different conditions, to achieve economical, efficient, and convenient virtual training and exercises, and to provide scientific, comprehensive, quantitative, and timely guidance and instruction for the optimization of physical force operations, training, command, and decision-making (Figure 6.1).

In the development of military parallel systems, there are three major difficulties. The first is to collect data on the blue force, which is difficult due to secrecy and other factors, and therefore requires an integrated information-mining and simulation approach. The second concern is optimizing models and algorithms, which includes describing the performance and behavior of physical forces, extrapolating and evaluating combat countermeasure simulations, and conducting actual exercises and operations. In addition to improving software systems, hardware systems such as computing chips and storage devices must be upgraded. The third is system validation. It is neither feasible nor realistic to rely solely on actual combat for verification,

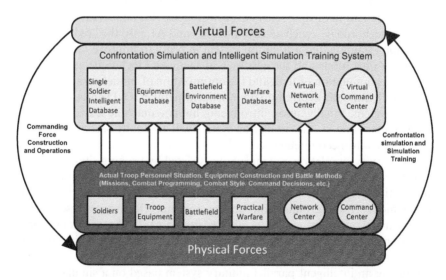

Figure 6.1 Intelligent military parallel system.

iteration, and optimization of systems. The only feasible approach is still to run simulations of semi-physical countermeasures in order to verify them experimentally.

Using digital software-defined models of troop personnel, weapons, and equipment, virtual battlefields, and warfare methods, the parallel system efficiently participates in simulation and confrontation calculations. Through continuous iteration and optimization of the military parallel system, it is possible to gradually establish an advanced intelligent parallel force with virtual–real interactive operations and machine-learning capabilities.

6.4 Virtual Battlefield Environment

Three features are present in this environment. The first is a simulated real-world environment, which includes mountains and rivers, highland forests, deserts, coastal islands, oceans, the atmosphere, space, and cities, including buildings, transportation networks, underground facilities, and architectural layouts. The second is a simulated meteorological and hydrological environment, including electromagnetic interference, radio/TV/network layout, communication pathways, nodes of information, etc. Finally, there is a simulation of a social environment, including national history, values, population groups, personnel practices, knowledge structures, and religious beliefs.

The advanced virtual battlefield environment should be able to accurately reflect the interactive relationship and qualitative and quantitative effects between humans, machines, and the environment. Among its contents are kinetic models of tanks, aircraft, artillery, warships, and other combat vehicles under complex geographical and meteorological conditions, and damage models for various weapons, firepower, and hard and soft means of killing or injuring. Only in this way can various situations be simulated, such as the movement speed, physical exertion, the mental state of personnel in different geographical environments, the mobility speed of equipment, ammunition, and fuel consumption, the adaptability and matching of battlefield road/sea conditions, and the striking accuracy, explosive power, and degree of destruction of various weapons against different targets at different distances in disturbed environments. It is therefore crucial to depict the operational capabilities and means of both red and blue forces, as well as the effectiveness of different operational formations and operational patterns, rather than simply producing a two-dimensional or three-dimensional electronic map and game environment.

Various technologies are needed to support the environment, such as remote sensing, detection of terrain, underground facility modeling, big data analysis, mining technology, social environment modeling, electromagnetic environment modeling, etc.

Also noteworthy is the use of holography in complex military training simulations. Holography allows combatants to master the battlefield environment from three-dimensional simulation images rather than maps and sandboxes. This technology breaks through the limitations of traditional flat display technology and provides commanders with intuitive perception and comprehensive information for decision-making and operational support.

6.5 Parallel Soldiers

Soldier systems are likely to take several directions in the future. How will they be integrated with humans, or will they be autonomous? Will they be human–machine hybrids? Accordingly, as long as humanity is involved in future warfare, the future soldier system will continue to evolve alongside operational practice and technological advancement. Under the strong impetus of information and networked intelligence technology, the future soldier system aims to enhance innovative combat capability in all aspects, and its connotation and functional composition will undergo significant changes. As a group-oriented system, the future soldier system integrates seamlessly with the OSoS. This system is expected to become an advanced special warfare equipment system with individual soldiers, machine soldiers, weapons, and sensors, capable of network communication, human–machine integration, intelligent interaction, coordinated attacks, comprehensive protection, human body enhancement, and other advanced capabilities.

Development of the future soldier system is likely to take place in three stages. In the first stage, with the development of big data, cloud computing, network communication, and other technologies, soldiers will be transformed into "transparent soldiers", supported by a networked information system with improved human–machine collaboration capabilities. In the second stage, with the support of distributed networks and clouds combined with the integration of physical assets and information, the soldiers on the battlefield will be transformed into "cloud soldiers" with numerous functions at their disposal and cloud support in the background. Individual soldiers' decision-making, human–machine cooperation, and swarm attack and defense capabilities will be greatly enhanced. In the third stage, through the continuous accumulation of experience and data from human–machine hybrid operations, as well as the optimization, upgrading, and perfection of models and algorithms, the soldiers will eventually become parallel soldiers with super capabilities in virtual–real interaction, human–machine integration, sensory control, and online and offline components. The operation units and systems will be intelligent and efficient.

In the 1980s, the US military proposed an equipment modernization program for soldiers. The US 2020 Future Warrior System concept incorporates weapons, head-to-toe soldier protection equipment, portable computer networks, soldier power sources, and advanced soldier performance equipment. Among them, the new body armor has a stronger ability to absorb kinetic energy from bullets when compared to existing body armor. With its physiological state-monitoring system, the soldier's internal and external temperatures, heart rate, body posture, and water consumption can be monitored. Upon detecting sickness, the computer can generate a map that directs the soldier to a medical facility. According to US military officials, operation clothing body armor will be integrated with a variety of microcomputers and will appear like traditional clothing. The material is normally soft, and hardens immediately to resist bullets after an attack, and then becomes soft again. In addition, the US military hopes that the "micro muscle fibers" embedded in the new system will realistically simulate muscles and provide soldiers with greater strength.

In 2010, the US Army Environmental Medicine Institute launched a virtual soldier research project to create a complete "Avatar" soldier. The US Army developed the first Avatar Warrior in April 2016 and has now successfully developed 250 Avatar virtual soldiers, known as "virtual cyber warriors," that have been used to test "vulnerabilities" during actual engagements. These warriors can be widely deployed for realistic, high-risk simulations as an alternative to real soldiers in real-world operation testing [4].

Parallel soldier R&D focuses on virtual soldiers, encompassing diversified simulation of soldier appearances, intelligent memory, capability differentiation, emotion calculation, path planning, perception, decision behavior, and operational behavior simulation.

Parallel soldier systems include simulations of the soldier's physical state, psychological quality, knowledge base, reaction speed, character traits, and command and decision level on a five-dimension scale encompassing physiology, psychology, cognitive ability, decision-making ability, and learning capability. With brain–computer technology and intelligent models, algorithms can be learned and improved through wartime, training, and daily life data. It will be possible to customize a virtual soldier for each member of the virtualized force, providing data for virtual operations, conducting potential mining analysis and growth judgments, and tailoring training and cultivation paths. Furthermore, it will be necessary to establish standardized processes for the collection, analysis, modeling, and optimization of soldier data in the future in order to automate the collection of five-dimensional data and build an intelligent repository of soldier models.

6.6 Parallel Equipment

Military equipment includes a range of complex and diverse weapons, weapon systems, information systems, security equipment, facilities, etc. For this reason, parallel equipment has unique characteristics. The purpose of parallel equipment R&D is to develop software-defined equipment systems with virtual brain AI capabilities to guide equipment development, operations, use, training, management, security, and improvement. It is mainly divided into three categories. One is a digital virtual equipment system, with models and data to characterize the performance of weapons and equipment for each combat technology, using characteristics and technical status. Another is an equipment simulation training system, focused on training troops' equipment operators, staff officers, and commanders. Third is an intelligent equipment AI system, which timely guides officers and soldiers to operate the usage of equipment and training, expanding the officers' and soldiers' operational and decision-making capabilities in wartime, and even replacing officers and soldiers to perform tasks when they lose their operational ability.

6.6.1 Digitalized Virtual Equipment Systems

This system includes a three-dimensional (3D) visualization model, basic performance data, combat effectiveness data, training data, daily mainten-ance data, etc. The core is to reflect the tactical, technical performance, and status indicators of the equipment. There are four categories of data in focus. First, the data of equipment categories, models, quantities, etc., including data of strategic deterrence weapons, army equipment, naval equipment, air force equipment, space equipment, network equipment, reconnaissance equipment, strike equipment, accusation equipment, electronic counter-measure equipment, and security equipment, and other quantitative scale and institutional structure. Second is the equipment operation effectiveness data, including voyage range, shooting range, surprise defense capability, hit accuracy, damage effectiveness, mobility, protection, environmental awareness, search and reconnaissance, target identification, charge communication, early warning, navigation, jamming/counter-jamming, interoperability, and other warfare technology indicators. The third is the general technical performance of the equipment, including environmental adaptability, transport adapt-ability, reliability, testability, maintainability, security, safety, damage resist-ance, concealability, economy, system contribution rate, etc. The fourth is the data and archives of the design, development, production, use, training, maintenance, security, current technical status, service life, and available times of the equipment.

In parallel equipment systems, it is necessary to gradually establish and improve model repositories and databases, to develop a regular system for collecting development, operation, training, repair, and maintenance data on weapons and equipment, and to continuously improve the accumulation of real-world data and virtual data. It is also necessary to conduct efficient management, updating, and analysis of big data, evaluate weapon equipment models and their effectiveness, and fully and accurately represent the actual installation performance and operational characteristics.

6.6.2 Equipment Simulation Training Systems

There are many types and models of equipment simulation training systems, including single equipment simulator trainers, integrated equipment simulator trainers, distributed simulation training systems, and human–computer inter-active simulation training systems. These simulators generally conduct training in a semi-physical manner. In terms of maneuver, firepower, reconnaissance,

Figure 6.2 The fighter jet flight simulator at the 2021 Zhuhai Airshow.

protection, and communication, the simulated equipment is consistent with real-world performance. Based on computer-generated force, equipment can be tested using physical and behavioral simulation requirements.

There is a substantial difference between equipment simulation training and real-world training. Large and intricate weapons and equipment systems are complex and expensive, live-load training affects the life span of the weapon and equipment, and it consumes a significant amount of time and money. Therefore, the simulation training system has been highly valued and generally welcomed (Figure 6.2). While traditional simulation training is used to train individuals, it does not use past training data, especially the real data from live loading exercises and the win–lose data from simulated training. Therefore, it is crucial to incorporate intelligent models and algorithms through machine learning so that software-defined equipment can be developed with machine brains that can assist users in their training and increase the efficiency of weapons, equipment, and operations.

6.6.3 Parallel Equipment AI Systems

The goal of parallel equipment systems is not only to demonstrate the capabilities and proficiency of equipment, but also to coach and guide soldiers and officers on how to use equipment in combat and missions as effectively as possible. In peacetime, the equipment AI allows officers and soldiers to familiarize themselves with the equipment and learn various operational skills and combat techniques. In wartime, with the assistance of optimized models and algorithms, it allows officers and soldiers to fight different enemies to the best of their abilities, and after the fight, improve their combat skills. Even after officers and soldiers have been transferred or discharged, the equipment AI continues to exist and guide the new officers and soldiers. Whenever an officer or soldier is injured, the equipment AI will assume their responsibilities and fight instead.

Simulated training data may be used as input for intelligent equipment systems. As the systems become more complex and intelligent, they can be equipped with more advanced models and algorithms, gradually approaching human intelligence and even exceeding it. The Alpha AI technology, developed by the University of Cincinnati, defeated the best human think tank team and pilots, indicating that such examples and situations will become more frequent.

6.7 Parallel Forces

Parallel forces are intelligent forces with virtual-reality interactive capabilities. Fundamental to this process is the development of breakthrough technologies such as efficient real-time confrontation simulations utilizing big data, soldiers, equipment, environment and warfare databases invocation and feedback, intelligent simulation training system development, force programming, warfare optimization algorithms, and self-adaptive mission planning and decision-making. The distributed intelligent simulation training system must be gradually established and improved. This system interacts with virtual battlefield data to undertake comprehensive calculation of simulation and confrontation to help ensure the efficient operation of the system. Indeed, the empirical data are augmented by tactical information on specific assaults conducted by the troops and with performance statistics for weapons and equipment. As a result of continuous accumulation and optimization, parallel forces with virtual–real interactive capabilities and machine–brain systems can gradually be created.

The core elements of parallel force construction consist of databases of blue forces, equipment system and operation process models, red–blue forces confrontation knowledge base, adaptation operation mission planning and dynamic adjustment algorithms, data collection and modeling of domestic and foreign troops, and so on.

If the complexity of an OSoS exceeds a certain threshold, the probability of uncertainty and unpredictability increases accordingly. As "accuracy" and "flexibility" of adversary simulation and "multiple objectives" are mutually exclusive, it is often difficult to find the exact and optimal solution. Through realistic scenario construction, comprehensive expert seminars, a priori design, virtual–real interaction, and simulation trial and error, it is possible to identify the optimal or suboptimal choice among the maximum, minimum, and most probable probabilities and reduce uncertainty and unpredictability.

From individual soldiers, squads, and battalions to regiments, brigades, and divisions, practical and virtual data are collected and advanced level by level, and can achieve higher-level battle and strategic simulation. At the same time, it is necessary to build a force confrontation simulation system based on multi-service joint operations, and this system helps to carry out joint warfare virtualization force building and virtual operation experiments, exercises, and training to effectively guide joint operations and forces building at all levels in the war zone.

Building an intelligent simulation training system can complete massive training for troops at a low time and economic cost and help force commanders, combatants, and security personnel form rapid response capabilities. It is necessary to combine AI algorithms and big data analysis technology to optimize troop programming and warfare to enhance troop operation effectiveness efficiently at a lower cost.

Intelligent simulation training systems can be used to solve data augmentation, reinforcement learning, adversarial training, and model optimization problems with a relatively small sample size of red and blue forces and operational data models through generative adversarial networks (GANs), Monte Carlo search, and Bayesian search for superiority.

GAN is a deep learning model that is one of the most promising approaches to unsupervised learning on complex distributions in recent years. The model learns by competing through (at least) two modules (the Generative Model and the Discriminative Model) to produce a good output. Discriminant models require input variables that some models predict. Generative models are given some implicit information to generate observations randomly. To give an example of generating pictures:

Discriminant model D: given a picture, determine whether the animal in this picture is a cat or a dog.

Generate model G: given a series of cat pictures, generate a new cat (not in the dataset).

The method of GAN is to let D and G play the game. During the training process, D and G compete to achieve model enhancement. During the training process, the generative model G continuously learns the probability distribution of the real data in the training set to transform the random input noise into images that can be faked. The more similar the generated images are to the images in the training set, the better. It is more useful to try to generate real images to deceive the discriminative network D. The goal of D is to try to separate the G-generated images from the real ones, with the real ones being 1 and the fake ones being 0. In this way, G and D constitute a dynamic "game process". What is the outcome of the game? In the best case, G can produce a picture for $G(z)$ that looks like the real one. For D, it is difficult to determine whether the image generated by G is real or not, so $D(G(z)) = 0.5$.

Thus, the goal is achieved: we get a generative model G which can be used to generate images.

In the discriminative model D, G is provided with the ability to approximate the real data well without a great deal of prior knowledge or prior distributions, making the generated data appear falsified, that is, such that D cannot distinguish real pictures from those generated by G.

GAN has become a hot topic in research, and many models have been proposed. The GAN algorithm is commonly used in image generation, compression, style conversion, and text-to-image conversion, among other applications.

The Monte Carlo method is a numerical calculation method that uses random numbers guided by probability and statistics theory. The general steps are:

1. For each operational countermeasure, enter the smallest, largest, and most likely estimated data and select an appropriate prior distribution model.
2. Using the above inputs, perform a sufficient and large scope of random sampling based on a set of operational rules.
3. Calculate the results based on a random sample of data.
4. Calculate the minimum, maximum, mathematical expectation, and unit standard deviation from the above results.
5. Based on the calculated statistical processing data, the computer automatically generates the probability distribution curve and the cumulative probability curve (normally based on the probability cumulative S curve of the normal distribution).
6. Use cumulative probability curves to perform a win–loss and win–risk analysis.

The more Monte Carlo method experiments are performed, the more accurate the results will be. Since the advent of computer technology, the Monte Carlo method has become increasingly popular. With modern Monte Carlo methods, it is no longer necessary to perform experiments manually. The high-speed capabilities of computers have revolutionized the experimental process, which was previously tedious and time-consuming. With the aid of computer technology, the Monte Carlo method is capable of both simple and robust execution, making it easily comprehensible and understandable by anyone. It is fast, and as long as the computational power and speed are sufficient, the results will converge quickly.

The Monte Carlo method offers a high level of adaptability, regardless of the complexity of the problem geometry. The convergence of this method refers to the convergence in a probabilistic sense, so that the increase in problem size does not impair the convergence rate, and the storage unit is very economical; these are the advantages of using this method to solve large and complex problems. Due to the development of electronic computers and the increasing complexity of scientific and technological problems, the Monte Carlo method has become increasingly popular.

Bayesian optimization is based on Bayes' Law for calculation and analysis. The Bayes' rule is a method of using observed phenomena to correct subjective judgments about probability distributions (e.g., prior probabilities). The mathematical formula is $P(B|A) = P(A|B) \times P(B) / P(A)$. The Bayes' Law can be expressed as follows in the context of a border dispute between two countries.

While country A is stronger than country B in terms of overall military power, B is stronger in terms of local border areas. A is constructing a national defense project close to the border. B plans to launch a small-scale para-militarized invasion of A but is unsure of the ease or likelihood that A will be able to counter the invasion with military means. On top of that, the B side has deployed a brigade at the border, whereas the A side only has a battalion on the border. Assuming that the probability of A being defeated is 70%, B estimates that when B enters A's border territory, the likelihood of A mounting a military counterattack is only 20%. Nevertheless, if A mobilizes his troops and has the ability to win absolutely, B believes there is a 100% chance of A launching a military counterattack in response to an invasion within A's borders.

Assuming that A will take a military counterattack with a 70% probability at the beginning of the game, B estimates that it will receive a military counterattack from A when it enters the border:

$$0.7 \times 0.2 + 0.3 \times 1 = 0.44$$

A military counterattack by A has a probability of 0.44, which does not exceed 50%, therefore it is considered unlikely.

As B approached the border, A used limited military force to prevent it from crossing. Even though there were no major clashes between the two sides, A promptly issued a diplomatic condemnation and reorganized and deployed its military, adding two battalions to the front lines. Based on these observations in diplomacy and military affairs, B concludes that the difficulty of taking military action against A has increased according to Bayes' Law. As B's actions have become more complex and difficult than originally anticipated, A's counterattack becomes less challenging, since 0.7 (A's prior probability) $\times$ 0.2 (A's probability of a counterattack under difficult circumstances) $\div$ 0.44 = 0.32.

Based on this new probability, B estimates the likelihood that it will be subject to A's military counterattack if it remains on A's territory as follows:

$$0.32 \times 0.2 + 0.68 \times 1 = 0.744$$

Consequently, the likelihood of a military counterattack increases significantly. A larger military deployment by A would result in a far greater number of troops than if B again redeployed its troops to stay in another country. In accordance with Bayes' rule, B believes that the likelihood of A launching a military counterattack is 0.32 (A's second prior probability calculation) × 0.2 (the probability of a counterattack under difficult conditions) ÷ 0.744 = 0.086.

As a result, B's assessment of the difficulty of A's counterattack gradually changes due to A's strategic adjustment. According to B, A is more likely to be superior to B in terms of military strength, and the probability of A winning in a conflict is 91.4%(1 – 0.086 = 0.914). The invading troops and facilities are ultimately withdrawn when B agrees to peace talks.

The above examples show that in a dynamic game with imperfect information, the actions taken by the participants have the effect of transmitting information. Although the likelihood of winning the military counterattack is low in the initial stages, A quickly redeploys his troops to change the disadvantages of the local confrontation and exerts strong military pressure on B, thereby halting B's invasion of the border.

Bayesian optimization uses information obtained from the previous sampling point to estimate the posterior distribution of the objective function before selecting a new sampling point. The Bayesian optimization theory can be applied to model high-dimensional military adversarial states and actions. Bayesian optimization is most effective when the computation is complex and with numerous iterations.

The use of GANs, Monte Carlo search, and Bayesian optimization can be combined. With GAN, we can obtain a large amount of data that approximates the real situation. A priori experiences, data, and probabilities of winning and losing can be obtained based on Monte Carlo search methods through combat simulation experiments and adversarial training. Combined with actual exercises, semi-physical simulations, or even actual combat training, Bayesian optimization can be applied to establish a probability distribution function that approximates the probability of winning and losing in a real-world scenario.

6.8 Virtual Staff Officers and Commanders

In the future, staff officers and commanders will be burdened by a great deal of work during complex and high-intensity confrontations, particularly the assessment of multiple threats, the sifting of vast quantities of information, and the deployment of suitable combat forces and methods. In the world of information and intelligent operations, this has become a major challenge.

At this point, developing technologies such as human–computer interaction and self-adaptive mission planning can significantly reduce the workload of staff and commanders. As a result of the continual improvement in the accuracy of human speech recognition, human–machine intelligent voice interaction technology and its products have developed rapidly upon entering the engineering and practical stages. The Echo, invented by Amazon, is considered one of the most significant technological advances since smartphones. This technology brings human beings into the era of voice Internet and voice control. Users find that speaking is the best method for receiving network information and services. Since 2017, millions of homes have been equipped with AI speakers that use voice interaction. In the first quarter of 2018, global manufacturers shipped an astounding 9 million smart speakers, more than doubling the growth of the previous year. Google, Amazon, Tmall, and Xiaomi accounted for four of the top smart speaker manufacturers. AI speakers have been widely used in environments such as smart home environment control, family education, information inquiry, air ticket ordering, and takeout ordering. In the near future, these technologies are expected to be widely used in the military, resulting in profound changes in military command and management.

The US Army launched its Commander's Virtual Staff (CVS) program at the end of 2016; this is a program that aims to create a commander-oriented virtual staff that combines cognitive computing, AI, and computer automation technologies to address massive datasets and complex battlefield postures. A CVS provides proactive recommendations, advanced analytics, and natural human–computer interactions tailored to individual needs. These capabilities assist commanders and their staff through every step of the tactical decision-making process, from planning and preparation to execution [5].

To ensure the compatibility of parallel military systems and forces, numerous virtual and real-time simulations and confrontations, simulation training, and necessary semi-physical and real military exercises should be conducted. Through this process, the level of capability of virtual staff officers and commanders can gradually approach or surpass that of human staff officers and commanders. Finally, auxiliary decision-making is transformed into AI-based hybrid and autonomous decision-making at the tactical and battle levels.

6.9 Personnel Recruitment and Training

An important aspect of intelligent training is the recruitment of personnel with good psychological and operational characteristics for the most appropriate positions. Personnel can be screened using brain–computer and interdisciplinary biology surveys, based on existing personnel recruitment procedures, training methods, and inspection methods. The pre-employment tests, on-the-job evaluations, and intensive training are conducted in compliance with the quality requirements of professionals, such as tank vehicle drivers, artillery operators, missile operators, aircraft pilots, radar operators, warship pilots, and operation duty officers, etc. Establishing the parallel soldier system described earlier in this chapter will permit a comprehensive assessment of personnel's basic qualities and professional skills. Additionally, through brain–computer technology, smart wear, and other technologies, the operational status and mental state of personnel may be monitored. Using images, sounds, and electrodes, it is possible to intervene and regulate personnel who have displayed symptoms of wartime stress. Training can be conducted according to their characteristics to minimize their physiological and psychological burdens. The entry of combat personnel can be well controlled through the assessment of intelligence, psychological quality, and mental state. It can further ensure and promote operation effectiveness through cognitive training and intervention.

Simulation, semi-ph simulations, and intelligent parallel system training based on sufficient battlefield data acquired from both parties (friends and foes) can enhance the intelligence of the combatants and their virtual brains. It is indeed challenging to replicate the requirements of different environments and tasks during actual operational training. The system will be capable of simulating situations with different opponents at different times and locations and solving problems encountered during actual operations through distributed, networked, virtualized, and intelligent parallel training. Then, when necessary, the simulated scenes can be examined through real-world exercises or actual operations, thereby improving training efficiency and achieving the improvement and advancement of both humans and machines.

As part of intelligent parallel training, it is also necessary to strengthen simulation training of virtual space and cross-domain operations in addition to physical space operations. The tracking and positioning of priority targets, the organization and execution of virtual and real decapitation attacks, cyber attack and defense, psychological warfare, public opinion management and psychological diversion, disaster emergency rescue, and action disposal are all training scenarios. The training can be conducted in simulated public locations such as government buildings, hotels, shopping malls, schools, airports, ports, and television stations in order to simulate possible counterterrorism operations. Radio and television, mobile telecommunications, network operation and management, data centers, communication base stations, power plants, substations, transportation systems, and financial systems may also be covered. This training may enhance the capability of cross-domain attack and defense, infrastructure control, and social stability maintenance.

References

[1]   Sun, W. 孙武. (n.d.). Sunzi bingfa ji pian _shici _baidu hanyu孙子兵法·计篇_诗词_百度汉语. Baidu百度. https://hanyu.baidu.com/shici/detail?pid=925ea831

[2]   Wang, F. Y. 王飞跃. (2015, April 4). *X5.0: pingxing shidai de pingxing zhineng tixi* X 5.0: 平行时代的平行智能体系. Kexue wang 科学网. as https://blog.sciencenet.cn/blog-2374-879754.html..

[3]   Popper, K. R. (1972). *Objective knowledge*. Oxford: Oxford University Press.

[4]   Cankao Xiaoxi 参考消息. (2016). *Meiguo lujun kaifa "Avatar" xvni shibing tidai shizhan ceshi* 美国陆军开发 "阿凡达" 虚拟士兵替代实战测试. Cankao xiaoxi wang 参考消息网. www.cankaoxiaoxi.com/mil/20160408/1121875.shtml.

[5]   *Xinxi xitong lingyu keji fazhan baogao* 信息系统领域科技发展报告. (2016). *Shijie guofang keji niandu fazhan baogao* 世界国防科技年度发展报告. Beijing: National Defense Industry Press.

7 The Evolution of OSoS

Since artificial intelligence (AI) practice has only begun recently, there are numerous uncertainties, making it impossible to offer a definite answer as to whether the Operation System of Systems (OSoS) can evolve or not. The general trend, however, is that with a new generation of AI technology, military intelligence, and AI-based battlefield ecosystems, single-tasking systems will exhibit life-like characteristics and functions, while multi-tasking systems will exhibit cyclic and evolving capabilities similar to those of a forest's ecological system. The future evolvable OSoS will be composed of several life-like task systems performing distinct functions that work in concert to form a unified ecosystem. Simultaneously, this evolvable system functions as a gaming system capable of self-competing, confronting, surviving, repairing, and evolving.

Theoretically, any system in which AI plays a dominant or important role can evolve. Overall, OSoS evolution is a complex and systematic process involving different categories, levels, and stages.

Generally speaking, OSoS evolution can be understood in terms of categories, levels, and stages. By category, it can be divided into the evolution of model algorithms, vehicles, swarms, task systems, and confrontation systems. By level, it can be divided into tactical, operational, and strategic levels. By stage, it can be divided into elementary, intermediate, and advanced stages. Evolution is a process that progresses from simple to difficult, inside to outside, and simple to complex. Given a finite amount of time and space, evolution is usually convergent. Model algorithms converge toward relative optimality; vehicles and swarms converge toward maximum capability and potential; task systems toward specific operational objectives; and OSoS toward missions, demands, and war winning.

The evolution of single-task systems is relatively simple, while the evolution of multi-task OSoS is more complex. Therefore, the evolution of the OSoS cannot rely entirely on its self-generation and self-progression but requires proactive planning and necessary environment and conditions. There are four basic principles to be followed.

The first is the bionic principle, which mainly refers to establishing a seamless connection between the AI brain and the information links that include information perception, acquisition, transmission, memory, storage, association,

DOI: 10.4324/b22974-8

analysis, decision-making, command and control, execution, and evaluation; and facilitating interaction, circulation, and feedback within the intelligent decision-making chain. As with the human body, the command-and-control center serves as the brain, the network serves as the nervous system, and the devices serve as limbs that the brain controls. The brain is intricately connected to and organically interacts with all body parts, performing self-adaptation, self-learning, self-coordination, self-repair, and self-evolutionary functions. As a necessary condition for evolution, the living-body functions are relatively simple to accomplish with a single vehicle or single-task system. They are, however, more challenging for the entire OSoS.

The second is the principle of survival of the fittest, which mainly means that operational systems and capability development comply with the law of the jungle, and they compete with each other. The advanced, efficient, and excellent systems and capabilities will be retained through virtual simulation training or actual combat, while the backward, incapable, and inefficient ones will be discarded.

The third is the principle of mutual reinforcement and neutralization, which indicates that a balanced system and an interactive mechanism of mutual reinforcement and neutralization should be established among the elements such as offense, defense, speed, strength, distance, and quantities.

The fourth is the principle of the whole system and full life cycle, which requires a large, rapidly iterated, and continuously optimized cycle in which components ranging from technology research to equipment development, production and procurement, application and training, comprehensive logistics, and operational capability generation ultimately meet operational demands via a closed loop with evolutionary functions.

If the above four aspects are strategically designed and planned for a future OSoS, it will gain the capability of continuous evolution.

7.1 Ecological Chain

Intelligent factors and the five "self" functions (self-adaption, self-learning, self-confrontation, self-repair, and self-evolution) are manifested throughout the OSoS – from the bottom to the top, throughout the entire loop – from the sensor to the shooter, and throughout the entire process – from demand to combat effectiveness generation – to create an intelligent and evolvable ecological chain. Numerous intelligent devices comprising materials, devices, power, structural components, sensors, software, and algorithms can compose intelligent operation task systems, forming intelligent OSoS at tactical, operational, and strategic levels. Intelligent perception, decision-making, offense and defense, logistics, and virtual–real interaction will coexist in an organic system that can be continuously advanced and optimized via human training or automatic iteration.

It should be emphasized that the process from the development of combat concepts, technology, applications, and modes to the generation of new combat effectiveness should feature human–machine interaction, rapid iteration, and continuous upgrading to promote OSoS evolution from low-level to high-level capabilities.

7.2 Distribution and Diversity

When confronted with different tasks, targets, and adversaries, an evolvable OSoS would exhibit distributed and diverse features. In terms of operational elements, the system data are connected via online and offline access; operational vehicles are organized by task; fire strikes and energy destruction are conducted based on vehicle and target distribution; and a machine AI system is distributed across clouds and terminals. In terms of operation space, networked, flattened, and distributed systems are the future development trend due to the emergence of network information and diverse AI technologies. The command echelons will become smaller, and humans will be required primarily in the pre-war period and backstage of war, while distributed AI and unmanned systems will carry out most in-war executions. Certain operational vehicles and task systems, such as intelligent ships, aircraft, ground combat vehicles, and unmanned vehicles, are directly operated and controlled by humans or AI. Some are decentralized and autonomously collaborative swarm systems, such as intelligent munitions and unmanned aerial vehicle (UAV) swarms; others are hybrid human–machine AI systems, primarily manned–unmanned cooperative operation systems, such as the "Loyal Wingman" project and hybrid human–machine special operation systems currently being developed by the US military.

7.3 Parallel Processing and Storage

The development of distributed clouds and diverse AI technologies enables the rapid processing of massive amounts of battlefield data and tasks in highly confrontational and dynamic situations that require rapid response. With the development of brain-like chips and systems, new machine learning technologies, and increased computing power, particularly through the application of mainframe and quantum computers, future systems will be able to perform parallel processing and storage in the same manner as the human brain neural network, but with lower power consumption and greater computing power. This is one of the fundamental conditions for OSoS evolution.

7.4 Network Connectivity and Mutual Feedback

It is also necessary to network AI brain systems and terminal execution systems for the evolution of OSoS and associated capabilities in order to establish interaction and mutual feedback between the internal systems and external battlefield environments. This lays the ground for the development of self-learning and self-evolving capabilities. Mutual feedback is enabled by the "brain" and "limbs" connected via networks. The OSoS interacts with task systems managed by the "brain" AI to achieve continuous optimization, iteration, and upgrading via the "limbs," i.e., mutual feedback, ensuring the network's cognitive communication capability, adequate bandwidth, anti-jamming capability, and the ability to combine online and offline processing as needed.

7.5 Self-Repairing

It is natural for an intelligent OSoS to sustain various types of damage. Given its constant exposure to high-intensity conflict and competition, the system should include self-repairing capabilities. When a component is missing, the system can be programmed to replace and enhance the component automatically. For example, if a few aircrafts are shot down in a swarm system, subsequent aircraft can step in and quickly complete the mission. Additionally, it is necessary to prepare in advance, fortifying the redundant backup. Once damage has occurred, the intelligent maintenance model can automatically detect, assess, repair, and replace the failure.

7.6 Self-Learning and Evolution

The military can conduct red–blue confrontations involving unmanned vehicles versus manned vehicles, unmanned versus unmanned, and manned versus manned, using generative adversarial networks, augmented learning, semi-physical simulation adversarial training systems, and mathematical modeling techniques such as Monte Carlo search, genetic algorithms, Bayesian optimization, particle swarm optimization (PSO), and blackboard systems. Through joint and self-exercise, the system can continuously accumulate data, revise models, optimize parameters, and thus simulate realistic combat situations. Assuming that the battlefield is sufficiently transparent, and that all pertinent information is available, the system's capacity for self-exercise and self-evolution will rapidly increase based on massive simulations, combat experiments, and self-learning practices. It will outperform the human brain in numerous areas, including information processing, target identification, knowledge accumulation, as well as rapidity, repeatability, accuracy, automation, and other professional skills, eventually surpassing commanders' capabilities and becoming battlefield "superheroes." If there is insufficient transparency and incomplete information on the battlefield, machine learning, and intelligent design can help predict upcoming operations and make adaptive adjustments, eventually achieving self-evolution. Although self-evolution at the strategic level is extremely difficult and may never occur, the rapid evolution of the system at the operational level, particularly at the tactical level, will gradually promote the system's overall progress.

7.7 Operation Rules

For the OSoS to be self-adaptive, self-learning, self-confrontational, self-repairing, and self-evolving, it is also necessary to establish adaptive mission planning and auxiliary or autonomous decision-making systems based on effective operational rules and processes, expert systems, knowledge, and experience for task systems.

For small unit operations, it is necessary to consider threats, the unit's capabilities and strength, as well as battlefield constraints and advantages, such as city streets, traffic, buildings, networks, and electromagnetic conditions, as well as hostile and friendly forces encircling the operation areas or operating beyond visual range. Then, various operation rules, specifying constraints and boundaries, should be formulated to guide pre-war training and education. Finally, it is essential to develop and optimize pertinent model algorithms to make scientific decisions and take precise actions.

In terms of operational command and control, it is critical to develop immediate plans for offensive and defensive operations, as well as military operations other than war, and then to conduct auxiliary or rapid decision-making using massive simulation calculations and prediction models of parallel systems that relate to the battlefield environment and the warring parties' strengths and capabilities. In this context, the system should be adjusted dynamically during the war and then optimized based on operational rules and knowledge. If the desired effect is achieved, the operation rules should be preserved while the model, parameters, and knowledge bases are modified and improved. If an undesirable effect occurs, the operation rules should be modified while the relevant models, parameters, and knowledge bases are re-established and improved. Operational support issues such as the supply, maintenance, and logistics of weapons and ammunition should be resolved quickly, efficiently, and automatically in response to wartime demand, battle damage, and troop deployment. The ammunition storage scheme, transportation capacity, technical experts, and maintenance capabilities should also be considered. In this context, the system should automatically select methods, tools, and modes to conduct targeted support and maintenance (Figure 7.1). To achieve intelligent, targeted support and maintenance, it is also necessary to build intelligent equipment and logistic network systems, remote visual maintenance, virtual reality (VR)/augmented reality (AR)-based targeted maintenance, three-dimensional (3D) and 4D on-site printing (Figure 7.2), distributed supply, intelligent logistic models and algorithms, and other hardware and software systems, as well as compliance with applicable operating rules and standards. These systems should be continuously optimized and evolved with sufficient simulations, combat training, and real-world operational experience.

Figure 7.1 Portable remote-control loading and unloading robot (2021 Zhuhai Airshow).

7.8 Self-Adaptive Factory

Seamless network connectivity and real-time data exchange will be established between the military and industry departments responsible for system planning, demand analysis, technology development, project research and development (R&D), and the subsequent application, training, maintenance, and support. In this context, the evolution of OSoS should be extended from the military to the defense industry's design and manufacturing fields, establishing a circular flow of systems from demand, capability, equipment, technology, research and development, and production to support, achieving iterative optimization (Figure 7.3). Virtual design and self-adaptive manufacturing are critical components of this cycle.

In 2010, Defense Advanced Research Projects Agency (DARPA) launched the Adaptive Vehicle Make (AVM) program to address military requirements for next-generation ground combat vehicles. Through crowdsourcing design and a meta-model approach, it established a "rapid adaptive factory". This new rapid adaptive R&D mode completed a vehicle's design in 2 months,

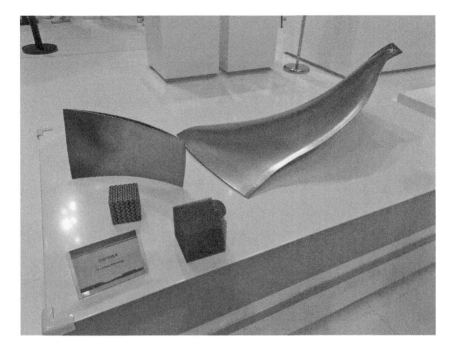

Figure 7.2 Three-dimensional (3D) printing products (2021 Zhuhai Airshow).

cutting the typical 2-year design time by 80%. This process resulted in the creation of the "democratic," collaborative, and innovative design, as well as the completion of the system design without a single error, with design data used to drive adaptive production. Adaptive manufacturing technology, it is estimated, can reduce the R&D cycle of a new armored vehicle from 13 to 2–3 years. Throughout 2014, the program was expanded and migrated to new areas, with several follow-up projects beginning in 2016.

At the 51st Paris Airshow in 2015, MBDA (Matra, BAe Dynamics, and Alenia) unveiled its latest fully modular air-launched missile concept, the CVW102 FlexiS. MBDA released a video demonstrating the future use of modular air-launched missiles. According to the video, when confronted with multiple air and ground threats, the aircraft carrier battle group could select various types of modular missiles and munition components from the equipment bay, including fuses, warheads, control parts, missile servos, and tail fins, assembling them into appropriate attack munitions before loading them on to a fighter aircraft, and finally conducting precise air-to-ground strikes.

On July 13, 2018, the US Army established a Future Command, aiming to achieve full operational capability within a year. With a responsive, flexible, and efficient Future Command, the US Army hopes to integrate engineering

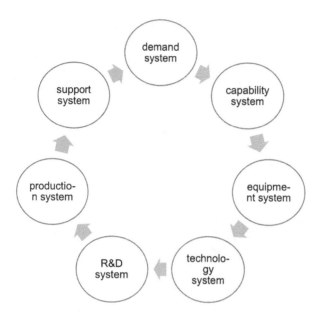

Figure 7.3 Cyclical military system extended to the industrial sector.

and acquisition capabilities, significantly shortening the US Army's modernization process. The Future Command will integrate the Army's concept of operations, operational requirements, acquisition, and field testing, and streamline the command structure to reduce the development cycle from 3 years to 1 year and the equipment delivery time from over a decade to a few years, thus completing a generational upgrade. By 2028, the US Army may be capable of defeating an adversary in joint, multi-domain, and high-intensity operations at any time or location; deterring potential adversaries; and sustaining unconventional warfare capabilities via the deployment of modern manned and unmanned combat vehicles, aircraft, support systems, and other weapons.

With digital parallel systems and parallel equipment manufacturing, people, machines, things, environments, and the supporting industry chain will become information-based, interconnected, and interoperable. Almost everything will be digitalized and modeled, from materials to components, subsystems, and equipment integration, completely changing the way weapons and equipment are developed and manufactured. The current single physical prototype demonstration can be upgraded to an integrated virtual prototype demonstration based on distributed, virtualized, and collaborative design methods, targeting operational demands and capability system requirements. Thus, virtualized integration and demonstration can be conducted from individual pieces of equipment to complete equipment systems, from demand

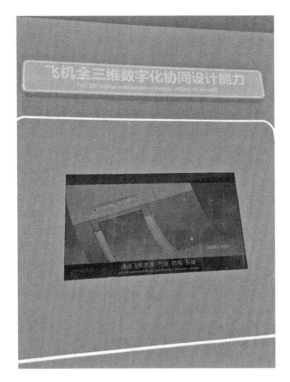

Figure 7.4 Three-dimensional (3D) digital collaborative design of aircraft (2021 Zhuhai Airshow).

analysis to logistics, from overall systems to supporting ends, and from self-development to collaboration and crowdfunding, significantly shortening the time required to complete the process from demand analysis to R&D. Updates and improvements may become simple when R&D, as well as manufacturing, become distributed, digitalized, modeled, and virtualized (Figure 7.4).

Theoretically, it is possible to transition from a manned- to an unmanned-oriented model in areas including design, manufacturing, R&D, and operations. However, in intelligent warfare, it is perhaps the most difficult task to make robots design equipment and operating systems in an innovative and creative manner. Given current technological advancements, robots continue to be superior at processing, producing, and manufacturing based on complete human-made designs. It is feasible for robots to produce and manufacture simple components and subsystems as well as complete pieces of equipment and systems. Many advanced factories worldwide now use industrial robots for welding, assembling, and inspection in unmanned workshops and assembly lines. Innovative design in general still requires manual labor. However, recently, AI has been observed in artistic fields such as music, painting, poetry, and others. It has begun to demonstrate creativity on a par with human work. In the future, as intelligent technology advances, virtual designers, and industrial robots will assume more complex tasks, ranging from manufacturing to design, and from single-piece equipment development to equipment system development. Their performance will be incrementally enhanced and optimized. When the intelligent era comes, many years later, anything will be possible.

In the age of AI, military and civilian sectors will collaborate on design, R&D, VR iteration, training, and upgrading, with concurrent research, experiments, applications, and iterations. This will become the guiding principle for the development of operational capability and equipment systems. The development of conventional mechanized vehicles and information technology only signifies the beginning. Military and civilian sectors will then proceed to co-analyze, design, and develop cognition-based operational concepts, combat simulations, and experiments via virtual and parallel systems, emphasizing military requirements. The two sectors may then coordinate intelligent adaptive manufacturing of software and hardware for equipment systems, conduct concurrent research, testing, development, and equipping, and optimize model algorithms, eventually forming civil–military integrated intelligent operation and support capabilities (Figure 7.5).

Due to the requirements for joint and multi-domain operations, as well as the demand for integrated development of mechanics, information technology, and intelligence, the emphasis of the future defense industry will shift from military service and platform construction toward cross-service

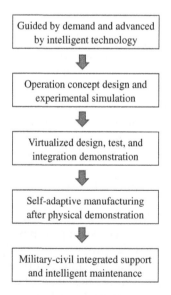

Figure 7.5 Intelligent design and manufacturing processes in the defense industry.

and cross-system integration. Currently closed, self-contained, mutually independent, physically based, and time-consuming research, design, and manufacturing processes will become open-sourced, virtual-based, democratically crowdfunded, self-adaptive, and more rapidly respond to military requirements. Thus, a new innovative and intelligent manufacturing system will be developed by combining hardware and software, virtual and physical worlds, as well as by establishing intelligent interactions between humans, machines, materials, and the environment, establishing effective connections of vertical industrial chains, and distributed collaboration between the government, industry, and academia.

7.9 Survival of the Fittest

In the age of AI, the OSoS evolution is governed by the law of the jungle, i.e., survival of the fittest. Naturally, the OSoS evolution is diametrically opposed to the evolution of an ecosystem. Natural selection is a lengthy process. The ecosystem must maintain a relatively stable state over time, where creatures interact and compete, forming a self-sustaining ecological chain. The OSoS evolution, on the other hand, is characterized by brief, intense, and violent conflicts between the warring parties. If an OSoS evolves at an advanced rate during peacetime, it will provide a better chance of surviving wars. During a regular warfare, if both warring parties are relatively equal in strength, the side with intelligence superiority has a better chance of victory. If the disparity between the two sides is significant, with the stronger side possessing superior intelligence, the outcome is self-evident. However, if the weaker party possesses superior intelligence, the outcome will be unpredictable, necessitating a meticulous calculation of both parties' relative advantages in all areas, followed by a thorough comparison. If both parties are evenly matched in terms of strength and intelligence, there will be numerous engagements. The whole war process, however, will be brief, characterized by high inertia, difficulty to control, and an unpredictable outcome. To summarize, the more sophisticated the AI, the more sophisticated the OSoS, and the greater the battlefield initiative, the greater the system's survival capability. Meanwhile, systems with less advanced AI will be gradually phased out.

Part II

Strategic and Tactical Applications

Intelligent warfare is more frequently manifested as artificial intelligence-based hybrid operations that combine mechanics, information technology, and intelligence. Typical patterns include unmanned operations in physical space, cognitive confrontation in virtual space, and parallel operations with virtual – real interaction.

DOI: 10.4324/B22974-9

8 Unmanned Operations

Unmanned operations refer to operational patterns based on unmanned vehicles, which, featuring mechanization, informatization, and intelligentization, are the culmination of human wisdom in the operation system of systems (OSoS). Unmanned vehicles can be defined in a narrow sense and a broad one. In a narrow sense, unmanned vehicles refer to unmanned aerial vehicles (UAVs), unmanned ground vehicles (UGVs), unmanned surface vehicles (USVs), unmanned underwater vehicles (UUVs), precision-guided weapons, and ammunition, etc. Broadly speaking, unmanned vehicles may include satellites, sensors, unattended guards, intelligent cyber attack and defense, and electronic confrontation systems. Unmanned vehicles are the basis and premise for unmanned operations. Without the development of unmanned vehicles, no unmanned operation would be possible. After a century of development, since the first UAV was explored in the twentieth century, great progress and achievements have been made, with UAVs, UGVs, UUVs, and bionic robots emerging in the last few decades. Many models have matured, laying a strong technological foundation for application in real unmanned operations. Predictably, with the wide application and rapid development of artificial intelligence (AI) and unmanned technology, direct confrontations between living forces in future battlefields will be significantly reduced. Instead, unmanned operations will dominate every battlefield, including land, sea, air, and space, at strategic, operational, tactical, and even individual levels. Unmanned operations will have a disruptive impact on the entire OSoS and become a basic form of intelligent warfare. In a primary stage, unmanned operations are led by humans and assisted by unmanned equipment. In an intermediate stage, unmanned operations are led by unmanned equipment and assisted by humans. In an advanced stage, unmanned operations are entirely performed by unmanned equipment under the guidance of rules made by humans. Unmanned operations involve single-system operations, manned–unmanned cooperative operations, swarm operations, and joint, multi-domain, cross-domain, and cross-media operations.

DOI: 10.4324/b22974-10

8.1 Development of Unmanned Technology

Due to their highly mobile and unmanned characteristics, unmanned vehicles are not limited by human physiological factors. They can undertake tasks that manned equipment cannot perform, such as dull, dirty, and dangerous tasks, while demonstrating greater capabilities than manned vehicles. Especially in recent local warfare, the operational advantages of unmanned combat vehicles, such as UAVs, have been demonstrated to the fullest.

Unmanned technology includes unmanned vehicles, environmental perception, intelligent control, system coordination, swarm networking, measurement and control communications, and other technologies. According to relevant information, military powers led by the US have been implementing unmanned operations based on the concept of "unmanned vehicles with a manned system" since the war in Afghanistan. This concept has also been tested in several local wars and has achieved outstanding operational results. *The Sunday Times* reported in June 2010 that the US military had deployed 7,000 UAVs in the wars in Iraq and Afghanistan [1], including 127 Predator reconnaissance aircraft, 31 Reaper reconnaissance aircraft, ten Global Hawk strategic reconnaissance aircraft, and other UAVs for various purposes. Among them, Predator and Reaper UAVs are capable of carrying missiles for ground attacks. The US Army has deployed more than 2,400 Talon robots, carrying cameras, motion and sound detectors that can work day and night, whose robotic arms have flexible rotating shoulders, wrists, and fingers with certain memory and learning capabilities for reconnaissance of the internal structures of buildings, courtyards, sewers, and caves, and for vehicle inspection, roadblock removal, and border security patrols, etc. In 2009, the *Guardian* reported that the US Air Force had trained more UAV operators than fighter and bomber pilots, with 240 UAV operators graduating by September, compared with 214 manned fighter pilots in the same period. The *Guardian* commented that this was "a sign that the era of fighter pilots is coming to an end" [2].

In 2015, Russia deployed its ready-framed unmanned forces for striking terrorist organization ISIS in the Syrian war and achieved remarkable results, giving the terrorists a confrontation with unmanned equipment deterrence. In Russia's fight against ISIS, six multi-purpose combat robots, four fire support combat robots, three UAVs, and the "Andromeda-D" automated command system formed an unmanned combat swarm that preceded manned combat forces to carry out reconnaissance and strikes. Syrian government forces then followed up with a clearing operation. The battle lasted 20 minutes, during which 70 enemy militants were eliminated in one fell swoop, while only four Syrian government troops were wounded. Among them, "Platform-M" and "Argo" combat robots participated in the battle, which was the first battle involving combat robots. During the battle, reconnaissance UAVs were responsible for monitoring the enemy's situation on the battlefield and transmitting information back through the "Andromeda-D" light automated command system. After receiving the request for fire support, the "Platform-M" and "Argo" fighter robots marched to 100–200 meters from the enemies' positions, under the remote control of operators in the rear, and launched an attack on the enemy, attracting and detecting enemy fire, directing artillery support from the rear. "Argo" adopts an 8 × 8 chassis, with a mass of 1.02 tons, a maximum speed of 20 km/h, and a maximum remote-control range of 5 km. It is equipped with a PTK 7.62 mm machine gun and three rocket-propelled grenade (RPG)-26s. Platform-M adopts a tracked chassis with a mass of 800 kg and is equipped with a 7.62 mm machine gun and four grenade launchers. The Iraq government has also purchased the CH-4 Rainbow reconnaissance and strike (R/S) UAVs, produced by China, which have completed several strikes against ISIS forces with excellent outcomes.

On August 5, 2018, Venezuelan President Nicolás Maduro was delivering a live television speech celebrating the 81st anniversary of the Venezuelan National Guard when a UAV exploded at the scene in front of a neatly lined parade of soldiers. Fortunately, the president was not injured, but seven soldiers suffered varying degrees of injury. This was the first terrorist UAV attack in public.

By the end of 2020, the exploration and application of unmanned technology had achieved a great leap forward. Unmanned equipment in developed countries has been delivered to the army in large quantities and put into operational application and has begun to undertake major assaults in addition to traditional reconnaissance, surveillance, and damage assessment tasks. The proportion of unmanned vehicles equipped in the US Army has exceeded one-third, of which the number of UAVs in the Air Force is well over one-third, with wide applications. Unmanned weapons and equipment and combat forces with manned and unmanned cooperative systems will lead the development direction.

In the future, vehicles, weapons, mission systems, and OSoS, during their process of developing unmanned and operational applications, will apply the following technologies related to intelligence.

First is global high-definition geographic information system (GIS) technology based on space–time benchmark, heterogeneous system cooperation, and big data association. This technology is mainly used for space–time synchronization of combat forces, efficient cooperation of manned–unmanned swarms, accurate positioning of battlefields, and precise target tracking.

Second is intelligent perception technology. Through convolutional neural network (CNN) image recognition, regression neural network (RNN) speech recognition, and other technologies, it will allow intelligent and rapid classification and recognition of information from multi-sources, including land, sea, air, and space, and information in multiple forms, including image, video, infrared, multi-spectrum, synthetic aperture radar (SAR), electronic spectrum, and voiceprint. Besides, targets can be quickly extracted from massive information through data association verification and satellite multi-spectral feature recognition.

Third is self-organizing network communication and navigation positioning technology. In an interference-free environment (under unified space–time reference coordinates), non-centered, self-organizing network, frequency-hopping communications, and mutual positioning between vehicles can be achieved using time, frequency, and code division in active multi-band communication modes. In an interference environment, the positions of swarm companions and the relative distance between them can also be calculated through passive detection means, including optical flow image recognition and tracking, binocular three-dimensional (3D) imaging and intelligent recognition, laser 3D active array imaging, and inertial navigation.

Fourth is human–machine integration and autonomous decision-making technology. Given the relative transparency of battlefield and enemy target information, autonomous machine decision-making can be achieved by fusing human decision-making with machine technologies such as intelligent recognition, situational awareness, and adaptive mission planning, with the help of simulation, adversarial training, and self-learning. Under conditions of incomplete information, machines can provide assisted decision-making or move from assisted and human–machine hybrid decision-making to an autonomous one through dynamic online learning.

Fifth is autonomous and cooperative operational technology based on network information. UAVs, UGVs, UUVs, and intelligent ammunition, with relatively independent development paths, in order to form a system and swarm operational capabilities in the future, need to execute tasks and assess the results in a self-adaptive manner based on the urgency and scale of battlefield threats and targets, pre-established operational rules, and standards, and information required for intelligent sensing and command and control (C&C), in which both autonomous single-system task execution and swarm coordinated task execution can be achieved. For example, suppose a swarm of 16 R/S UAVs flying over a battlefield detects four types of target consisting of command posts, radar communication stations, tanks, and squad personnel. In that case, two UAVs closest to the command post will attack the post, four will attack the radar communication stations, six will attack the tanks, two will attack the squad personnel, and the remaining two will be used for battlefield assessment, communication relay, and backup, etc., according to the pre-established operational rules.

At present, as unmanned technology develops, unmanned vehicles, with increasingly cheaper production costs, will be deployed in large numbers in operations, realizing the tactical concept of "quantity over quality." Among the operations, swarm and anti-swarm operations have attracted great attention from all over the world.

The US Army has coordinated unmanned systems development through regular publication of documents such as Unmanned Systems Integrated Roadmap and Unmanned Ground Systems Roadmap. In August 2018, the US Department of Defense (DoD) issued the Unmanned Systems Integrated Roadmap in the Fiscal Year 2017–2042, which proposed four "global themes," including interoperability, autonomy, secure network, and human–computer cooperation. The US DoD sorted out 14 supporting factors, 17 related challenges, 11 future development directions, and 19 key technologies based on these themes. For the first time, this roadmap lists AI and machine learning as important factors for developing unmanned systems and abandons the normal acquisition process for unmanned systems, with a renewed emphasis on multi-path agile acquisition.

8.2 UAVs

Controlled by radio remote control or its program, the UAV comprises air-frame, flight control and management, reconnaissance payload, weapon payload, ground measurement and control, data communication, and other systems. Towards the twenty-first century, driven by military and civilian demand and inspired by the outstanding performance of UAVs in several local wars, many countries have paid unprecedented attention to the research and development (R&D) of UAVs, and the industry has seen unprecedented explosive growth. According to limited statistics and analysis, since 2010 more than 80 countries worldwide have developed and produced UAVs, and dozens of series, hundreds of models, and varieties of UAVs have been developed. The world's leading national armed forces are all equipped with UAV systems.

In the military sector, UAVs can be classified into penetration, endurance, tactical, mini, and micro tactical uses according to their operational missions, as shown in Table 8.1. The penetration, endurance, and tactical UAVs are generally multi-functional.

According to flight distance and function, military UAVs can be divided into five categories ((1) long-range; (2) medium-range; (3) short-range; (4) ultra-short-range; and (5) unmanned helicopters) and 11 types ((1) ultra-high-speed in near space; (2) high-altitude, high-speed, and long-endurance; (3) medium- and high-altitude, and long-endurance; (4) ground/sea striker; (5) medium-range, high-speed; (6) medium-range, low-speed; (7) close-range; (8) ultra-close-range; (9) large, (10) medium, and (11) small unmanned helicopters).

Table 8.1 Unmanned aerial vehicles (UAVs) classified by missions

Group 1	Micro tactical	Mass 0–9 kg, flight altitude < 350 m, flight speed < 185 km/h
Group 2	Mini tactical	Mass 10–25 kg, flight altitude < 1,000 m, flight speed < 460 km/h
Group 3	Tactical	Mass < 600 kg, flight altitude < 5,500 m, flight speed < 460 km/h
Group 4	Endurance	Mass > 600 kg, flight altitude < 5,500 m
Group 5	Penetration	Mass > 600 kg, flight altitude > 5,500 m

Guided by global warfare strategy and supported by strong demand and financial resources, the US Army has formed a complete UAV development system, with multiple UAV series forming squadrons. By 2018, the US military was equipped with a series of small, medium, and large UAVs as well as short-, medium-, and long-range UAVs, including more than 10,000 micro and small UAVs (under 20 kg) and 1,139 large UAVs (over 20 kg) [3].

Israel's UAV equipment was developed in wars in the Middle East in the twentieth century. In response to frequent armed conflicts, the Israeli army urgently needed weapons to reduce the cost of fighting, reduce casualties, and facilitate victory. With their low cost, high efficiency-to-cost ratio, and ability to reduce casualties, UAVs are undoubtedly the best option for dealing with skirmishes and battles. Therefore, the Israeli army has had experience proving the conception that "UAVs are better than manned aircraft," which has also been the main guiding ideology and driving force for the development of the UAV industry in Israel.

Israel is closely behind the US and far ahead of other developed countries in terms of UAV technology. Israeli UAV equipment covers all operational applications, including reconnaissance, surveillance, searching, electronic warfare, anti-radiation, decoys, communications, and strike. Israel has formed a series of UAV equipment covering micro and small UAVs, medium-range tactical UAVs, and strategic UAVs, with a complete system of more than 20 series, typically including: "Heron," "Helms," "Seeker," "Harpy," and i-View series, as shown in Table 8.2.

The Israeli Air Force plans to build a new air force fleet with more than 50% UAVs by 2030. Israel has developed and produced ten types of tactical drones, all of which are exported, among which the largest export volume is the "Harpy" UAV with armed strike capability. The UAVs have been frequently exported to the US, Europe, South America, Asia Pacific, Africa, and other countries or regions.

In addition to the US, European countries have not been shy in accelerating the development of UAVs. European UAV technology also leads the world in terms of key performance indicators such as payload, endurance, and ceiling,

Table 8.2 Key performance indexes for typical European unmanned aerial vehicle (UAV) equipment

Country	Model	Payload (kg)	Speed (km/h)	Flight time (h)	Ceiling (m)
France	Sparrow	45	235	6	5,000
	Neuron	1,000	900	3	5,000
Germany	KZO	35	220	3.5	4,000
	Typhoon	50	120	4	4,000
Russia	Dozor-4	12.5	120	8	3,000
	BLA-05	10	200	3	3,000

Typical UAVs developed by European countries are shown in Table 8.2. Currently, European countries have installed mainly tactical small and medium unmanned aircraft systems (UAS). The European Union is also stepping up the development of its own high-altitude, high-speed, high-payload stealth unmanned attack aircraft, "Neuron." On February 5, 2014, the British military announced that it had completed the flight test of the "Raytheon" supersonic combat UAV.

Despite being a major aviation and nuclear power, Russia has been conservative in developing its UAV equipment and has under-explored UAVs' potential military value, resulting in a relative lag in developing UAV technology and its tactical applications. However, following the 2008 Russo-Georgian war in South Ossetia, the Russian army found that its previously highly efficient anti-personnel weaponry was rendered "blind" by weak reconnaissance and surveillance technology. Since then, the Russian army has changed its mindset and increased its investment in the R&D of medium and small UAVs and launched a project to develop a series of heavy stealth UAVs (the MiG "Mikoyan Skat").

India is in the midst of a comprehensive equipment upgrade, of which enhancing its UAV combat capability is a key component, with plans to create one of the most powerful UAV formations in the Asia-Pacific region by 2020. India is currently one of the largest buyers of UAVs in the Asia-Pacific region and is heavily equipped with advanced UAVs such as the Heron TP UAVs manufactured by Israel Aircraft Industries (IAI). Meanwhile, based on its technological strengths, India is increasing the R&D of local drones by absorbing foreign UAV technology.

The rapid development of UAVs has been driven by political factors and war demands, but more importantly, by the enormous operational successes achieved by UAVs in modern warfare. In 2019, many countries, including the US, China, Israel, Canada, Germany, Britain, France, and Russia, have developed UAVs, and more than 80 countries have been equipped with UAVs. Among them, the US owned the largest share (nearly half) of the military UAV market, ranking first in the world, while Israel, Europe, Japan, South Korea, and China ranked second, with a relatively complete industrial chain and a certain scale of production, developing rapidly.

According to the data released by the Stockholm International Peace Research Institute (SIPRI), 12 countries worldwide exported 35 types of military UAVs, totaling 963 UAVs (excluding micro-UAVs) from 2010 to 2019. The US accounted for 41%, China 26%, Israel 15%, and other countries 18%. For China, the UAVs currently exported are mainly the Pterosaur series (Figure 8.1), Rainbow series (Figures 8.2 and 8.3), and new R/S unmanned helicopters such as Jingdiao CR500, to be launched (Figure 8.4).

At present, high altitude, high speed, long endurance, stealth, multifunctionality, and autonomy have become the main development trends for UAV equipment of countries worldwide. For the US military, there are six key technical areas for UAS development: interoperability and modularity,

Figure 8.1 China's Wing Loong 2 unmanned aerial vehicle (UAV) (2021 Zhuhai Airshow).

Figure 8.2 China's Rainbow CH-6 unmanned aerial vehicle (2021 Zhuhai Airshow).

Figure 8.3 China's Rainbow-7 high-altitude long-endurance stealth unmanned aerial vehicle (UAV) (2018 Zhuhai Airshow).

Figure 8.4 China's Golden Eagle CR500A search and fight integrated unmanned helicopter (2021 Zhuhai Airshow).

communications capabilities, security (intelligence/technology protection), sustainable self-recovery, autonomy and cognitive behavior, and weapon systems. In the future, there will be four major development trends for UAVs.

First, flight time and speed improvement. This mainly involves low altitude, low speed to high altitude, high speed, and hypersonic speed in near space. With their high survivability and efficient reconnaissance capabilities, high-altitude and long-endurance UAVs will continue to see their applications expanded (Figure 8.5). With the widespread use of UAVs on the battlefield,

Figure 8.5 China's anti-unmanned aerial vehicle (UAV) system (2021 Zhuhai Airshow).

emerging countries are intensifying research into countermeasures against UAVs, leading to the creation of counter-unmanned aircraft systems (C-UAS) and other interception systems. Improving the speed of a UAV is one of the primary means to reduce the probability of interception by C-UAS.

Second, stealth and miniaturization. The airframe of UAVs is developing towards stealth and miniaturization to improve mobility and battlefield survivability further. Micro aerial vehicles (MAVs), because of their small size and light weight, can imitated various insects and thus are highly stealthy. MAVs can shuttle between obstacles or buildings in complex terrain or urban operations to obtain detailed real-time battlefield information. Figure 8.6 shows China's series of small UAVs at the Zhuhai Airshow in 2021. The smallest UAV is only 249 g, with only half an hour of flying time, but a flight control distance of more than 10 km.

Third, intelligent manipulation and decision-making. At present, most UAVs are remotely controlled by human operators, which requires high manipulation skills and limits the machine's ability to grasp battlefield information more accurately than humans. UAV intelligence is the ability of UAVs to make autonomous decisions, enabling the UAV to follow instructions or pre-programmed missions, react promptly and autonomously to known threat targets, and respond and act quickly to emergencies that arise at any time.

Figure 8.6 China's small unmanned aerial vehicles (UAVs) (2021 Zhuhai Airshow).

Fourth, integrated sensing and swarm cooperation. In the future, UAVs will also develop towards system integration, integrated sensing, and swarm operations, enhancing their versatility, system support, and learning capabilities. For example, the US military Predator UAV is equipped with an observer, a zoom color camera, a laser rangefinder, a third-generation infrared sensor, a charge-coupled device (CCD) camera (capable of imaging in both visible and mid-infrared bands), and a synthetic aperture radar.

8.3 UGVs

Since the 1980s, major countries worldwide, including the US, Israel, Britain, France, Germany, Russia, Japan, and other countries, have launched research programs on unmanned ground systems (UGS) and increased investment. Since then, the R&D of UGVs for military uses has achieved numerous outcomes, some of which are applied in real operations. UGVs can be divided into semi-autonomous, autonomous, and fully autonomous. Among them, full autonomy is the ultimate goal for the development of UGVs. Depending on different demands and purposes, UGVs have a variety of missions. UGVs undertake reconnaissance and surveillance, patrol and vigilance, fire guidance, fire strike, communication relay, mine hunting, barrier-breaking, anti-terrorism, stability maintenance, and rear-loading support. UGS technology involves multiple disciplines and expertise involving mechanics, electronics, dynamics, automatic control, computers, AI, environmental perception, path planning, behavioral control, and human–machine interaction. Manufacturing UGVs is a complex and systematic project. Compared to the aerial, surface, and unmanned underwater systems, UGVs operate in complex geographical environments, with winding roads and numerous obstacles, placing higher demands on their environmental adaptability, battlefield mobility, controllability, and interoperability, as well as communication and C&C.

The US military has prioritized top-level planning for UGS, with the regular publication of documents including Unmanned Systems Integrated Roadmap and Unmanned Ground Systems Roadmap, guiding the development of military UGS and related technologies. As early as 2001, the US military set the ambitious goal of "achieving 1/3 of all ground combat vehicles being UGVs by 2015" [4]. Although this goal has not yet been achieved, driven by the Future Combat Systems program and military operations in Iraq and Afghanistan, the US autonomous UGS technology has been rapidly developing, with an increasing proportion of UGVs in the US military's weaponry system, among which semi-autonomous UGVs have already been in service. Apart from autonomous UGS, the US military has also been vigorously developing various forms of advanced UGS, including quadruped robots, humanoid robots, climbing robots, robotic exoskeletons, and exploring cutting-edge technologies in manned–unmanned cooperation, unmanned swarms, and brain–computer integration. The US military predicts that its UGS will be able to perform fully autonomous missions by 2034.

In the Future Combat Systems project, the US has put forward ten levels of evaluation criteria for the autonomy and intelligence of unmanned systems:

Level 1: the system relies on remote control with no decision-making capability, where the remote operator commands the system to operate in a relatively simple stationary environment.

Level 2: the system relies on remote control with no decision-making capability, where the remote operator, according to the status displayed by the dashboard sensor and depth images, commands the system to operate in a relatively complex static environment.

Level 3: the system, based on pre-planned tasks, may avoid simple obstacles, with the basic ability to track paths with the help of the operator.

Level 4: the system, based on local environmental knowledge through processing sensor images, can follow a navigator with the operator's help.

Level 5: the system can detect, avoid, or overcome simple obstacles/hazards and predict paths in real time based on assessment, with semi-autonomous navigation and basic off-road capability.

Level 6: the system can detect complex obstacles and analyze terrain, avoid, or climb over complex terrain and obstacles with the operator's help.

Level 7: the system can detect and track moving objects by processing sensor information. In on- and off-road environments it can autonomously avoid or overcome complex terrain and environmental conditions, hazards, and obstacles, basically without the operator's help.

Level 8: the system can cooperate and escort, with advanced decision-making capability based on shared data from other vehicles, and drive quickly and efficiently with minimal operator input.

Level 9: the system can perform complex cooperative tasks through cooperative planning and execution under the operator's supervision.

Level 10: the system is fully autonomous. It can integrate data from all combat equipment without the operator's surveillance and accomplish mission objectives through cooperative planning and execution.

UGVs can be classified in many ways, for example, by size: micro, small, and large; by weight: light, medium, and heavy; by nature: mechanical and bionic. Micro and small UGVs mainly assist individual soldiers and teams in operations, with a small range of movement (tens to hundreds of meters) and slow travel speeds (generally under 10 km/h), focusing on reconnaissance and surveillance, explosive detection and disposal, route clearance, and toxic agent/radiation detection. For example, in the 2021 Zhuhai Airshow, China Northern Company's UGV carried by individual soldiers enabled personnel to complete operation and disposal at a safe distance (Figure 8.7).

Figure 8.7 An unmanned ground vehicle (UGV) carried by individual soldiers.

Figure 8.8 China's team unmanned ground vehicles (UGVs) (2021 Zhuhai Airshow).

Light UGVs, with a rather small mass (generally 1,000–2,000 kg), a certain load capacity, large range of movement, and relatively fast speed (generally above 50 km/h or even 100 km/h) can assist teams in border patrols, armed guards, follow-up operations, and service support. Typical light UGVs include, for instance, the Squad Mission Support System (SMSS) transporter UGV of the US, the AvantGuard and Guardium UGV of Israel, and the Team UGVs of China (Figure 8.8). The SMSS transporter UGV is based on a 6 × 6 chassis, with a mass of 1,960 kg, a payload of 540 kg (provided by the cargo floor and folding rails on both sides), a maximum on-road range of 160 km, a maximum off-road range of 80 km, a vertical obstacle height of 0.6 m, and a trench width of 0.7 m.

The SMSS transporter UGV has two control modes: remote control mode, in which the operator controls the UGV through the handheld universal control unit; and surveillance-based autonomous mode, in which the sensor component of the UGV can follow a soldier by identifying, locking on to, and tracking the 3D image of the soldier, and return to the mission starting point by tracking the electronic markers left on the route of travel with the aid of global positioning system (GPS) navigation.

Medium UGVs, generally with a mass of several tons and strong off-road and protective performance, can perform assault, transportation, reconnaissance, and patrol missions alone or in support. Figure 8.9 shows medium UGVs at an unmanned vehicle challenge in China. The Israeli medium UGV "Crusher," with a mass of 6,000 kg and a maximum load capacity of

Figure 8.9 A medium unmanned ground vehicle (UGV) at the Unmanned Vehicle Challenge in China.

more than 3,600 kg, features a 6 × 6 highly adaptive travel system driven by hydro-pneumatic suspension and hub motor, which enables differential steering in situ. The UGV can overcome vertical obstacles of less than 1.2 m and ditches of 2 m and climb slopes less than 40°. The UGV's hybrid engine supports purely silent electric driving with a top speed of approximately 42 km/h.

Heavy UGVs, with a large payload (generally around 10 tons), can be heavily equipped with weapons, highly protected, and thus particularly suitable for independent reconnaissance, assault, and fire support missions. The "Black Knight" heavy UGV, developed by British Aerospace, with a mass of 9.5 tons, is based on a scaled-down Bradley chassis and the power, protection, and firepower technology from the M2 combat fighting vehicle. The UGV is available in both reconnaissance and assault types. For the reconnaissance type, the weapon system mainly consists of a 30 mm machine gun. The weapon is replaced by larger guns, medium over-the-horizon missiles, and modular remote-controlled self-defense weapon stations for the assault type. Figure 8.10 shows China's heavy unmanned platform at the 2021 Zhuhai Airshow.

The key technologies of UGVs include UGS, intelligent perception, environmental understanding, autonomous planning and decision-making, highly adaptive driving, core devices and typical mission payloads, human–computer interaction, and C&C technologies.

Figure 8.10 China's heavy unmanned ground vehicles (UGVs).

Countries including the US, Britain, Russia, Canada, Israel, France, and Spain are developing a new generation of vehicle-mounted and self-propelled UGVs.

Defense Advanced Research Projects Agency's (DARPA's) Ground X Vehicle Technology (GXV-T) program demonstrates advances in UGV technology for fast travel over different terrain, as well as higher situational awareness and operability. The GXV-T project envisages a future in which UGVs will traverse the most complex terrain, including various slopes and grounds. UGVs will feature revolutionary wheel track and suspension technology, enabling on- and off-road access and faster travel than existing ground vehicles. The GXV-T seeks a variety of in-vehicle sensors and technology solutions designed to provide high-resolution, 360° situational awareness while maintaining vehicle containment.

In the future, the development of UGV will focus on full autonomy and swarm operation.

First, autonomous perception technology. The UGS, based on information about environmental features detected by its sensors and through information fusion algorithms, detects, identifies, and calculates terrain, buildings, and moving targets in operational scenarios.

Second, autonomous navigation technology. The UGS features an autonomous perception of the environment and its state, by which it plans a path to the target. The technology enables UGS to determine its own friendly and target positions concerning the global coordinate system, autonomously plan the optimal path between the starting point and the target off obstacles, and finally correct the path to reach the target.

Third, the wireless ad hoc network technology. The wireless ad hoc network technology is a specially structured wireless communication network technology that relies on the mutual collaboration of nodes to communicate in a wireless multi-hop manner, independent of any fixed facilities, with self-organizing and self-managing characteristics. The technology originated from the packet radio network project initiated by the US in 1972. After that, DARPA initiated the Survivable Radio Adaptive Networks (SURAN) and the Global Mobile Information Systems (GloMo) programs to develop a self-organizing, peer-to-peer, multi-hop ad hoc network for mobile communication. At present, the ad hoc network, with high flexibility and survivability, is the focus of developing wireless self-organizing network technology, mainly involving routing, security, and other key technologies. The key to routing technology is to develop a dynamic protocol that efficiently finds routes between network nodes. Such a protocol will sense changes in the network topology, maintain network topology connectivity, and be highly adaptive.

Fourth, autonomous control technology. Autonomous control technology enables control of power, payload, navigation, communication, and weapon firing of UGVs with the help of human–machine interfaces, communication systems, and various algorithms, enabling UGVs to operate autonomously or form swarms with other unmanned and manned systems.

Full autonomous control technology allows UGVs to operate autonomously without the intervention of an operator, or to be transferred to the operator's control in case of emergency, and to operate in formations and swarms with manned and unmanned systems. The technology will have a disruptive impact on future equipment development and operational patterns.

8.4 USVs and UUVs

The new century has seen the rapid development of USVs and UUVs. The US military has developed a series of USVs for reconnaissance, anti-submarine, and mine hunting and is also developing armed ones, demonstrating and validating USVs for swarm attacks.

In January 2015, the US showcased the Sensor Hosting Autonomous Remote Craft (SHARC), jointly developed by Boeing and Liquid Robotics. SHARC is an automatic long-range remote control USV powered by wave and solar energy. The USV, including surface and underwater components, is mainly used for data acquisition and acoustic monitoring. The USV features advanced situational awareness above and below the water, operating in a swarm with other ground, surface, and aerial vehicles. It is equipped with towed array sonar and can sail for up to a year at sea without crew and maintenance.

In March 2015, DARPA issued an information consultation document to solicit new detection technology solutions for long-endurance anti-submarine USVs, mainly including three technical fields.

In March 2015, DARPA issued a call for proposals for novel detection technologies in long-range anti-submarine USVs, covering three main technology fields. First, maritime sensing sensors, capable of detecting and imaging ships around the clock and in all weathers, with an action distance of 4–15 km; priority is given to passive detection means, including but not limited to passive visible and infrared imagers, laser range finders, lidar, etc. Second, maritime perception software, i.e., algorithms capable of detecting, tracking, and identifying targets (e.g., ships) for passive optical or active non-radar imagers. Third, algorithms and software used to detect, track, and identify ships' signal flags and navigation lights; these are standard tools used to communicate a ship's heading, position, and status. The purpose of the DARPA call for proposals is to improve the detection and identification capabilities of the Anti-Submarine Warfare Continuous Tracking Unmanned Vessel (ACTUV). Primarily intended for continuous, autonomous tracking of quiet conventional submarines, ACTUV has three main objectives: to outpace conventional submarines at a significantly lower cost than in-service systems, navigate safely in the oceans following maritime laws, and accurately track diesel-electric submarines in any position.

The mine-hunting USV (MHU) being developed by the US Navy is expected to provide a new means of countering mines. The USV has a sonar detection accuracy of up to 7.6 cm and laser detection accuracy of up to 2.5 cm and is capable of automatic target identification, classification, and optical imaging, based on several proven technologies such as the Navy's 11 m long USV technology and the AQS-24A mine detection system.

For the MHU, the operational uses are as follows: the MHU navigates remotely to the mission area and drops the mine detection system; after detecting the target, it transmits information such as situational awareness and sonar detection data to the C&C station via a tactical data link; at the end of the mission, it retrieves the mine detection system and returns. The vehicle is used to detect not only mines but also other dangerous objects underwater.

USV technology in China is also developing rapidly, with a few private companies entering the field. At the 2018 Zhuhai Airshow, Zhuhai Yunzhou Intelligent Technology Co., Ltd. debuted the "Lookout II" (Figure 8.11), an R/S USV that has passed a missile flight test, being the first missile USV in China and the second missile USV in the world that can launch missiles. The company also completed the world's largest USV swarm test at Zhuhai Wanshan Offshore Test Site (Figure 8.12).

At present, USVs are showing the development trends of larger hulls, smaller weapons, more diversified functions, and swarmed formations. UUVs, likewise, are developing to be more serialized, swarmed, and submarine-like,

Figure 8.11 Lookout II reconnaissance and strike unmanned surface vehicle (R/S USV).

Figure 8.12 The world's largest unmanned surface vehicle (USV) swarm test.

Figure 8.13 China's high-speed unmanned surface vehicles (USVs) (2021 Zhuhai Airshow).

with longer-range and higher deep-sea operability. Figure 8.13 shows China's high-speed USV at the 2021 Zhuhai Airshow, with a maximum speed of 80 knots and an endurance of 300 nautical miles. Figure 8.14 shows China's UUVs at the 2021 Zhuhai Airshow.

Unmanned combat swarm in deep sea is one of the key points for future unmanned maritime combat. DARPA has proposed projects involving the underwater carrier "Hydra" (an unmanned aircraft carrier that carries UAVs and UUVs) and the upliftable underwater preset weapon system, which remains underwater on standby, and surfaces when receiving remote orders, significantly enhancing concealed strike capability. In May 2016, the DARPA Distributed Agile Anti-Submarine Warfare (ASW) project had completed sea trials of all subsystems. In the future, it will deploy dozens of UUVs at 6 km submersible depth to detect all submarines within a sea area of up to 180,000 km² (Figure 8.15).

8.5 Bionic Robots

The bionic robot is a recently rapidly developing unmanned vehicle, the product integrating biotechnology, information technology, mechanical technology, AI,

Figure 8.14 China's unmanned underwater vehicles (UUVs) (2021 Zhuhai Airshow).

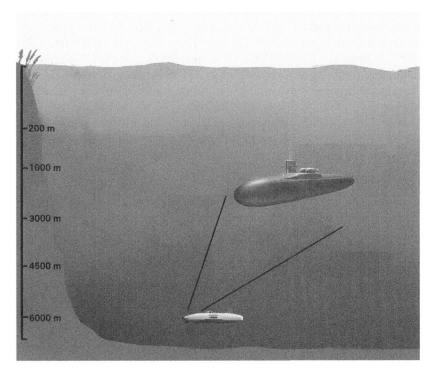

Figure 8.15 Concept map of unmanned underwater vehicle (UUV) navigation and anti-submarine detection.

and other fields, with significant autonomous and intelligent characteristics. Bionic robots, of which there are many varieties, are constantly developing, including humanoid, dog-like, fish-like, and bird-like robots. The most famous of these include the humanoid robot ATLAS, the dog-like robot BigDog, and their upgraded version and series of models, developed by Boston Dynamics, and Raven, a robotic bird developed by the US Army.

ATLAS features striding, single-leg standing, jumping, obstacle avoidance, and fall protection, with a human body shape and a standing height of 1.88 meters. ATLAS is equipped with 28 hydraulic drive joints for closed-loop position and force control, as well as a computer for real-time motion control, a hydraulic pump, and a thermal management system. Its wrist can be equipped with application program Interfaces (APIs) for different functions, as well as fiberoptic Ethernet access. Humanoid robots are currently being developed in the US, Russia, India, Japan, and South Korea.

The BigDog is a four-legged bionic robot, successfully developed by Boston Dynamics in early 2006 with funding from DARPA. The robot can follow soldiers over hilly terrain. It is equipped with a stereo camera and laser scanner in its head, relying on a stereo vision system or remote control to confirm its path. Since 2013, the BigDog bionic robot has introduced an upgraded version, Alpha Dog, which, in addition to carrying objects over long distances in the field, featured avoiding obstacles autonomously and would be equipped with the ability to interact with people. The Alpha Dog project was another innovative project undertaken by Boston Dynamics in collaboration with DARPA. Meanwhile, Boston Dynamics is also developing the Cheetah robot project, which will run at speeds of up to 113 km/h. Once developed, it can resolve the problem of slow-moving quadrupedal walking robots, with cross-country speeds far exceeding those of tracked and wheeled vehicles.

In June 2013, a British website reported that the US Army had developed a bionic robot bird, Raven, with a mass of only 9.7 grams and a 34.3 cm wingspan. In flight, the bird surprisingly attracted the attention of nearby crows, which thought it was one of their kind and approached, accompanied, and called out to it.

Bionic robots in China are also developing rapidly, including humanoid, four-legged, multi-legged, and bird robots, and robotic fish, etc. Figures 8.16 and 8.17 show China's robot dog and dexterous two-armed robot at the 2021 Zhuhai Airshow.

Bionic ammunition technology refers to the development of ammunition with a biologically distinctive form, structure, function, and behavior. This technology can be divided into principle, functional, and structural bionic technologies. Bionic ammunition features a more concealed shape, accurate damage, flexible movement, and clearer reconnaissance.

Since the 1990s, bionic micro weapon technology integrating bionic and microsystem technologies has developed rapidly. In 2007, the US took a research initiative in bionic ammunition, completing a design of a sparrow-sized swarm bionic micro ammunition and in-building flight tests. After

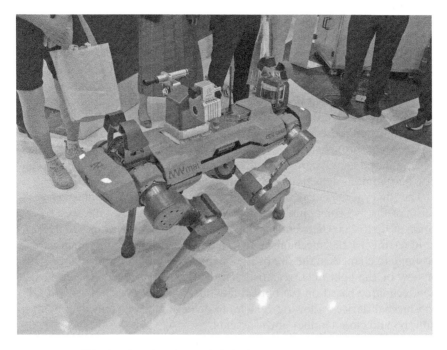

Figure 8.16 China's robot dog (2021 Zhuhai Airshow).

Figure 8.17 China's dexterous two-arm robot (2021 Zhuhai Airshow).

2009, the US began to focus research on swarm synergistic bionic ammunition. In 2011, AeroEnvironment completed flight testing of the prototype Hummingbird reconnaissance bionic munition. The German company Festo has been developing various bionic aerial vehicles, e.g., bird-like and dragonfly-like vehicles; the former were first exhibited and flight tests were completed in 2011, whereas the latter have been under development since 2013, currently without destructive capability. This kind of bionic ammunition, which "flies and walks" like living creatures, is considered one of the most important pieces of military equipment in the post-2025 era.

Key breakthroughs in bionic ammunition systems include overall design, structural and aerodynamic design, materials applications, bionic flight control, autonomous detection, and control and attack technologies.

The "fly-eye" camera, developed by US scientists imitating the imaging mechanism of the fly's compound eye, can take 1,329 pictures at a time, with a resolution of up to 4,000 lines/cm, making it an effective reconnaissance tool [5].

Referring to the mechanism by which bees navigate autonomously in response to changes in the size of the surrounding groundscape and the speed of flow of the light field, scientists have also developed a precision navigation technique based on the optical flow method that does not rely on GPS and inertial devices, allowing vehicles to achieve precise, controlled flight in subways and closed indoor environments.

Since the 1980s, bionic robotics based on the theory of bionics and biophysics has become of concern in the world, and at present, both aspects have made great progress with many practical achievements. Bionic robots developed by the US, Russia, Japan, and Germany are rich in variety and realistic in appearance, with functions close to or even surpassing those of natural creatures, facilitating activities such as detonation, reconnaissance, search and rescue, and transport of goods and casualties on the battlefield. The robots mainly include humanoid robots, quadruped/hexapod robots, crawling robots, bionic UAVs or ammunition, and bionic fishes (Table 8.3), many of which have been put into use in the War of Afghanistan, with their technology becoming increasingly mature.

Bionic robots mainly involve power, walking, and control technologies. Driven by the development of autonomous and interoperable technologies, future bionic robots, with increasing autonomous combat capabilities and inter-system synergy, will take on more and more combat tasks, expanding from the current auxiliary tasks of reconnaissance, surveillance, and material transportation to the main combat tasks of firepower strikes and mobile assaults.

8.6 Intelligent Ammunition

Intelligent ammunition is the core of intelligent weapons and one of the ultimate solutions to efficient strikes. In the future, the development of intelligent ammunition will focus on intelligent perception, mobile orbiting,

Table 8.3 Development of bionic robots in the US, Russia, and other countries

Serial no.	Robot	Country	Category	Bionic features	Note
1	ATLAS	US	Humanoid	A variety of fine movements and 360° backflips	
2	Fidor	Russia	Humanoid	Flexible fingers to complete fine motor skills, even capable of holding a gun	
3	HRP-2	Japan	Humanoid	Quick and smooth movement	
4	BigDog	US	Dog-like	Quadruped robot capable of walking, running, carrying ammunition, food, and other items for soldiers in inaccessible areas; obstacle-crossing capabilities	Used in War of Afghanistan
5	Bobcat	Russia	Dog-like	Similar to BigDog	
6	Lex	US	Walking animal-like	Hexapod robot	Used in War of Afghanistan
7	Guardian	US	Reptile-like	Serpentine robot with life-like appearance	
8	Omnitd	US	Reptile-like	Serpentine robot with outstanding crawling ability	
9	Scorpion	US	Reptile-like	Smooth walking on complex terrain; transmission of pictures via the camera on its tail; obstacle avoidance	
10	War Eagle	US	Bird-like	Flexible wings that can be folded in the packaging tube, with advanced lift-drag characteristics and stronger wind resistance capacity	
11	Hummingbird	US	Bird-like	Small, palm-sized, flapping-wing design with a realistic appearance	Abu Dhabi Defense Exhibition
12	Sniper System	US	Flock-like	Swarm coordination for urban warfare	
13	Octopus	US	Fish-like	The world's first fully soft-bodied autonomous robot resembling an octopus, manufactured using three-dimensional (3D) printing technology	
14	Robotic fish AIRACUDA	Germany	Fish-like	Artificial muscles, used in swing tail movement; water-free style manipulation by remote control	

high-speed cruise, inter-bullet networking, precision control, swarm attack and defense, and efficient damage. Among these, hypersonic swarm attacks and defense are important developments. Hypersonic can be subdivided into hypersonic boost-slip long-range guided ammunition, hypersonic ramming cruise long-range guided ammunition; swarm attack and defense can be subdivided into long-range artillery-launched UAVs, patrol flight ammunition swarm, and long-range end-sensitive bomb swarm combat system. As hypersonic weapons are covered exclusively in Chapter 10, the focus is on basic intelligent ammunition technology.

8.6.1 Fully Modularized Flexible Ammunition

Fully modular flexible ammunition refers to generic components, functional components, and auxiliary devices that adopt a modular design concept and open system architecture, stored in various component forms in peacetime and assembled into various ammunition families for different operational scenarios in wartime. Whereas "modularization" was previously concerned with functional components, the "full modularization" proposed here covers generic components, functional components, and auxiliary devices, extending the improvement of intelligent ammunition systems to the component level, where new and updated technologies can be embedded at any time.

At the 2015 Paris Airshow, Matra, BAe Dynamics, and Alenia (MBDA) presented the ICVW101 airborne flexible missile concept, dividing airborne missiles into three diameter families: 180 mm missiles (1.8 m and 3 m in length), 350 mm missiles (3.5 m), and 450 mm missiles (5.5 m). MBDA has completed top-level design and concept studies for the three missiles, focusing primarily on the full modularization of an 180 mm missile, which is planned to be fielded in 2035.

In 2014, the US Army launched the Modularized Missile Technology Project to study the full modularization design technology and open system architecture of helicopter/UAV air-to-ground missiles. It has developed the modularized six-degrees-of-freedom simulation software and new guidance and control algorithms for missile-oriented modularized open system architecture and carried out simulation and live ammunition verification. It is confirmed that adopting an open system architecture for missiles will promote technological innovation, improve procurement mechanisms and storage methods, and lay a technical foundation for the interchangeability and integration of missiles of various mission types.

8.6.2 Intelligent and Controllable Ammunition

With the systematization and networking of operations, the entire OSoS is becoming increasingly large and complex, raising demands for weapons and equipment with dexterous strike capabilities for future multi-domain combat and autonomous operations. In areas of urban operations, mountain

Figure 8.18 China's intelligent and controllable ammunition (2021 Zhuhai Airshow).

operations, local airborne and landing operations, and operations in special confined space, with numerous blind spots, the rapid removal of multiple small targets such as enemy personnel and fire points requires autonomous, low-cost, dexterous ammunition or micro missiles that can precisely seek out the positions and weaknesses of the enemy (Figure 8.18).

In the field of controllable ammunition, there are five key issues to be addressed. First, circular error probable (CEP), the core way to improve equipment damage effect and efficiency ratio, to be controlled within 1 m, with the efficiency ratio to be at least doubled compared to ordinary munitions. Second, 90% improvement to target identification probability. Third, precise identification and strike capability of target weakness and vitals. Fourth, adaptability of multi-platform delivery and launch. Fifth, adjustable damage effect.

Ammunition with adjustable damage effect is a kind of ammunition that produces the appropriate energy and controlled elements required to destroy a target depending on the type of target detected, combining precision detonation control technology, warhead charge construction, and an onboard sensor. This kind of ammunition can be "roughly loaded" before firing depending on the target type, as well as "finely controlled" after firing depending on the nature and location of the target. Compared with traditional ones, the ammunition with adjustable damage effect reduces the logistics burden of the army and improves mission flexibility. Simultaneously, it can control damage effects according to specific targets, reduce collateral and environmental damage, and reduce the cost of post-war reconstruction. This kind of ammunition is an important stage for intelligent ammunition evolving from "controlling the hit point" to "controlling the damage power and pattern," and represents an important trend for the future development of intelligent damage.

Adjustable damage effects mainly include adjustable damage power and pattern. Adjustable damage power includes damage radius and result (detonation or deflagration). For this part, key issues involving innovative structural design, explosive charge design, and detonation control are solved by realizing control of the warhead's output energy and damage pattern, improving the combat cost-effectiveness ratio, and reducing collateral damage. For adjustable damage patterns, key issues involve making damage patterns and equivalent decided by the warfighter or even the ammunition itself, based on adequate target area information at the end of the trajectory.

8.6.3 Guided Ammunition

Guided ammunition is a kind of bullet that, fired from firearms, can correct its trajectory to improve shooting accuracy. Like simple guided artillery shells, the guided bullets are flown to the target based on sensor signals, with instructions generated from the onboard chip and transmitted to the actuator, which repeatedly corrects the trajectory according to the instructions until the bullet hits the target. Guided ammunition can significantly improve the shooter's accuracy at long range, making hits independent of factors such as target movement and wind speed, as well as allowing the shooter to attack moving targets from more positions.

Research into guided gunshot technology began in the US in the early 1990s, and the project was officially launched in the early twenty-first century. A breakthrough in US-guided ammunition research resulted in two 12.7 mm guided gun cartridges: laser-guided ammunition by Sandia National Laboratory and optically guided ammunition by DARPA. In 2012, Sandia National Laboratory completed the simulation and prototype test of laser-guided ammunition. In 2014, during a live firing test conducted by DARPA on a 12.7 mm optically guided ammunition developed by Trinity Science and Imaging, the shot autonomously hit a target that had drifted from where it was aimed. In 2015, in another live firing test, the guided ammunition hit a moving target, again striking a moving target.

8.7 Unmanned Ground Operations

Unmanned ground operations are relatively complex among all operations because of undulating terrain and numerous obstacles involved, resulting in numerous variables and blind spots faced in mobile assaults, firepower strikes, and communication links, as well as formation and coordinated operations. They also involve challenges in urban, landing, mountain, border and coastal defense, as well as overseas military operations. Therefore, unmanned ground operations are characterized by complex patterns, forces, and difficulties. There are many patterns for unmanned ground operations, and the following is an analysis of nine typical patterns.

8.7.1 Single Unmanned System Operation

Single unmanned system operation is mainly divided into two forms: R/S UAV and UGV operations. Among them, the main task of R/S UAV operations is

Figure 8.19 An unmanned ground vehicle (UGV) at the unmanned vehicle challenge in China.

reconnaissance or strike at short and medium range, low and medium altitude. For short- and medium-range operations, the US military has deployed reconnaissance (R/S UAVs), including the Shadow 200, Grey Eagle, and Predator, and is developing new models such as unmanned helicopters and vertical takeoff and landing UAVs. Single unmanned ground system operations are mainly aimed at reconnaissance, search, and indicating targets within visual range to better provide accurate local battlefield information in fire systems and command, with an activity range of several tens of kilometers and a vehicle mass and load of tens to hundreds of kilograms. There are numerous UGV models, ranging from small to large, from short- to long-range, from reconnaissance and patrol models to search and strike models, mostly for urban alley warfare, explosive and drug detection, indoor operations, underground trench warfare, mountain operations, cave operations, etc. Figure 8.19 shows the UGVs at the unmanned vehicle challenge in China.

8.7.2 Manned–Unmanned Cooperative Ground Operations

In this kind of operation, operative formations are mainly composed of tanks, armored vehicles, artillery and missile vehicles, and small and medium UAVs and UGVs, which are connected into a whole through a special network.

Figure 8.20 China's unmanned ground vehicle (UGV) launcher for coordinated ground operations (2021 Zhuhai Airshow).

Figure 8.20 shows the coordination between UGVs on the China Challenge. Before the battle, the vehicles collectively move towards the target area based on a situational awareness map of the battlefield beyond a 40 km line-of-sight provided by the command post or space-based satellites and medium- and long-range UAVs. During the process of moving, the UAV carries out careful detection and camouflage identification at short and medium range (10–30 km), sends threat targets and suspicious targets back to the command post for screening, and then carries out attacks through accompanying fire vehicles and guided weapons to clear threats beyond the visual range of tanks and armored vehicles. When approaching enemy target areas or dangerous areas such as minefields, UGVs, ahead of manned vehicles, take the initiative to move forward for reconnaissance, raid, target indication, and obstacle clearance, and guide gun-launched missiles or other firepower of tanks to carry out precision attacks.

8.7.3 Low- and Medium-Altitude Manned–Unmanned Cooperative Operations

In this kind of operation, operation formations are mainly composed of reconnaissance (R/S UAVs), unmanned helicopters, manned helicopter gunships,

Figure 8.21 Manned–unmanned cooperative operations at low and medium altitudes.

manned transport helicopters, manned ground attack aircraft, ground command posts, and remote-control stations. The main operational process includes the UAVs carrying out long-range reconnaissance or R/S tasks based on the long-range battlefield information provided by the command post or satellites, then unmanned and manned helicopter gunships carrying out mobile or long-range attacks, and finally, transport helicopters carrying out personnel or equipment delivery to occupy the target area. This operation can be either a small-scale operation or a large-scale swarmed one (Figure 8.21).

8.7.4 Air–Ground Integrated Unmanned Operations

There are various patterns for air–ground integrated unmanned operations. There are two main unmanned cooperative operational patterns: medium- and long-range UAVs + ground fire and UGVs + accompanying UAVs. UAVs are fast in mobility, with a wide reconnaissance range but a limited number of loads; at the same time, UGVs are slow in mobility but have many loads. A feasible operational pattern is using UAVs, with air superiority, to carry out mid- and long-range battlefield reconnaissance while guiding medium- and long-range fire from UGVs to perform saturation precision attacks on target areas (Figure 8.22).

Similarly, another good option would be using small to medium UAVs to accompany UGVs, carrying out limited-range forward detection or limited strike, and providing battlefield and target information to strong ground fires behind, forming an integrated and coordinated air–ground operation at short to medium range. Figure 8.23 shows the small and medium-sized UAVs and unmanned vehicles specially used for air–ground cooperative operations at the 2021 Zhuhai Airshow.

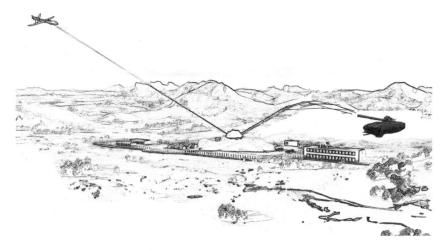

Figure 8.22 Air–ground integrated unmanned operations with air guidance as the key.

Figure 8.23 China's air–ground cooperative warfare unmanned system (2021 Zhuhai
Airshow).

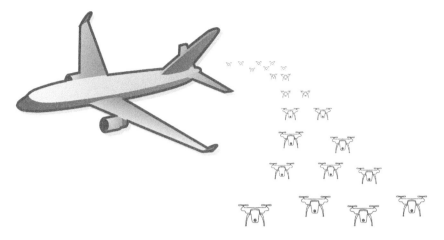

Figure 8.24 Low- and medium-altitude unmanned aerial vehicle (UAV) swarm operations.

8.7.5 Low- and Medium-Altitude UAV Swarm Operations

In this kind of operation, dozens, hundreds, or even more small and medium UAVs are launched using ground launching, airdrops, and gun launching to carry out multi-batch, swarmed, and saturation precision attacks on the target area (Figure 8.24). The formation and damage can be in many patterns, such as "many against one," "many against few," "many against many," etc., or multiple waves and directions of attack against the same target, or distributed and coordinated strikes against multiple targets in a similar timeframe (e.g., 1 or 5 minutes).

8.7.6 Medium- and Long-Range Loitering/End-sensitive Missile Swarm Operations

In this kind of operation, cruise missile groups at different distances are launched through ground rocket launchers or air-launched vehicles. Terminal-sensitive missile groups are launched to form a combat area covering tens to thousands of kilometers to carry out uninterrupted reconnaissance, surveillance, and intelligent rapid identification. After finding the target, loitering munition swarms and many terminal-sensitive missiles will be used according to the type and quantity of the target to perform swarmed and coordinated attacks against large-scale ground mobile targets (Figure 8.25). The operation is suitable for targeting swarms of tank armor vehicles, such as the tops of tank armor vehicles, which are the weakest part and are vulnerable to missile attacks. Figure 8.26 shows China Northern Corporation's missile patrol swarm launching system at the 2021 Zhuhai Airshow.

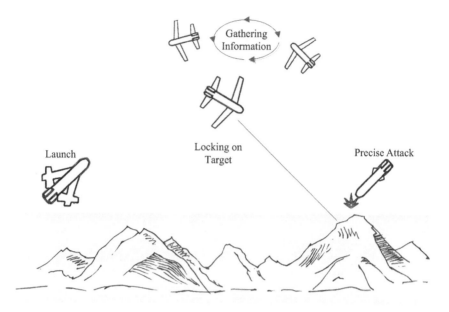

Figure 8.25 Coordinated attack by swarm of loitering missiles.

Figure 8.26 China's loitering munition swarm launch system (2021 Zhuhai Airshow).

Figure 8.27 China's unmanned ground vehicle (UGV) series (2021 Zhuhai Airshow).

8.7.7 *UGV Swarm Coordinated Operations*

The operation is suitable for environments such as cities, plains, hills, landing islands, and military operations other than war (MOOTWs). The formation comprises unmanned light tanks, heavy unmanned vehicles, bionic unmanned vehicles (e.g., BigDog), and humanoid robots (Figure 8.27).

In urban operations, for example, the main operational process includes the following: gaining air supremacy, based on the image information returned from forward positions, including the direction, location, and depth of the target area, scientific formations are established, with heavy tanks with protection capabilities as the main force, supplemented by short- and medium-range R/S UAVs; after the tanks and UAVs have broken through the city's outer defenses, light vehicles plus small and medium accompanying UAVs will conduct alleyway battles, followed by bionic robots (such as BigDog), humanoid robots and hand-thrown or micro UAVs to carry out in-building operations and underground clearance until the end of the battle (Figure 8.28).

8.7.8 *Human–Robot Mixed Formation Operations*

There are mainly three patterns in this kind of operation: individual soldier + portable (generally hand-thrown) UAVs, individual soldier + dog-like robot (BigDog), and individual soldier + humanoid soldier. In the first pattern,

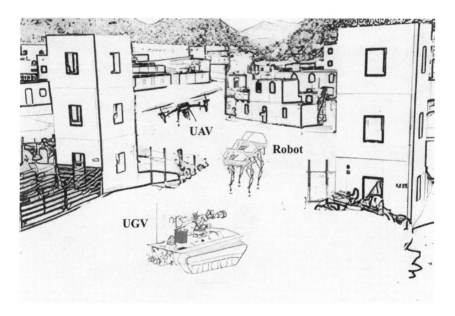

Figure 8.28 Unmanned ground vehicles (UGVs) and robot swarm operations.

MAVs carried by individual soldiers or teams scout and detect blind corners and dead ends in alleys, partition walls, building floors, and ground covers to provide accurate information on the battlefield prepared for coordinated operations. In the second pattern, "BigDogs," being able to carry heavy loads, transport weapons, ammunition, and equipment in complex terrains, as well as undertake dangerous tasks such as detecting dangerous explosives, removing obstacles, and reconnaissance. In the third pattern, human soldiers are completely or partly replaced by robots on the battlefield, with humans operating in the background or performing human–robot cooperative combat in a narrow space, with robots in the front and soldiers in the back (Figure 8.29).

8.7.9 Unattended Ground Systems

Unattended ground systems are mainly used for policing and defending borders and coasts, radar stations, barracks, ammunition depots, field command posts, and troop quarters. The systems are usually composed of radar detection, photoelectric reconnaissance, optical fiber detection, intelligent mines, and denial weapons. The main operational process that involves this kind of system proceeds as the system detects multi-source signals such as radar, visible light, infrared, and vibration, carries out intelligent and rapid identification of targets including people, animals, weapons, and equipment in the vicinity, and automatically correlates image and location information with

Figure 8.29 Human–robot mixed formation operations.

weapons to take destructive attacks or non-lethal strikes through manned or unmanned control and other modes for security protection (Figure 8.30).

8.8 Unmanned Marine Operations

At present, military powers are implementing independent ocean network plans, constituting a 3D ocean observation network through satellite remote sensing, aerial detection, surface buoys, naval towing, underwater submersibles and self-sinking buoys, shore-based radar systems, accumulating basic data to establish a "digital ocean," laying the foundation for unmanned marine operations. Unmanned vehicles at sea mainly include UAVs, anti-ship missiles, USVs, UUVs, and torpedoes. From a strategic point of view, manned vehicles such as aircraft carriers and large destroyers will remain manned for a long time and will not be completely unmanned, but the trend has been towards unmanned combat vehicles and escort vehicles. From an operational and tactical point of view, the proportion of unmanned equipment will gradually increase in ground defense, ground boarding operations, regional operations, and overseas military operations, and these operations may even be completely unmanned. There are three major unmanned marine operations: unmanned operations based on manned warships, including aircraft carriers, large destroyers, and amphibious assault ships; unmanned swarm operations against ships, submarines, and ground

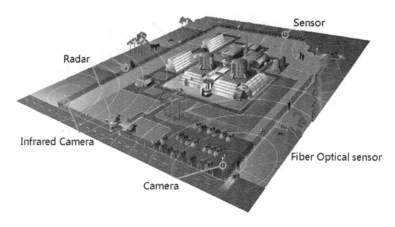

Radar

Sensor

Infrared Camera

Fiber Optical sensor

Camera

Figure 8.30 Unattended ground systems.

vehicles; and overseas unmanned ground operations. The following is an analysis of ten typical unmanned operational patterns at sea.

8.8.1 Single Unmanned System Operation at Sea

This kind of an operation mainly relies on large UAVs to carry out medium and long-distance attacks against ships, aircraft, and ground vehicles; USVs, to carry out detection and attack against ships, ground targets, and submarines (Figure 8.31); UUVs and torpedoes, to attack submarines and surface ships; and unmanned anti-submarine aircraft, to detect and attack submarines.

8.8.2 Manned–Unmanned Aerial Vehicle Cooperative Operations at Sea

This kind of operation relies on large ships such as aircraft carriers, amphibious assault ships, and mixed formations of unmanned and manned UAVs to carry out long-range strikes up to thousands of kilometers, both against high-value fixed and mobile targets on the ground and against combat formations at sea (Figure 8.32), as well as carrying out medium- and long-range air-to-air operations.

8.8.3 3D Unmanned Sea-to-Ground Attacks

First, this kind of operation mainly relies on aircraft carriers, large destroyers, amphibious assault ships, submarines, and carrier-based aircraft to launch long-range 3D missile-to-ground attacks outside the defense area or air-to-ground attacks by carrier-based aircraft when conditions permit. For example, on April 14, 2018, the US, the UK, and France launched Tomahawk

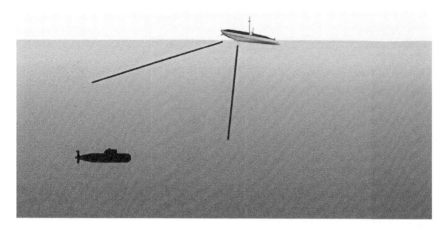

Figure 8.31 Anti-submarine unmanned surface vehicles (USVs).

Figure 8.32 Swarm formations at sea.

Figure 8.33 Missile patrol unmanned surface vehicles (USVs) by Zhuhai Yunzhou
Intelligent Technology Co., Ltd.

cruise missiles from destroyers, missiles from Ohio-class submarines, and air-launched cruise missiles from naval aircraft, with a total of 105 missile strikes against Syrian military and civilian facilities, achieving a 3D strike on the ground by the combined maritime forces.

Second, in the future, for this kind of operation, it will be possible to launch UAVs and missiles directly from USVs for sea-to-ground attacks (Figure 8.33), air-to-ground missiles from UAVs for sea-to-ground attacks; and at closer distances, missiles from UUVs for sea-to-ground attacks; or UUVs for cross-media sea-to-ground attacks from underwater.

8.8.4   3D Unmanned Attacks Against Ships

Primarily, this kind of operation mainly relies on manned vehicles such as aircraft carriers, large destroyers, amphibious assault ships, submarines, and carrier-based aircraft to carry out anti-ship missile and torpedo attacks (Figure 8.34), or relies on unmanned carrier-based aircraft to carry out air-to-ship attacks when conditions permit. Additionally, in the future, similar to 3D unmanned sea-to-ground attacks, it will be possible to launch UAVs and missiles directly from USVs for attacks against ships; air-to-ground missiles from UAVs for attacks against ships; and at closer distances, missiles from UUVs for attacks against ships; or UUVs for cross-media attacks from underwater attacks against ships. Third, in the future, in joint multi-domain operations, it will also be possible to launch ground-based medium- and long-range missiles or cruise weapons to attack ships.

Figure 8.34 Precision attacks by stealth anti-ship missiles.

8.8.5 3D Unmanned Attacks Against Submarines

Submarine attacks are carried out primarily by anti-submarine USVs detecting and releasing torpedoes and guided depth bombs by UUVs targeting submarines or releasing torpedoes; naval guns firing guided artillery shells; and, in a joint multi-domain operation, ground-based medium- and long-range missiles (Figure 8.35).

8.8.6 Unmanned Air-to-Sea Operations

First, in terms of short and medium range, this kind of operation mainly relies on air defense and interception such as air-to-sea missiles, phalanx, high-power microwave weapons, laser weapons, and electromagnetic cannons, combining point and surface strikes as well as soft skilling and physical damage (Figure 8.36). Second, in terms of long range, air combat is carried out through maritime early-warning aircraft, carrier-based UAVs, and carrier-based long-range missiles. Third, in joint multi-domain warfare, it is also possible to rely on ground and air forces for unmanned, high-speed, stealth air fire support within the combat radius.

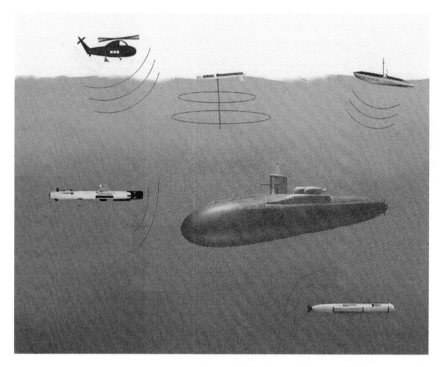

Figure 8.35 Three-dimensional (3D) unmanned attack operation against submarines.

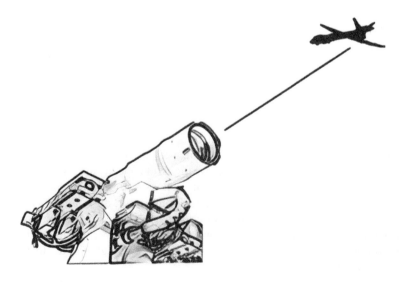

Figure 8.36 Shipborne laser weapon air defense.

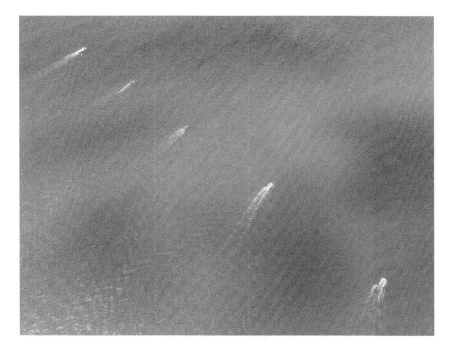

Figure 8.37 Swarm cooperation between unmanned underwater vehicles (UUVs).

8.8.7 Maritime Unmanned Swarm Operations

Maritime unmanned swarm operations mainly rely on shipborne manned vehicles that release UAV swarm for attacks against ships and ground vehicles outside the defense area or in the target area; USVs and small and medium UAVs, carrying out swarm attacks against ships and ground vehicles (Figure 8.37); shipborne precision-guided weapons and anti-ship missiles, carrying out saturation swarm attacks; and finally, UUVs and torpedoes, carrying out swarm attacks on submarines and surface ships.

8.8.8 Sea–Air Integrated Unmanned Operations

For this kind of operation, there are mainly three modes: sea-based medium- and long-range cruise missiles + long-endurance UAV for ground-to-sea attacks; shipborne short- and medium-range missile + R/S UAVs for sea-to-ground attacks (Figure 8.38); short- and medium-range USV missiles + small and medium UAVs for sea-to-ground attacks.

8.8.9 Unmanned Marine Operation System

This system relies on amphibious assault ships and landing ships at sea, forming a new marine operation system consisting of R/S UAVs, short- and

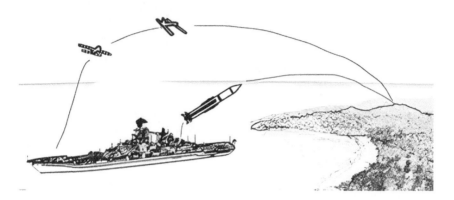

Figure 8.38 Air–sea integrated unmanned operations.

medium-range precision-guided firepower, USVs, unmanned amphibious vehicles, UUVs, UGVs, and man–machine mixed individual formations. It is designed mainly for unmanned landing, ground attack, coastal city attack, ground defense and expulsion, maritime anti-terrorism, sea rescue, and other MOOTWs (Figure 8.39).

8.8.10 Cross-Media Unmanned Maritime Operations

In the future, in joint multi-domain operations, supported by network and target detection, this kind of operation mainly relies on submarine-launched missiles to carry out cross-media ground and ship attacks (Figure 8.40); ground-based medium-range missiles and airborne weapons to carry out cross-domain and cross-media attacks on submarines, surface ships, and other targets; and large maritime vehicles to carry out anti-missile and anti-satellite operations.

For example, the US has conducted the Standard three missile tests against satellites and has completed several mid-course anti-missile tests. On June 6, 2015, the US Missile Defense Agency announced that the US Navy, in conjunction with the Technical Research Headquarters under the Japanese Ministry of Defense, had completed the first flight test of the Standard-3 Block IIA anti-missile.

In July 2018, foreign media published an article saying that a new submarine-launched missile system code-named "Sea Dragon" of the US Navy gives submarines "breakthrough offensive capability" and could be a new multi-purpose missile system with both anti-ship and anti-aircraft capabilities. The secret weapon was likely to modify the latest US Standard 6 anti-aircraft missile, according to the analysis. If the description of the "Sea

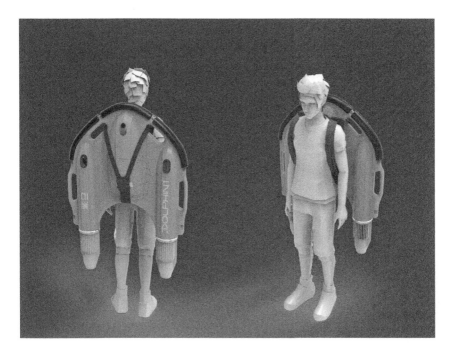

Figure 8.39 "Dolphin 1" unmanned lifeboat.

Dragon" missile is true, then submarine air defense may usher in a new era, from "hiding" to "engaging."

8.9 Unmanned Aerial Operations

In the future, manned air combat vehicles may maintain their presence but in reduced numbers, while UAVs will increase. Air operations will be predominantly based on manned–unmanned cooperation early and may shift to unmanned operations later. Unmanned aerial operations mainly involve long-range unmanned bombers, unmanned stealth combat aircraft, a series of unmanned reconnaissance UAVs and R/S UAVs, diversified airborne weapons, and air defense and anti-missile weapons. The following focuses on seven typical unmanned aerial operational patterns.

8.9.1 Single UAV Operations

As air forces worldwide are strategic military services, their UAVs are mainly tasked with strategic reconnaissance, campaign operations, and tactical strikes. Therefore, single UAV operations mainly include medium- or

Figure 8.40 Submarine-launched missile weapon system by China (2021 Zhuhai
 Airshow).

long-range ground reconnaissance and strikes, air reconnaissance, and early
warning. Typical UAVs include the US Army's Global Hawk and the new
series of SR72s under development. The US Air Force's Predator UAVs
have been used extensively in Afghanistan, Iraq, and Syria. Although these
UAVs have caused accidental injuries to civilians and civilian facilities, they
are highly cost-effective, with almost no combat casualties. Figure 8.41 shows
China's large, high-altitude, stealth, unmanned attack aircraft at the 2021
Zhuhai Airshow.

It is said that most of the US Air Force UAV operators work not in front-
line field conditions but at Nevada Air Force Base or other airbases thousands
of miles away from the battlefield. US military UAV operators usually arrive
at the base in business attire, change into their uniforms, operate various types
of drones according to their commanders' operational orders, finish combat
missions and conduct real-time battlefield assessments, and then change back
into their suits before going home and reuniting with their families. The process is like a live version of online war games, with the operators strategizing
and winning thousands of miles away.

Figure 8.41 China's large, high-altitude, stealth, unmanned attack aircraft (2021 Zhuhai Airshow).

8.9.2 Manned–Unmanned Cooperative Aerial Operations Supported by Loyal Wingman

The Loyal Wingman project, with UAVs accompanying manned aircraft, is being developed by DARPA. According to the project, reconnaissance and early warning of aerial threats are carried out by UAVs in advance of manned aircraft, and enemies are attacked by UAVs beforehand under the command of manned aircraft. It can greatly improve the detection, early warning, and attack distance of manned aircraft, forming air-to-air and air-to-ground superiority without the need to improve the combat performance of manned aircraft (Figure 8.42).

8.9.3 Early-Warning Aircraft + Mixed Formation Aerial Operations

The main purpose of this operation is to invest in the advantages of early-warning aircraft over long distances, multiple directions, and multiple targets in the air, and to command manned or unmanned aircraft in

Figure 8.42 Manned–unmanned cooperative aerial operations supported by Loyal Wingman.

air-to-air operations. The main process of its operation is early-warning aircraft carrying out rapid detection, intelligent identification, correlation printing of incoming targets in multiple directions, and automatically carrying out mission planning according to the spatial location, trajectory, and characteristics of the targets following the priorities. There are many ways of mission planning, including direct first-wave attacks by UAVs and second-wave attacks by manned aircraft; coordinated attacks by manned aircraft with "loyal wingmen"; or simultaneous coordinated attacks by manned and unmanned aircraft. The mission-planning process and various operational patterns can be simulated and trained before a war through the self-adaptive mission-planning system, i.e., the machine brain, that conducts confrontation exercises, self-learning, and self-improvement. In the early stage, exhaustive methods can be adopted to rehearse and verify possible confrontation modes in advance. When various operational experiences and data are accumulated in the later stage, operational rules could be formulated, and an intelligent decision-making system based on operational rules could be established, so that manned and unmanned vehicles can make and execute decisions autonomously (Figure 8.43).

8.9.4 Manned–Unmanned Coordinated Ground Operations

In the future, manned and unmanned combat aircraft with stealth, high-speed, and even hypersonic functions will still be the first choice for striking high-value targets and key areas on the ground. In recent local wars, the B-2 strategic stealth bomber has become the first choice for the US military to carry out the first wave of attacks. Especially after the emergence of UAVs, manned and unmanned coordinated ground attacks will become common, with the

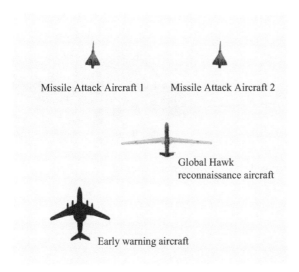

Figure 8.43 Early-warning aircraft + mixed formation aerial operations.

following typical operational scenario and process. The space-based reconnaissance satellite and the high-altitude stealth unmanned reconnaissance
aircraft similar to the Global Hawk provide the battlefield environment and
target information of the operational area. After being verified by large data
correlation and various military and civilian means, the data are handed over
to the mission-planning system and the commander for decision-making. The
main purpose of correlation verification is to confirm the nature of the target,
such as distinguishing whether it is a command post, a radar station, or a
troop-gathering area; whether it is a military, civilian, or dual-use target; fixed
or moving, as well as the geographical location and movement speed of these
targets, to select the optimal means of attack and damage mode in mission
planning. It is possible that after the mission is determined, the stealth UAV
will be used to carry out the first round of attack on the enemy, destroying the
enemy's command post, air defense radar, and weapon system, and then the
manned combat aircraft will carry out a large-area saturation ground attack
(Figure 8.44).

8.9.5 *Air-Launched UAV Swarm Operations*

This kind of operation relies on long-range bombers, large transport aircraft,
or unmanned aircraft carriers to release UAVs for air-to-ground or air-to-sea
attacks outside the defense area or in the target area. On October 25, 2016,
the US military demonstrated an air-launched MAV swarm attack, setting
a record for the largest number of UAVs deployed at that time. During this

Figure 8.44 Manned–unmanned cooperative ground combat.

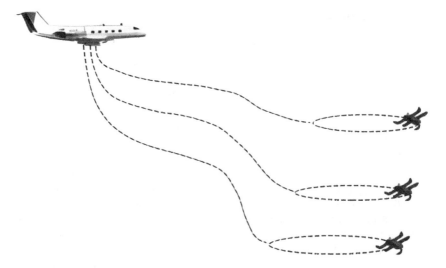

Figure 8.45 Air-launched unmanned aerial vehicle (UAV) swarm operation.

demonstration, three "Super Hornets" successively launched 103 "Perdix" MAVs to form a swarm, which completed four operational tasks that had previously been assigned (Figure 8.45). When the Perdix grounded, it passed through a series of navigation points with high precision and reached the predetermined target position. Each Perdix can communicate with each other and collectively fly towards the target as decoys to confuse the enemy's air defense system or be equipped with electronic transmitters that perform jamming on enemy radars, ensuring the safety of friendly fighters. At present, the US DoD has incorporated the Perdix into the Future Combat Network, which may soon carry out operational missions. Through this demonstration, the US military has taken a big step in applying swarms, revealing the future of small and medium UAV swarm operations as a new operational pattern. Swarms can be used not only for ground combat but also for air and sea combat and even for striking aircraft carriers. Small UAVs with high-explosive bombs flying all over the sky can carry out irregular, targeted mobile raids or distributed strikes, which would be difficult even for the best defense system to protect against. No matter how good a defense system is, there will always be a weak point.

In 2018, the DARPA "Gremlins" project planned to launch a swarm of UAVs with rapid networking and cooperative capabilities for reconnaissance, surveillance, suppression, and strike missions against the enemy. The UAVs were planned to be launched on C-130 transport aircraft, strategic bombers, and fighters.

The basic idea is to use C-130 transport aircraft as an aircraft carrier to launch a swarm of Gremlin UAVs with rapid networking and coordination capabilities outside of the enemy's defense radius. Gremlin UAVs can be equipped with reconnaissance, surveillance, electronic warfare, and even warhead payloads. After the UAV swarm penetrates the enemy's air defense zone, it can perform intelligence surveillance and reconnaissance, electronic attacks, fire strikes, psychological warfare, and other combat tasks.

After the mission, the Gremlin UAVs that are not shot down will be recycled one by one in the air by the C-130 that launched them. The Gremlins UAVs are designed to be reusable up to 20 times in adversarial airspace, and personnel on the ground will complete the reset and repair within 24 hours after each recycle, making the UAVs fully prepared for the next use. For the Gremlin project, each aircraft carrier is designed to launch 8–20 Gremlin UAVs. Since the UAV swarm will inevitably face a certain proportion of unre-coverable damage, it is required that 4–8 UAVs be recycled within 30 minutes after the launch.

The Gremlin UAVs are similar to a cruise missile: slender fuselage, small wingspan, foldable wings, and a small turbofan engine. Its range and combat radius are designed to be relatively large, and the minimum performance indicators include a combat radius of 555 km, an air cruise of 1 hour in the combat zone, a payload of 27.3 kg, and a maximum speed of 0.7 Mach. Optimal performance indicators include a combat radius of 926 km, 3 hours of air patrol, a payload of up to 54.5 kg, and a maximum speed of no less than 0.8 Mach.

The Gremlins project was initiated in three phases, with the first starting in March 2016 and the third starting in April 2018. According to planning, the launch and safe recycling of several UAVs by C-130 transport aircraft were conducted in 2019. The whole Gremlins project lasted 43 months and cost US$64 million.

8.9.6 Air–Ground Integrated Unmanned Ground Attacks

In the future, air–ground integrated unmanned ground attacks will be achieved in joint multi-domain operations, based on target information sensed by air-to-ground integrated networked information and control

Figure 8.46 China's unmanned armed helicopters with air-to-ground missile series (2021 Zhuhai Airshow).

systems, reconnaissance UAVs, and manned–unmanned systems on the battlefield front, with the support of R/S UAVs, networked patrol missiles, and unmanned helicopter gunships deployed in formation or swarms by air, gunfire, catapult, and direct ground launch and takeoff (Figure 8.46).

8.9.7 Anti-Stealth, Anti-Hypersonic, and Anti-Swarm Operations

Where there is a spear, there must be a shield. In addition to intercepting traditional incoming targets, air defense and anti-missile systems will face three new types of priority threats in the future: stealth multiple targets, swarm attacks, and hypersonic weapons and munitions. As hypersonic attack and defense are covered in Chapter 10, the focus is on operations against the first two types of threat. For the defense of stealth multi-targets, it is important to address the core issue of long-range detection. At present, the radar cross-section of advanced stealth combat aircraft or UAVs is within the range of 0.01 square meters or even lower. Targets that could be detected from hundreds of kilometers away now can only be detected within tens of kilometers due to the increasing speed of new aircraft and missiles, leaving less time for the early warning of air defense systems, and even if the targets are detected, they may not be intercepted effectively in time. Therefore, effective defense must be implemented using close reconnaissance, coordinated interception, and multi-layer protection. To this end, it is necessary to adopt a cooperative air–space sensing and detection approach to detect incoming targets in advance using low-orbit grouping small satellites, near-space grouping detection systems, and high-altitude long-range solar-powered UAV grouping detection systems, in combination with ground-based air defense detection systems. At the same time, radar, photoelectricity, and infrared means are used to implement detection, trajectory tracking, and targeting, and locking of stealth incoming multi-targets in a staged manner and with a combination of high and low accuracy.

If necessary, it is also feasible to launch unmanned reconnaissance aircraft and patrol reconnaissance missiles to form a detection network in key directions and critical areas, carry out close reconnaissance and early warning, and hand over the detected target to the air defense system for interception. This is equivalent to adding a temporary mobile relay detection network over the traditional air defense system to carry out tasks such as early-warning reconnaissance, real-time perception, high-precision measurement, real-time multi-target cataloging and tracking, and electronic confrontation, to make up for the problem of short detection range and difficulty in attacking stealth multi-targets in the existing air defense system. It is also a good option to adopt multi-missile cooperative interception or directional energy weapon interception to deal with stealth multi-targets.

Anti-hypersonic targets pose challenges to higher detection systems, especially interception, as traditional air defense missiles cannot catch up with hypersonic targets. There are examples of successful deployments of anti-drone swarms by Russian forces in the Syrian war. The detection system of a UAV swarm can be shared with low, small, and slow target detection systems. The interception system mainly includes electronic jamming, high-power microwave weapons, and anti-aircraft gun systems. Electronic jamming is mainly aimed at jamming GPS navigation and data linking. High-power microwave weapons are mainly adopted to do physical damage to front-end radio frequencies, power amplifier systems, and back-end control chips of UAV systems. The anti-aircraft gun system relies primarily on kinetic energy and fragmentation kills for physical damage (Figure 8.47).

8.10 Space and Cross-Domain Unmanned Operations

Space remains an unknown terrain for humanity, and international conventions still govern space warfare. Technologically, however, the world's military powers have invested enormous resources and funds in exploring options for accessing, using, and controlling space, although secretly, the space arms race has already begun, and the militarization process is inevitable.

8.10.1 Space-Based Controllable Unmanned Operations

Space-based vehicles mainly consist of satellites, manned aircraft, space capsules, and space laboratories. These space-based vehicles are predominantly civilian and only partially military in peacetime, while the vast

Figure 8.47 Anti-swarm measures.

majority are available for military usage in wartime. At present, space-based vehicles and information have become the main support for US military operations, and the applications of space-based resources in major countries have been extensive in both military and civilian fields. In the future, space-based vehicles, used in both military and civilian fields, will become more prominent, and the networking application of space-based vehicles, competition over space-based resources, and a certain degree of offensive and defensive space-based operations will become inevitable trends. Physical damage caused in space operations will result in potential disasters such as space debris and irreparable losses to the Earth, the common home for humanity. Therefore, space operations are mainly embodied in controllable unmanned operations, emphasizing three aspects: deception and jamming of space information links, non-explosive soft and hard killing through directed energy weapons and other means, and capture and dismantling of satellites in orbit through space robots and manipulators. By October 27, 2019, the US X-37B orbital test vehicle had made five space trips, finishing multiple experiments, research, and exploration in nearly 2,865 days in space. This kind of vehicle will become a major weapon for space networking and attack and defense in the future.

8.10.2 *Land–Sea–Air–Space Integrated Defense*

The US missile defense system, which evolved from the Star Wars program, comprised the National Missile Defense system and the Theater Missile Defense (TMD) system. The core concept is the integrated defense of land,

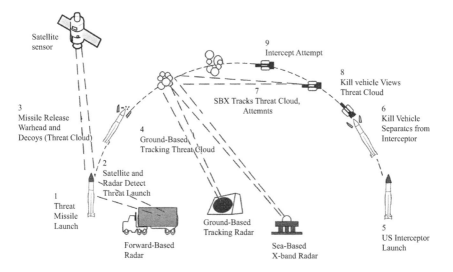

Figure 8.48 Ground-based midcourse anti-missile interception.

sea, air, and space domains. It detects and intercepts ballistic missiles from their launch stage, the out-of-atmosphere stage, and their re-entry into the atmosphere. The Terminal High Altitude Area Defense system, deployed in South Korea in 2017, is an important part of the TMD system. Anti-missile operations are divided into stages of detection, tracking, and interception. The detection system is composed of strategic and early-warning satellites, long-range warning radar, and tracking radar. Once Elon Musk's Star Chain plan is complete, the detection capability of ballistic and hypersonic missiles will be greatly improved in the future, as electronic and optical detection functions will be integrated into space-based Internet, and targets, with photoelectric and electronic features exposed, will be identified intelligently on or off satellite. At present, the interception system mainly uses kinetic kill vehicles, air-defense missiles, and other means for interception. In the future, it may use high-power microwave weapons, laser weapons, high-speed electromagnetic cannons, and other means (Figure 8.48).

8.10.3 Suborbital High-Speed Space-to-Ground/Sea Strike

"Suborbital space" refers to the space between 100 and 300 km above sea level because vehicles in this range are in an erratic, progressively descending orbit around the Earth. They can usually remain in space for up to a month before crashing into the atmosphere. Taking advantage of this feature, suborbital vehicles can be launched weeks before a war, orbit the Earth, and deorbit as required before entering the atmosphere to deliver hypersonic strikes against high-value targets on the ground and at sea (Figure 8.49).

Figure 8.49 Suborbital high-speed ground/sea attack.

8.10.4 Near-Space Information Support and Strike

At present, there is no unified definition for "near space," which generally refers to the airspace 30–100 km above sea level. The near-space area is higher than the flight altitude of general aircraft and lower than the orbital altitude of general spacecraft, where the air is thin and air resistance is low, providing a certain amount of lift and lateral mobile control for aerial vehicles, which is conducive to hypersonic flight and of great military values. At present, air defense and anti-missile weapons are powerless against hypersonic aircraft flying in this area. Near-space vehicles are usually composed of airships, balloons, and other aerostats. Such vehicles can be loaded with radar, photoelectric detection equipment, and communication relay equipment, as well as information-jamming and strike weapons. Once an aerostat detects a target, it can be loaded with weapons to carry out information jamming and attacks against the target in the air, on land, or at sea, or it can transmit target information to the ground or airborne hypersonic weapons to carry out rapid strikes (Figure 8.50).

8.11 Autonomous Swarm Operations

Autonomous swarm operations are based on individual autonomous capabilities. The individuals in the swarm cooperate to complete complex tasks, creating an emergence effect of higher intelligence. Each individual in the swarm

Figure 8.50 A model of a suborbital and near-space hypersonic vehicle (2021 Zhuhai Airshow).

has a high degree of self-organization and self-adaptability. The overall system exhibits non-linear and emergent characteristics.

8.11.1 Swarm Effects and Advantages

Birds, fish, frogs, ants, and bacteria do not possess advanced intelligence such as the complex logical reasoning and comprehensive judgment of human beings. However, through adaptation to the environment and group collaboration and self-organization, motivated by the same goal, e.g., food, they exhibit powerful group intelligence, which provides new ideas for human beings to solve complex problems. Swarm unmanned combat systems use numerous micro and small unmanned vehicles to accomplish combat tasks based on a swarm mechanism simulating natural swarms' behavior and coordination mechanisms, including ant colonies, bird flocks, and fish schools. Without being controlled by a unified external system, each unmanned combat vehicle forms an understanding of the battlefield situation based on its state, detection of local environmental information, and interaction with other unmanned vehicles in the swarm, and autonomously makes and executes decisions to achieve mission objectives, coordinate swarm behavior, and improve operational efficiency.

From a macro point of view, the autonomous swarm operation system presents a spontaneous and organized mission behavior process. Compared with unmanned vehicle control stations, autonomous swarm operation systems have the following advantages.

First, quicker response to changes in the battlefield situation. Unmanned vehicle controllers can quickly access environmental and state information by sensors, understand battlefield situations by combining information by interaction with other unmanned vehicles, and effectuate a rapid response to local situation changes through autonomous control.

Second, low cost of decentralized decision-making. In this system, unmanned vehicles only need to plan their behavior for achieving mission objectives, and the control software only needs to be designed for a single vehicle at a time.

Third, low dependence on communication conditions. With their autonomy, unmanned vehicles quickly complete a large amount of information processing at the front end, carrying out only cooperation-oriented interactions, reducing dependence on communication through a combination of active and passive communication.

Fourth, high security of the system. The system is highly secure because if any unmanned vehicle malfunctions or loses functionality, there is no impact on the system's overall performance, and the other unmanned vehicles can still function normally.

Realizing the importance of developing autonomous swarm C&C technology, countries like the US and the UK have researched self-organizing technology for unmanned vehicles. The US Joint Forces Command's "Project Alpha" Laboratory has studied UAVs' effectiveness of autonomous swarm operations. In the simulation, 100 UAVs equipped with sensors and weapons destroyed 63 targets and detected 91% of simulated enemy forces compared to the existing basic deployable units that destroyed only 11 targets and detected less than 33% of enemy forces.

8.11.2 R&D Goals of Swarm Systems

The basic principle for the R&D of autonomous swarm operation systems is 1 + 1 > 2. Under the conditions of uncertain or unknown environmental factors involving terrain, meteorology, enemy targets, and communication in the mission area, the swarm system needs to achieve the transfer of reconnaissance and search results of the battlefield, the location and trajectory of enemy targets, and the unmanned vehicles' status through network communication and data links between unmanned vehicles – in other words, to realize the sharing of battlefield situational information and autonomous search, confirmation, attack, and damage assessment of targets between unmanned vehicles, obtaining a far higher mission execution capability than that of a single vehicle or non-coordinated multi-vehicles.

8.11.3 Multi-Vehicle and Multi-Task Cooperation

Due to the limitations of sensor detection accuracy, detection range, vehicle flight, and mobility, a single unmanned vehicle cannot complete the search coverage of a wide range of mission areas in a short time. Therefore, to find as many targets as possible quickly, multiple unmanned vehicles must search targets together. At the same time, due to the highly dynamic nature of the battlefield environment, a mission area that has been cleared may become uncertain due to factors such as the movement of enemy targets and the possible addition of new targets to the mission area. Under these conditions, unmanned vehicles are required to conduct intermittent and repeated reconnaissance of the searched area.

Unmanned combat vehicles have different mission execution capabilities due to different payload capabilities. For example, reconnaissance UAVs have stronger reconnaissance and search capabilities and target identification and damage assessment capabilities than attack UAVs, while the latter have target damage capabilities that the former lack. For prompt effective damage to enemy targets, a combination of unmanned vehicles is required to perform a continuous search, identification, strike, and damage assessment in the target area; for example, to assess the damaging effect to the target in time after completing one phase of a strike and confirm the need for a second strike.

8.11.4 *Communication Between Swarm Vehicles*

The self-organization behavior of unmanned combat vehicles depends on the information exchange of data links and active and passive network communication, which has become a determining factor for self-organization. In future battlefields, communication conditions are characterized by strong interference, high confrontation, and low bandwidth. The potential for packet loss and time delays in information transmission will undoubtedly significantly impact the self-organizing behavior of unmanned combat vehicles that rely on communications. To ensure the survival of unmanned combat vehicles and effective mission execution, the impact of communication conditions on their behavior should be minimized, and the robustness of their communication conditions should be ensured. Inertial navigation, passive 3D imaging, optical flow navigation, and frequency hopping can aid contact and sense when necessary.

8.11.5 *Swarm Control Theory and Algorithm*

At present, the research on unmanned combat vehicle swarms control theory mainly focuses on multi-agent systems, such as multi-UAV, multi-robot autonomous cooperation, and other fields. Most of them are based on the self-organization method of a bionic algorithm, to use self-organization of groups by simulating the behavior of biological groups and referring to the information transmission mode of biological individuals, so that individuals with low intelligence can coordinate through information interaction and the system can show collective intelligence through continuous evolution. At present, the representative bionic algorithms include ant colony algorithm or optimization (ACA/ACO), particle swarm optimization (PSO), bee colony optimization (BCO), fish swarm algorithm (FSA), and artificial shuffled frog leaping algorithm (SFLA) that simulates the mechanism of information sharing and communication in the process of frog foraging. Some scholars proposed avant-garde swarm intelligence algorithms concerning dolphin, rat, monkey, and wolf swarms. In addition to bionic algorithms, there are self-organizing algorithms involving potential artificial fields and agent negotiation.

The autonomous swarm operation system is essentially a self-organizing network system with the following characteristics:

1. The system is composed of several lower-level individuals who interact dynamically and equally.
2. In the system, individual behaviors depend entirely on the practical information, state, and interaction with other individuals and are not controlled by an external entity.
3. The system gains swarm behaviors through evolution.
4. The system can achieve one or more goals that are superior to the individual.

Despite no coordinator for the individuals, the whole system is in a coordinated, orderly, and intelligent state for many biological groups in nature, such as ants, bees, and fish.

Ants building nests: when building a nest, ants initially pile up the carried material in random piles in various places. The material piled up by one ant in one place may be carried away by another ant, and then it may be carried back to the same place by other ants, with each ant repeating the process independently. As time goes by, more and more material accumulates in a designated place, and after reaching a certain density, the accumulated material forms a source, attracting other termites to pile material in this place.

Ant foraging: the ant colony forages from the nest and disperses around the nest at the beginning. When an ant finds food, it will inform other ants by releasing pheromones. Ants will then flock to the food, and the path they take is often the shortest path between the nest and the food.

Honeybee foraging: in some species of bees, worker bees perform a "dance" when they return to the hive after foraging, thus sending a message to their companions. The other worker bees can tell the dance movements' direction, distance, and abundance of the food source. The better the food source, the more bees will participate in this dance.

In a stretch of water, fish can often find nutrient-rich areas on their own or by following other fish. When fireflies gather together, the rhythm of their luminescence can be observed from dispersion and disharmony at the beginning to gradual convergence after grouping. These biological groups take cooperation as their survival strategy, in which each individual performs only simple and limited actions, interacting with each other through different media while exhibiting intelligent behaviors such as coordination, concerted aggression, defense, etc.

The above self-organization phenomenon originates from the feedback mechanism of biological systems, which exists in two forms: positive and negative. Positive feedback reinforces a change in the same direction, intensifies it, and causes the system to take on a macroscopic form in this direction eventually. The phenomenon of self-organization arises mainly as a result of positive feedback. In contrast to positive feedback, negative feedback weakens the original change, causing the system to move in the opposite direction of the macroscopic form and maintain the original form. Negative feedback has the effect of stabilizing a system. The formation of self-organizing phenomena is also constantly influenced by negative feedback. When bees find a food source, they will transmit the information about the food source to their companions through "dancing" that guides other bees to the food source. As the number of bees participating in the "dance" increases, more and more bees will fly to the food source. This is positive feedback. However, not all bees will fly to the food source they have found, with a small number of bees flying to other places or bad food sources. This is negative feedback. Ants release pheromones on the foraging path, marking the path from the nest to the food source. With the increase in the number of ants passing through this path, more pheromones will be left, further attracting more ants to the food source. That is positive feedback. With an increasing number of ants gathered at the food source, some moved away from the food source to search for other food sources. That is negative feedback.

The mechanism and principle of the autonomous swarm operation system combine ACO, PSO, BCO, artificial potential field algorithm, agent negotiation algorithm, and other algorithms to achieve self-organizing swarm combat.

ACA is a pheromone-based individual behavior coordination mechanism. As stated above, when an individual ant finds a food source by searching, it will release pheromones on the path from the nest to the food source to inform other ants to move towards the food source. Based on this idea, the unmanned vehicle swarm can be compared to an ant colony, with each unmanned vehicle regarded as an ant. When it finds the target, the unmanned vehicle adjusts the pheromone distribution to the environment, such as enhancing the pheromone concentration at the enemy target, which affects the behavior of other unmanned vehicles.

PSO is an evolutionary algorithm based on population search inspired by the law of birds' swarming activity. This algorithm improves the natural selection mechanism by featuring social organization behavior and searching optimal solutions through individual cooperation among groups. The basic principle is to randomly initialize a swarm of particles, determine the velocity and position of each particle, the movement direction and relative position to the target area, and the fitness value (determined by an optimization function) of the particle that is closer to the target. If the current fitness value is higher than the historical best position (p_{Best}), it is updated to p_{Best}. Otherwise, it is eliminated. The particle's p_{Best} is then compared with the global best position of the particle swarm (g_{Best}) and updated to g_{Best} if it is higher, or otherwise eliminated. Then, the position and velocity of the particle swarm are substituted into the following equation.

$$V_{iD}^{k+1} = \omega V_{iD}^{k} + c_1 r_1 \left(p_{iD}^{k} - x_{iD}^{k} \right) + c_2 r_2 \left(p_{gD}^{k} - x_{iD}^{k} \right)$$
$$x_{iD}^{k+1} = x_{iD}^{k} + V_{iD}^{k+1}$$

The velocity and position of each particle are repeatedly updated until a convergence or a preset termination condition is realized. In the formula, V_{iD}^{k} is the velocity of the Dth variable of the ith particle at time K; ω is the inertia factor; C_1 is the learning step of self-record; C_2 is the learning step of group record; r_1 and r_2 are random numbers in the interval [0, 1]; p_{iD}^{k} is the value of the Dth variable at the position where the maximum fitness value of the ith particle is located at time K; P_{gD}^{k} is the value of the Dth variable at the position of the maximum fitness value in all records before time K, and x_{iD}^{k} is the value of the Dth variable of the ith particle at time K. In fact, the trajectory of each particle at the next moment is pulled by three forces at the same time: the original direction of action, its record of the highest point, and the global record of the highest point.

PSO has the advantages of simplicity, fast convergence, and fewer parameters. With UAVs and other vehicles mapped as particles in PSO and keeping a safe distance between particles, the whole swarm can become a formation when the speed of particles is dynamically adjusted by their own and their companions' flight, coordinating the flight direction of the swarm. When an enemy target is found, the particles may reach the target by dynamically adjusting the movement direction and minimizing the distance between the particles and the target. Because PSO particles follow the best particles in the solving space, they will eventually form swarm behaviors such as UAV launching swarm attacks on targets.

BCO is a non-pheromone-based algorithm inspired by bees gathering. The collective intelligent behavior of a honey swarm contains three types of individuals: honey source, honey-collecting bees, and bees waiting for work. It involves three modes of behavior: searching for honey sources, recruiting for honey sources, and abandoning honey sources. In BCO, the UAV swarm can be mapped as a bee colony, and the threat information can be mapped as a honey source. By referring to honeybees searching for honey sources and individual interaction, the course planning, enemy target search, and reconnaissance of UAV swarm operations can be achieved. The autonomous intelligent allocation of multi-target attacks can also be realized by mapping the enemy target to the honey source.

The self-organization method based on the potential artificial fields is to construct potential fields with opposite polarities: attractive potential field and repulsive potential field. The attractive field has an attractive effect on a moving subject and moves it closer to the field. The repulsive field has a repulsive effect on the subject and moves it in a direction away from that field. According to this idea, an attractive field is created at the enemy target, while a repulsive field is created at the enemy threat and other danger factors. Influenced by these two fields, the UAV can move towards the target and launch an attack while ensuring safety. When the target is destroyed, the attractive field disappears, and the repulsive field causes the UAV to exit the target area.

The self-organization method based on agent negotiation is achieved through various negotiation mechanisms and algorithms: for example, negotiation based on contract netting. When a UAV finds that it is not capable of completing certain tasks or it is too costly to complete them, these tasks can be handed over to other UAVs by way of an auction.

8.11.6 Swarm Attack Vehicles and Ammunition

Swarm attack vehicles and ammunition constitute a system that operates in a coordinated manner of multi-vehicle and multi-bullet networking. Each bullet can be equipped with various payloads, with communication and autonomous decision-making, and can fly in formation over a target area to perform multiple operational tasks such as vehicle reconnaissance, attack, and damage assessment. Figure 8.51 shows China's UAV swarm combat system at Zhuhai Airshow 2021. It is composed of three to four launching units; each launching unit can launch 48 UAVs, and one time can launch nearly 200 UAVs to form swarm combat.

Over the years, the US Army, Navy, Air Force, and DARPA have been researching swarm attack systems, including the Army's swarm attack ammunition carrying smart submunitions, the Navy's low-cost UAV swarm technology, the Air Force's swarm attack munitions, and DARPA's Gremlin UAVs. DARPA's Gremlins project involves key air launch/recovery technologies and high-speed digital flight control of small UAV swarms and it completed the first phase of R&D in 2017. In 2018, it launched the "Offensive Swarm Tactics" project to study the ability of swarms with 100 ground and low-altitude unmanned systems to perform autonomous missions in urban environments [6]. In August 2014, the US Navy used a swarm of 13 USVs, adjusting its route autonomously in real time, and successfully surrounded and intercepted "suspicious vessels." In 2016, the US Navy verified the orderly flight capability of the "Coyote," loitering up to 30 munitions on the ground; this solved the technical problems of single-missile positioning and autonomous flight. The US Navy then conducted a sea launch swarm demonstration test to study more complex autonomous capabilities.

Since 2010, networked cooperative attack munitions have entered a multiway rapid development stage, emphasizing autonomous intelligent networking and decentralized control to improve the confrontation capability in complex

Figure 8.51 Swarm unmanned aerial vehicle (UAV) launch system (2021 Zhuhai Airshow).

combat environments and the cooperative combat capability of multi-level weapons and sensors (Table 8.4).

The key technologies of swarm attack munitions include autonomous dynamic mission planning and coordination, inter-swarm networking communication and transmission, adaptive control, new power, micro-mission payload, and others. Further objectives to be achieved include reducing the dependence of unmanned vehicles on mission control stations, improving autonomous combat capabilities, enhancing situational awareness and autonomous decision-making, reducing dependence on operators, and reducing interaction with operators. A possible way to achieve these goals is to use a swarm self-organizing technology based on bionic mechanisms, which integrates numerous decentralized unmanned vehicles.

8.12 Human–Computer Intelligent Interaction

With the rapid development of AI, brain–computer interface, virtual reality/augmented reality (VR/AR), and other technologies, human control over weapons and equipment will undergo subversive changes. Humans

Table 8.4 Typical US networked swarm combat vehicles and munitions research projects

Name of weapon/item	Developer	Function description	Performance Index
"Air Dominator" Loitering munition swarm	US Air Force	"Gateway" ammunition control, wireless link networking coordination, autonomous reconnaissance, identification, attack target, long-term control of the target area	F-22 and C-17 transport aircraft can release it; the loitering time is more than 12 hours
Low-cost unmanned aerial vehicle (UAV) swarm technology	US Navy	Distributed use, wireless ad hoc network, swarm flight, situational awareness, disposable use	30 fighters launched at a time; with reconnaissance and strike payload modules; $15,000 for a single fighter, with a target price of $5,000–7,000
"Gremlins"	DARPA	Distributed use, wireless ad hoc network, low cost, partially recyclable, reconnaissance and electronic warfare swarm operation	Mass 320 kg; air drop and recovery; reusable at least 20 times; factory price less than $700,000
"UAV" "Payload"	Strategic Capability Office	Decentralized control, autonomous group coordination, saturation attack	In-flight stability at a wind speed of 30 m/s Disposable, non-recyclable
OFFensive Swarm-Enabled Tactics project	DARPA	Decentralized control, autonomous human–machine formation coordination, capable of carrying out complex swarm tactical operations	250 munitions, capable of operating in eight city blocks in 6 hours; with open systems architecture and software and distributed perception

DARPA, Defense Advanced Research Projects Agency.

and equipment can merge as an organic whole, i.e., to accomplish human–machine integration. In the future, it is necessary to establish a new relationship between human and intelligent systems, evolving from passive control of traditional unmanned equipment to the autonomy of unmanned systems, giving full play to the advantages of machine intelligence. The evolution will free humans from complex control, drawing their attention to battlefield situational awareness, mission decision-making, and evaluation, improving combat effectiveness.

In 2014, the Tactical Technology Office (TTO) of DARPA issued Military Mission Innovation System Collection Guidelines, which emphasized strategic and tactical assault to the enemy in land, sea, air, and space battlefields, with four key areas of concern: human unmanned system interaction innovation, unmanned autonomy, rapid global deployment, and strategic confrontation cost.

Technological advances in human cognition will create intelligent formation systems that surpass machines and human beings. The intelligent formation comprises human soldiers and physical or virtual artificial agents, connected by brain–brain and brain–computer interaction, forming a network that operates semi- or fully autonomously. Participants of this formation can be defined based on each force's mission, posture, and status, incorporating new combatants or machine agents free from extensive training. In intelligent formations, information exchange can be actualized between humans or AI without voice communication, improving the accuracy and timeliness of transmission in extreme operational environments.

The technology of human–machine intelligent interaction and integration focuses on dividing tasks between human and AI systems, making machines quickly and accurately understand the complex intentions of human beings and realizing complex machine behavior with minimal human manipulation.

First, how can a single combatant cooperate with multiple intelligent systems, such as UAVs and battlefield robots, to complete the whole process of reconnaissance, C&C, strike, and evaluation? Different from present human–machine interactions, the technology needs to carry out cooperation between human and intelligent systems in terms of battlefield perception, judgment, and decision-making, action and strike, support and evaluation, and to establish a model for the evaluation of cooperation and operational effectiveness among multiple personnel and intelligent vehicles.

Second, how can an unmanned system quickly identify the intention of combat personnel in extreme battlefield environments? Traditional unmanned systems mostly adopt one-to-one remote-control mode, which, with rather slow feedback and high personnel consumption, is not suitable for a high-intensity unmanned swarm combat environment in the future. Therefore, it is necessary to have a look into advanced interactive recognition technology that realizes quick judgment of human intention from action, expression, voice, electroencephalogram (EEG), and other cognitive means, and accurate understanding of fuzzy and random human behaviors based on situation, task state, operation habits, and other factors.

Third, how to effectively control a large swarm of unmanned systems with limited control ability of the combat personnel? In response to the sparsity of control, research on intelligent matching technology must be carried out to build an interaction system between a small number of combatants and many unmanned systems, achieving perfect human–machine integration in situational awareness, decision-making, and mission cooperation so that a combatant controls an intelligent unmanned system like controlling his body, releasing huge combat potential.

8.13　UAV Warfare in Turkish–Syrian and Nagorno-Karabakh Conflicts

Since the beginning of the new century, with the delivery of numerous UAVs to the armies of various countries, unmanned operations have appeared in large-scale regular operations.

From the end of February to the beginning of March in 2020, Turkey launched the "Spring Shield" operation against Syria, dispatching several UAVs that severely damaged Syrian ground forces and directly affected the result of the war. In this operation, the Turkish army took the domestic "Anka" -S and "Bayraktar" TB2 unmanned combat aerial vehicles (UCAVs) as the main combat force, together with some long-range artillery, and, under cover of E-737 early-warning aircraft and F-16 fighters, carried out large-scale and high-intensity fire strikes on Idlib, Syria, causing heavy losses to the Syrian army. The Turkish army has deployed dozens of UAVs and launched hundreds of sorties for airstrikes in Aleppo, Hama, and other places in Syria. The UAVs also carried out decapitation operations against senior generals of the Syrian army and its allies, with impressive outcomes.

The "Anka"-S large medium-altitude long-endurance R/S UAV developed by Turkey has a length of 8.6 meters, a wingspan of 17.5 meters, a maximum level-flight speed of 217 km/h, endurance time of more than 24 hours, a practical ceiling of 9,000 meters, a maximum takeoff weight of 1,600 kg, and a maximum mission load of 200 kg.

The "Bayraktar" TB2 UCAV developed by Turkey is a medium medium-altitude long-endurance R/S UAV with a length of 6.5 meters, a wingspan of 12 meters, a maximum level-flight speed of 220 km/h, a practical ceiling of 8,200 meters, an endurance of more than 24 hours, a maximum takeoff weight of 630 kg, and a maximum mission load of 55 kg. It can carry four Turkish MAM-C and MAM-L miniature precision-guided munitions.

The MAM-C and MAM-L small laser-guided bombs have a mass of 6.5 kg and a range of 8 km, using semi-active laser guidance, while the latter has a mass of 22 kg and a range of 14 km, using GPS/inertial plus semi-active laser guidance.

In the later stage of the operation, the Syrian/Russian army shot down 23 Turkish UAVs, including a TB2 UAV signed by Erdogan, through the combination of "soft killings" such as electronic jamming suppression and "physical damage" of air defense missile interception, which successfully contained the Turkish UAV attacks and quickly turned the tide of the battle. The combat results show that the anti-UAV operations carried out by the Syrian and Russian armies are very effective and have played an immediate and important role in the battle. The Russian army's electronic warfare equipment has performed well. Major equipment includes the latest "Krasuha"-4 ground electronic warfare system, the "Krasuha -2 system, the Tu-214R electronic reconnaissance aircraft, and the Il-20 PP electronic warfare aircraft, which participated in the operation. Besides, the "Armor"-S1 missile and artillery integrated air defense system and the "Buk"-M2 medium-range ground-to-air missile also played a certain role in defense.

Russia's "Krasukha-4" ground-based mobile electronic warfare system, with broad-spectrum and strong noise-jamming capability, can jam various radar, communication systems, UAV-controlled links, and navigation systems, and is mainly used to suppress spy satellites, ground radar, early-warning aircraft, UAV, and other sky–ground-based detection systems. It is capable of countering electronic information and unmanned equipment such as US E-8C early-warning aircraft, "Predator" unmanned reconnaissance attack aircraft, "Global Hawk" unmanned strategic reconnaissance aircraft, and "Lacrosse" series reconnaissance satellites.

From September to November 2020, the role of UAVs became more prominent in the Nagorno-Karabakh conflict between Azerbaijan and Armenia. At the beginning of the conflict, it appeared to be traditional ground warfare involving airstrikes and counter-air strikes, long-range artillery/rocket attacks, tank and infantry combat vehicle confrontations, anti-tank missile launchers, and light weapons engagements, with casualties on both sides. However, in late September, the Azerbaijan Army highlighted the role of systematic UAVs, turning the tide of the war and taking the strategic initiative.

Before the war, the Azerbaijan Army purchased numerous UAVs from Turkey and Israel, including TB2 R/S UAVs, Hermes long-range reconnaissance UAVs, Harop suicide UAVs, Seeker medium-range reconnaissance UAVs, Orbiter series short-range reconnaissance UAVs, and Azerbaijani-modified An-2 UAVs.

The Azerbaijan Army first used An-2 UAVs for the strike, which deceived Armenia's air defense radars. Then it used Harop suicide UAVs, followed by TB2 R/S UAVs, to carry out large-scale ground strikes, achieving good results. Among them, TB2 and Harop UAVs made great contributions to their victory. It is reported that more than 75% of the Armenian weapons and equipment destroyed in operation were attributed to attacks by these two types of UAVs (Figure 8.52).

Harop anti-radiation suicide attack UAV has attracted great attention for its successful damage of Armenia's C-300 air defense missile system in this

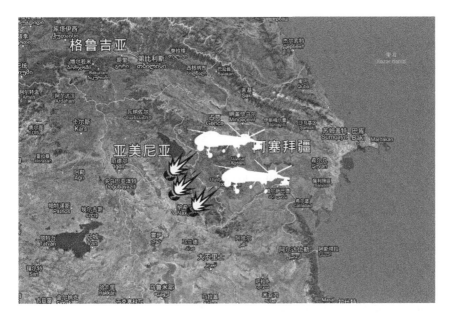

Figure 8.52 Screenshot of a live battlefield video shot by a Turkish unmanned aerial
vehicle (UAV).

conflict. The UAV is designed to deal with radar and air defense systems. It is
launched by a vehicle-mounted launcher and can fly for 6 hours. Once a radiation source is found, it can automatically/manually select a route towards
the target and destroy it by detonating the warhead. The UAV can also attack
ground targets without radiation sources, such as light-armored vehicles and
soldiers.

The Armenian Army has fewer UAVs, mainly indigenously developed
MAVs, such as the "Crane" series of reconnaissance UAVs, mainly used for
close-range reconnaissance missions. The "Beast" suicide drone, designed as
a loitering munition, weighs only 7 kg and has a warhead mass of 1.6 kg. It
can attack tanks and armored vehicles 20 km away and has destroyed several
tanks of the Azerbaijani Army.

The Armenian Army mainly takes three measures against the enemy: electronic jamming, surface-to-air missile interception, and false-target deployment. However, the measures turned out to be very limited in terms of
operational effectiveness. The traditional air defense radar and surface-to-
air missile system, which filters the radar reflection of many small UAVs
as clutter, cannot guide surface-to-air missiles, resulting in the damage of
numerous 9k33 surface-to-air missile systems.

Overall, in the Nagorno-Karabakh conflict, according to the relevant data
released by the Ministry of Defense of both sides by the end of 2020, the

Azerbaijan Army had the upper hand, for which the role of its UAVs was indispensable.

Cumulative losses of Armenia: 118 armored vehicles; 66 artillery pieces; 15 anti-aircraft missiles; one fixed-wing combat aircraft; 172 vehicles of various types; 12 anti-tank missiles; eight military installations (two bases, two ammunition depots, one airfield, two command posts, one warehouse); 3,330 casualties.

Cumulative losses of Azerbaijan: 40 armored vehicles; seven An-2 long-range UAVs; one helicopter; 16 other UAVs; 11 vehicles of various types; 2,783 casualties.

The UAV attack and defense wars between the Turkish–Syrian border and in the Nagorno-Karabakh conflict have been very enlightening in terms of:

First, UAV/anti-UAV operations will become an important operational pattern for future warfare.

Second, the effectiveness of UAV is closely related to the OSoS.

Third, the "combination of hardware and software" is an effective way to implement anti-UAV operations.

Fourth, traditional air defense weapons, such as anti-aircraft guns and anti-aircraft missiles, are ineffective in anti-UAV, especially UAV swarms. In contrast, electronic warfare equipment and high-power microwave weapons with strong, soft killing capability will play a more prominent role.

8.14 Intelligence in Israeli–Palestinian Conflict

From May 10 to 20, 2021, the most serious military conflict since 2014 between Israel and Palestine commenced. The armed forces of Hamas and the "Al-Quds Brigade" fired about 4,070 rockets into Israel. Israel used the "Iron Dome" terminal interception system to intercept the rockets. While fighter planes and attack helicopters were dispatched to retaliate against Palestinian airstrikes, special forces and teams were dispatched to decapitation strikes and key raids. Heavy ground troops were assembled on the Gaza border to carry out ground attacks. Among them, Israel has implemented a noteworthy series of intelligent operational means.

According to the Arab Defense website, Hamas had about 14 types of unguided rocket-propelled grenades (RPGs) in its arsenal, with a range of 8–180 km. In this attack, Hamas mainly used the new A-120 RPG, with a maximum range of 120 km. The A-120 RPG weapon system had one 8-unit rocket launcher with a diameter of 333 mm, a length of about 5 meters, and a weight of about 300 kg. There was no crane: six soldiers were needed to assist reloading, resulting in low reloading efficiency after launching and a high risk of being destroyed as the position could not be transferred in time. The guidance system of the A-120 rocket was simple and crude, with low hit accuracy.

Israel's Iron Dome terminal interception system played an important role. The Iron Dome system does not intercept all incoming rockets but only launches Tamir interceptor missiles when the impact point of an incoming rocket is determined to be a threat. In particular, the system's phased-array radar can detect and target 200 airborne targets simultaneously while searching, identifying, and tracking them within 5 seconds, then calculating their trajectory and predicting their impact point. The Iron Dome terminal interception system consists of a radar system, a fire control system, and three launching units, each of which is equipped with 20 "Tamir" interceptor missiles equipped with active radar terminal guidance seekers. Tamir interceptor missiles, with active radar guidance at the end, intelligently engage with multiple targets and have proved to be able to counter better the "saturation attack" of Palestinian armed forces. Tamir intercepted more than 1,200 threatening rockets out of more than 4,000, while most of the rest crashed by themselves. It did not intercept the rockets that posed no threat, which greatly reduced the damage of rockets to cities and towns.

The Iron Dome system has also been improving in the face of changing threats such as Hamas militants in the Gaza Strip and the emergence of suicide drone swarms. In March 2021, the Israeli Ministry of Defense announced that the Israel Missile Defense Organization and Raphael Defense Systems had completed a series of flight tests of the upgraded Iron Dome system. In testing, the Iron Dome successfully intercepted UAVs, salvos, and rockets simultaneously, emulating various ground and sea threats that the "Iron Dome" system may face in the future and demonstrating its intelligent defense capability to deal with a variety of complex threats.

In December 2020, the US and Israel jointly completed the first test of the Iron Dome system to intercept cruise missiles and UAVs, evaluated the comprehensive interception capability of Israel's multi-layer air defense and anti-missile system, and verified the coordinated interception and interoperability among the "Iron Dome," "David's Sling," "Arrow"-2, and "Arrow"-3 systems.

The "Iron Dome" system initially intercepts and destroys short-range rockets, and artillery shells launched 4–70 km away, and after upgrading, it can intercept cruise missiles and different types of UAV within 40–300 km. By the intelligent upgrade of C&C, the system has gained stronger target identification and multi-target-tracking capabilities. The software of interceptors has also been upgraded accordingly to deal with targets that require different damage patterns. In addition, the deep integration of the Iron Dome system into Israel's air defense and anti-missile system has been the best way to deal with multi-targets. In the field of national air defense, the Israeli army has also carried out intelligent transformation and upgrading of the adaptive joint defense early-warning, joint mission-planning, autonomous decision-making, and multi-mode automatic interception fire control systems. Without the upgrading, it would be impossible for the Israeli army to have a 90% interception probability in the face of all-weather, multi-wave, and large-scale rocket attacks.

During the 11-day military conflict, Israel launched multiple decapitation operations against Hamas. Dozens of Hamas heavyweight militants were killed, including Bassem, commander of Hamas's "Qassam Brigades," Hassan, deputy commander of Hamas's "Al-Quds Brigades," and Kaogi, director of the Hamas intelligence agency.

Decapitation strikes and targeted killing operations were carried out jointly by Israeli special forces, weapons and equipment department, and intelligence department. They were advanced operations based on the deep integration of big data association algorithm + human behavior computing model + multi-source detection and perception + AI.

In this large-scale conflict, the Israeli military used advanced AI to sift through huge amounts of data intercepted and collected from Gaza: telephone calls, text messages, surveillance footage, satellite images, and a large number of various sensors. AI transformed this data into usable intelligence, such as the time and location of an enemy commander. In order to understand the scale and accuracy of the amount of data collected, any given location in the Gaza Strip was photographed at least ten times a day during the conflict.

A senior Israeli Defense Forces intelligence official told the media that this was:

> the first time AI has become a key component and power amplifier in fighting the enemy, which is unprecedented for the Israeli military ... The combination of various intelligence sources with artificial intelligence and the forces in this field have transformed the mode of cooperation between intelligence personnel and front-line personnel.

According to the disclosed information, Israel has formed an elite team code-named 8200, specializing in developing algorithms and writing software, and has hatched at least three programs, Alchemist, Gospel, and Depth of Wisdom, all of which were used in war activities. In this conflict, Israel has carried out more than 1,500 precision attacks on Hamas. The targets include rocket launchers, rocket manufacturing, production and storage sites, military intelligence offices, UAVs, command headquarters, and so on.

References

[1] Smith, M. (2010). *BigDog to lead US robot army*. www.thetimes.co.uk/article/big dog-to-lead-us-robot-army-jg063q2hpsk
[2] *Global Times* 环球时报. (2009). Mei cheng zhanji feixingyuan shidai xingjiang jieshu 美称战机飞行员时代行将结束. FX361 参考网. www.fx361.com/page/ 2009/0825/5651765.shtml
[3] 360 Geren Tushuguan 360个人图书馆. (2020, February 21). *Meijun wurenji zhuangbei qingkuang ji yingyong qianxi* 美军无人机装备情况及应用浅析. 360 geren tushuguan 360个人图书馆. www.360doc.com.
[4] US Congress, 107th Congress. (2001–2002). *National Defense Authorization Act for fiscal year 2001*. www.congress.gov/

[5] 360 Wenda 360问答. (2019, April 9). *Yingyan zhaoxiangji you shenme yongchu?* 蝇眼照相机有什么用处? 360 wenda 360问答. https://wenda.so.com/q/155613278 4211862.

[6] Baidu Wenku 百度文库. (2018, October 3). *DARPA "xiaojingling" (Gremlins) xiangmu jinru disan jieduan* DARPA "小精灵" (Gremlins) 项目进入第三阶段. Baidu wenku 百度文库. https://wenku.baidu.com/view/ac9b4811326c1eb91a37f 111f18583d048640f6d.html.

9　Cyberspace Operations

Cyberspace operations are a subset of virtual space operations and a novel type of military operation that places a greater emphasis on deterrence than on combat. With the advancement and development of the Internet and the Internet of Things (IoT), the network will become ubiquitous, with almost everyone and everything connected. Cyberspace operations are becoming increasingly important as a demonstration of genuine combat capability and as a deterrent force against interstate strategic competition. In the event of a cyber attack against the US domestic infrastructure, which includes the Internet, telecommunications networks, power grids, financial networks, and military networks, the US military will respond in various ways, including but not limited to nuclear strikes. In June 2011, as part of the launch of its first cyber strategy, the US Department of Defense (DoD) stated unequivocally that it would treat high-level cyber attacks as acts of war and would consider using military means to retaliate: "If the other side uses a computer network to disrupt our power grid system, we may fire a missile at them." Numerous senior military officials in the United States believe that:

> Strong retaliatory measures may be necessary if a serious attack results in the failure of critical infrastructures, such as military or financial systems ... Countries considering launching a high-end attack should be fully aware that the US will respond in kind and that the attack will not be limited to cyber facilities [1].

Since the inception of the military Advanced Research Projects Agency Network (ARPANET) in 1969, cyber technology has had a profound impact on human society, affecting nearly every facet of politics, economics, technology, culture, and security and ultimately contributing to the formation of cyberspace. The world's major military powers have identified cyberspace as a critical area of military competition and have committed substantial human and material resources to researching and preparing for cyberspace operations. Cyberspace has become inextricably linked to all facets of national security, serving as a new frontier for national security, a new arena for great power

DOI: 10.4324/b22974-11

competition, and a primary theater of ideological conflict. It has ushered in a new era of human society's saber-rattling.

9.1 Cyberspace

Fundamentally, cyberspace is a space created by humans or a global domain within the information environment defined by the electronic and electromagnetic spectrum to establish interdependent and connected networks for creating, storing, exchanging, and using information [1].

Apart from the Internet, cyberspace encompasses a variety of networks, including telephone, cellular, cable, corporate and government networks, and military network systems. Cyberspace consists essentially of four domains.

The first is the system domain, which includes cyberspace's technical foundation, infrastructure, architecture, and software and hardware.

The second domain is the content and application domain, which includes the cyberspace-based information base and the mechanisms for accessing and processing data.

Third, there is the human and social domain, which encompasses human-to-human communication and human-to-information interaction. Business, consumers, political activities, and social movements all fall under this domain.

The fourth is the governance domain which encompasses all facets of cyberspace, including technical specifications for the system domain, exchange regulations for content and application domains, and legal systems for human and social domains [1].

9.1.1 The System Domain

The system domain is comprised of infrastructure that transports, stores, and processes data. In cyberspace, hundreds of millions of computer and network systems communicate and transport the data generated by billions of people.

9.1.1.1 Network module and structure

The Internet is a globally connected, publicly accessible network that utilizes Internet protocol (IP) technology to connect the world. It comprises a network of backbone routers operated and managed by the world's largest Internet service providers (ISPs). The network module of the system domain consists primarily of a local area network (LAN), an interconnected LAN, and a global backbone network managed by ISPs [1].

Initially, the Internet was designed for computer-to-computer communication, primarily for bidirectional data transmission. With the development of technology, it was gradually adopted for telephony, television, and dedicated data communications. Later, it was seamlessly integrated with wireless networks, including WiFi and the IoT. Notable is the interdependence of the power network, the communication network, and the data acquisition and monitoring system: the communication infrastructure relies on the power network, which relies on the data acquisition and monitoring system. Convergence of the Internet and the IoT increases the risk of cascade failure [1].

9.1.1.2 Protocols and data packets

The various computers in the system domain are connected via protocols and data packets to facilitate data communication. While today's Internet is largely based on the International Organization for Standardization's (IOS's) Open Systems Interconnection (OSI) model, a seven-layer protocol stack defined in 1980, it does not strictly adhere to the OSI's layers. The OSI's seven layers are as follows: layer 7: application, which includes browsers, servers, and email servers; layer 6: presentation, which includes the representation of digital time bits and byte arrangement orders; layer 5: session, which coordinates communication between two machines; layer 4: transport, which includes, but is not limited to, packet loss retransmission, data sorting, and error checking; layer 3: network, which is in charge of data communication between networks via routers; layer 2: data link, which enables data transmission between systems and the nearest router or target machine; and layer 1: physical, which is responsible for actual data transmission over physical links such as copper, fiber, radio transmitters, and receivers.

In terms of practical applications and data transmission, the Internet comprises four major layers of transport protocols. The application layer, which includes OSI layers 7, 6, and 5, is the first layer; the protocol layer for the transmission control protocol (TCP)/IP protocol suite is the second layer; the data link layer is the third layer; and the physical layer is the fourth layer [1]. The data transmission procedure is as follows. The applications generate data packets, which may take the form of a web request or a fragment of an email message. The transport layer prefixes the data with headers indicating what information to send and where it should be sent. The network layer augments the data with IP headers, which function similarly to the return address on an envelope. The packet is then transmitted to the data link and physical layers.

End-to-end communication on the Internet is based on IPV4, and each data packet transmitted via the IPV4 protocol consists of a header and a data load. The header contains a 32-bit source IP address and a destination IP address. The router routes the packet based on the address information. The message structure of the IPV4 protocol is depicted in Figure 9.1 [1]. The data packet leaves the sending source bit by bit, and upon arrival at the router, the router examines each field of the received packet and forwards it to the network interface corresponding to the destination or another hop route near the destination. This is the protocol used to transmit and address data packets across a network.

Vers	Hlen	Type of Service	Length of data packet	
Identifier			Flags	Fragment offset
Time to Live		Protocol	Header Checksum	
Source Address				
Destination Address				
IP address (if applicable)			Padding	
Data				
Data				
Data				
Data				

Vers=Version
Hlen=Header length

Figure 9.1 Message structure of the IPV4 protocol.

9.1.2 The Content and Application Domain

While the system domain is concerned with the network infrastructure and support, the content and application domain is concerned with the available information applications and the data processed.

9.1.2.1 Content Storage

Content is primarily stored in two types of file system on the Internet: hierarchical file systems and relational databases. A hierarchical file system uses small sectors of hard drives or small blocks of memory to store files separately on one or more computers. Typical computer files contain the operating system, such as Windows, and executable programs, such as web browsers, as well as configuration files that store program settings and associated data. The files are organized in directories containing subdirectories, creating a hierarchical structure (Figure 9.2) [1]. Today, a large number of computer file systems are structured in this manner. The hierarchical file system is extremely advantageous because it provides a retrieval method that is consistent with the relationship between files. Meanwhile, it acts as a means of file navigation, as files within the same directory share certain characteristics.

Another popular method for content storage is using a relational database (Figure 9.3) [1]. This database stores the information in a series of tables that define several fields, each defining and storing a particular data type. While the fields are somewhat related, they all exist as distinct records in the database. A relationship may exist between two or more tables that share the same fields. Consider the database depicted in the figure as being related to

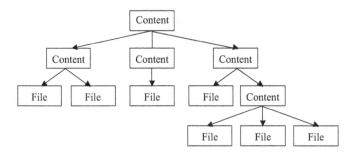

Figure 9.2 File system of a hierarchical structure.

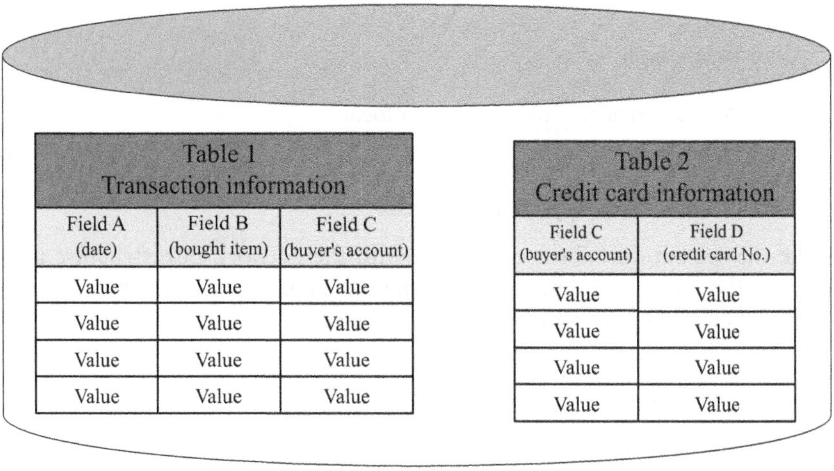

Table 1 Transaction information				Table 2 Credit card information	
Field A (date)	Field B (bought item)	Field C (buyer's account)		Field C (buyer's account)	Field D (credit card No.)
Value	Value	Value		Value	Value
Value	Value	Value		Value	Value
Value	Value	Value		Value	Value
Value	Value	Value		Value	Value

Figure 9.3 Relational database.

e-commerce. Table 1 may contain customer transaction data, which record the items purchased by customers in an online store. The customer's credit card information may be contained in Table 2. Both tables may contain information about the buyer's account or credit card.

9.1.2.2 Application Architecture

The term "application architecture" refers to the pattern of consumer and business applications frequently used on today's enterprise networks and the Internet. The most prevalent pattern is the client/server application architecture. The client application is a desktop or laptop application that serves as the user interface and connects to the server via a web browser, which later evolved into a ubiquitous client for programming, displaying, and processing

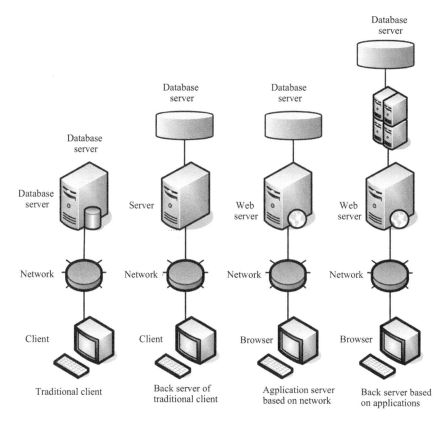

Figure 9.4 Common application architecture.

user data. Generic web browsers establish an Internet connection between the client and the server, thus the back-end database, defining the modern era's dominant application architecture (Figure 9.4) [1]. A third component, referred to as an application server, is used in certain applications.

Since 2005, a new service-oriented architecture (SOA) has emerged (Figure 9.5) [1]. While the client still communicates with the web server via browsers in this model, the communication is typically conducted via the Extensible Markup Language (XML). The web server can communicate with other servers via XML. The servers share and collaborate on data processing, providing users with network services. This SOA model is also referred to as "cloud computing". The model enables the distributed execution of functions such as geographical mapping, product searches, price calculations, and comparisons across multiple computation servers, with data flowing seamlessly between them. This enables users to purchase items more efficiently through querying and calculation, creating a shopping list, and selecting

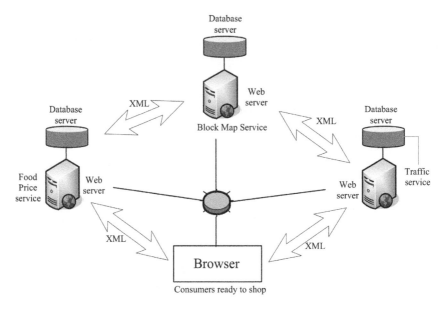

Figure 9.5 Service-oriented architecture.

nearby stores and courier services. It enables users to save money and time on travel.

Email, instant messaging, and search engines are all examples of wide-spread cyber applications. While email relies on store-and-forward technology, with messages waiting to be accessed by users, instant messaging technology emphasizes real-time information exchange between users, with such "chats" occurring between one-to-one or multi-to-multi users. Another common type of cyber application is search engines, which use specialized browsing programs called "crawlers" to connect pages. The "crawlers" can detect new pages and links, enabling the collection of massive amounts of data. Search engine companies have developed software that compiles the pages captured by "crawlers" into a searchable index to provide users with web-based front-end services. Web services such as e-commerce, instant messaging, and search engines are experiencing explosive growth at the moment, with Amazon, Alibaba, Facebook, Tencent, Google, and Baidu being just a few examples.

9.1.3 The Human and Social Domain

Cyberspace is a human-created domain that enables communication and information sharing between humans and machines. With the increasing integration of cyberspace into contemporary life, a new human and social domain emerges. This domain is exemplified perfectly by the online community. People from all over the world can connect through online communities to discuss business, politics, and various other topics. These communities include blogs, mailing lists, video sites, chat rooms, social networking sites, and virtual cities that cover a wide variety of topics such as news, healthcare, religion, chess, cooking, and take-out. Cyber or online communities are extremely diverse and dynamic, with new communities forming all the time and established communities contracting.

9.1.4 The Governance Domain

Global cyberspace governance is extremely complicated and involves several institutions and organizations, including the Internet Corporation for Assigned Names and Numbers (ICANN), the Internet Society, the International Telecommunication Union (ITU), the Organization for Economic Cooperation and Development (OECD), the IOS, the International Electrotechnical Commission, and the Institute of Electrical and Electronics Engineers (IEEE). Technology, standards, protocols, security, legislation, services, and management are all entwined and distinct at the same time. Whether cyber attacks qualify as acts of war is one of the most contentious issues in cyberspace.

The Internet is a complex megasystem, and the network we use every day is frequently referred to as the "surface web." However, the surface web is only the tip of the iceberg compared to the Deep Web and Dark Web (Figure 9.6), which are significantly larger [2].

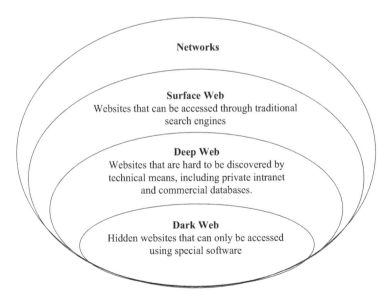

Figure 9.6 US Congressional Research Service's classification of the Internet.

The surface web is the web that users are accustomed to, and that can be easily crawled and accessed by any search engine.

The Deep Web refers to all networks that are not part of the surface web and are therefore inaccessible to search engines. It is not entirely hidden. However, standard search engines are incapable of detecting its tracks.

The Dark Web is a subset of the Deep Web. The Dark Web conceals everything, including websites, user identities, and IP addresses, enabling Internet users to truly vanish without a trace.

Thus, while the Internet has facilitated human interaction and created a virtual society of human–computer interaction, it is inherently insecure and destined to become a new space for exploitation, confrontation, and competition.

Cyberspace operations are primarily concerned with network-centric warfare that involves the civilian Internet, open-source information, human–computer interaction systems, and their derivatives such as public opinion, intelligence, and information warfare. Countries with advanced militaries are well prepared to defend against cyber attacks. They have already conducted several live-fire tests of their defense capabilities and are investigating cyber attack methods.

9.2 Cyber Information

The Internet is a veritable goldmine of open-source knowledge. Military use of open-source data is critical for improving intelligence operation capabilities and is a critical strategic factor affecting future operations. Since its inception, the Internet has contained the most political, economic, technological, military, social, livelihood, religious, and cultural information, demonstrating an increasing trend toward omnipresence and omnipotence. Since 2007, the advent of cloud computing and big data technology has been an impetus for the rapid development of the intelligent networked industry. Utilizing the Internet and other open-source information to the fullest extent is a critical and timely strategic issue for military construction and future operations.

As of January 2021, Internet World Statistics (IWS) estimates that there are 4.66 billion Internet users, 5.22 billion mobile phone users, and 4.2 billion social media users worldwide, with 99% of social media users accessing the platform through their mobile devices.

The Internet is all-pervasive. The US military, according to reports, routinely obtains between 50% and 80% of intelligence through open sources, including the Internet. In short, military combat forces can be multiplied through the use of open-source information from the Internet.

1. Open-source data can be used to improve macro- and micro-environments, such as geographic, meteorological, traffic, and other battlefield environments.

2. Open-source data enable the study of customs, religious beliefs, lifestyles, and other social conditions prevalent in war-torn countries and regions and their historical evolution and origins.

3. Through the use of open-source data, it is possible to obtain a holistic view of targets, including the shape and structure of target buildings from the outside and the distribution of incoming agencies and personnel from the inside.

4. The use of open-source data not only provides insight into combat adversaries' scientific and technological development but also enables prediction and early warning of their war preparations, mobilization, and operational intentions, primarily through data on power consumption, material procurement, product delivery, overtime work, and general movement of military personnel in core enterprises.

The army's future operations will rely on Internet-based open-source information as critical strategic assets, operational support, and critical battlegrounds, establishing a new domain and method of occupation control and effective utilization.

9.3 Cyber Battlefield

Cyber warfare is at the heart of cyberspace operations. In 1991, the National Academy of Sciences already warned that "future terrorists will use keyboards to inflict far greater damage than bombs" in a report on computer security [3]. Since the 1990s, the United States has increased its investment in cyber warfare, conducting exhaustive and systematic research and validation of operational concepts, doctrines, pathways, and approaches, to the point where it took the lead in establishing the world's first cyber warfare force.

Edward Snowden, a former US National Security Agency's defense contractor employee, made global headlines in early June 2013. He revealed a list of 49 "cyberspace" surveillance and detection programs in the United States, shocking the US government and the general public worldwide and causing powerful ripples in international politics. His disclosures about a series of US cyber-surveillance and intrusion programs codenamed "Prism" have heightened global political and military competition in cyberspace, exposing the virtual digital battlefield.

From as early as 2007, the US National Security Agency and other intelligence agencies have monitored the leaders of dozens of countries in real time, according to the secret documents disclosed. US intelligence agencies have conducted covert surveillance on former United Nations (UN) Secretary-General Ban Ki-moon, German Chancellor Angela Merkel, Brazilian President Rousseff, and other dignitaries. Apart from these, significant institutions such as the UN headquarters and the European Union's permanent mission to the UN have been subjected to prolonged US surveillance. The US has gained a significant informational advantage due to this maneuver at the international political negotiating table. When the UN Security Council debated whether to sanction Iran in 2010, Susan Rice, then-US Ambassador to the UN, in order to take targeted measures in advance, reportedly requested the intelligence division to monitor member countries whose voting intentions were unknown. By hacking into the world's mobile networks and major Internet companies, the National Security Agency illegally obtained massive amounts of personal information, call records, short messages, emails, stored data, transmitted files, video conferencing, and other types of data and information. It could surveil almost all the online activities of a particular target Internet user.

In May 2008, the United States began constructing the National Cyber Range (NCR), completed in 2012. The NCR establishes a realistic operational environment conducive to conducting cyber warfare exercises and weapon evaluations. From 2006 to the present, the US has conducted a series of cross-border, cross-nation, and cross-region "cyber-storm" exercises, each directly targeting the Internet and modeled after real-world cyber attack and defense events. The exercises, sponsored by the Department of Homeland Security, were similar to the biennial "Schriever" series of space security exercises conducted by the US DoD. Along with Homeland Security, the DoD, Departments of Commerce, Energy, Justice, Treasury, and Transportation participated in the elite member exercises. The exercise series brought together technical personnel from several allied nations, paving the way for the US to conduct cyber warfare operations in collaboration with other countries. The US military conducted an annual exercise called "Cyber-guard" to evaluate security cooperation between governments and military sectors, as well as collaboration between governments and non-governmental organizations in response to cyber attacks. The North Atlantic Treaty Organization (NATO) conducted the world's largest cyberspace operational simulation in 2014, with over 670 soldiers and civilians from 28 countries and 80 organizations reportedly participating.

President Obama launched a 60-day cyberspace security assessment at the start of his presidency and established an independent Cyber Command to integrate the Navy, Army, and Air Force cyber warfare forces. Since then, the development of the US cyber force has been accelerated, and the US Navy, Air Force, and Army have all established their cyber warfare units. Meanwhile, the United States has released a series of strategic plans, including a cyberspace policy review, a national cyber strategy, an international cyber strategy, and a strategy for cyberspace operations, which explicitly link cyber attack and defense to acts of war.

In 2012, the US DoD launched the Foundational Cyberwarfare (Plan X) research initiative to develop cyber attack and defense strategies from a warfare perspective. It attempted to construct a kill chain capable of effectively controlling the entire process of cyberspace attacks by incorporating operational planning, coordination, and execution, and damage assessment.

Table 9.1 Specific distribution of US Cyber Mission Force

Service/ mission	National mission	National support	Battle mission	Battle support	Cyber defense	Total teams/ members
Navy	4	3	8	5	20	40/1,860
Air Force	4	2	8	5	20	39/1,821
Army	4	3	8	6	20	41/1,899
Marine Corps	1	0	3	1	8	13/607
Total	13	8	27	17	68	133/6,187

The US Army explicitly proposed in the 2014 *Quadrennial Defense Review* to invest in enhanced cyber capabilities and establish a new Cyber Mission Force comprised of 133 teams [4]. By 2017, all 133 teams, including 41 in the Army, 40 in the Navy, 39 in the Air Force, and 13 in the Marine Corps, were combat-capable (Table 9.1).

With approximately 21,000 personnel, the US Army Cyber Force was formed to support expeditionary, special operations, and counter-terrorism operations. Simultaneously, the concept of a "2050 Cyber Army" has been proposed to address cyber attack and defense requirements and spur cyber technology development. The cyber force of the US Army is integrated in terms of strategy and tactics, offensive and defensive capabilities, cyber attack and defense, electronic confrontation, intelligence signals, and electronic warfare.

The US Defense Advanced Research Projects Agency (DARPA) recently launched the MEMEX "Dark Web Search Engine" project — a new cyber attack and defense initiative focused on searching and tracing "completely invisible" websites, identities, and IP addresses on the Dark Web in order to detect and combat drug trafficking, arms smuggling, and the activities of cult and terrorist organizations. The Dark Web may also be used to steal covert intelligence, recruit cyber talent, and disseminate information about cyber security. In addition, the US Army has formed a joint cyber-space security research alliance with academia, business, and government researchers, establishing a military-oriented, integrated civil–military power dynamic.

In 2013, Russia established cyber defense forces and a cyber warfare command. The latest version of the Russian Military Doctrine makes a compelling case for enhancing the strategic status and role of "non-nuclear deterrence", the centerpiece of which is information and cyber warfare to bolster the deterrent capability.

In 2014, Japan's Ministry of Defense (MoD) established a separate "Cyber Defense Unit". Moreover, Japan has been reportedly actively developing cyber warfare weapons, including one capable of rapidly identifying the source of a cyber attack and launching a retaliatory strike directly at the source.

NATO has established a cyber warfare center in Estonia to conduct periodic cyber warfare exercises. It has issued legal guidelines for cyber warfare, the *Tallinn Manual*, which serves as a legal foundation for using cyberspace by the United States and other Western countries. In short, a cyber arms race is on the horizon, rattling the nerves of major powers.

In 2016, global attention was focused on artificial intelligence applications in cyber attacks and defense. Artificial intelligence is being used in the US military for various purposes, including autonomous network encryption, vulnerability detection, and malware behavior learning. A new cyber defense technology for power grid infrastructure was developed. In October 2016, the Google Brain Team successfully enabled its neural systems to create their algorithms via launching mutual attacks between three neural networks. Two neural systems, "Bob" and "Alice", generated a shared secret key for sending and receiving messages, while a third neural system, "Eve", made an unsuccessful attempt to steal and decode the message. This demonstrates the feasibility of achieving autonomous encryption through the use of machine learning and neural networks.

In August 2016, DARPA held the Cyberspace Grand Challenge, during which numerous "robot hackers" demonstrated their ability to detect and fix software vulnerabilities using artificial intelligence methods automatically. However, prolonged dormancy and accidental damage to the protected system have occurred in practice. The "robot hackers" also demonstrated attack capabilities by exploiting the vulnerabilities of network systems. These robot hackers, powered by artificial intelligence, have the potential to significantly improve the efficiency and versatility of a military's cyber attacks.

In terms of malware behavior learning, at the August 2016 "Black Hat" security conference, a company called Spark Cognition announced the release of an anti-virus product called "Deep Armor" that, based on automatic modeling algorithms and other artificial intelligence technologies, can understand the behavior of new malware and identify the process by which viruses attempt to bypass the security system.

In July 2016, PFP Cybersecurity launched an IoT security solution based on electricity usage analytics, machine learning, and cloud technology, which could detect supply chain anomalies, internal tampering, and persistent attacks to protect any IoT-connected device.

In 2020, the US Army developed a radio that conserves power in sleep mode while monitoring communication traffic and ensuring the mesh network operates normally. It can also automatically adjust the rate and power of data transmission in response to interference and noise, demonstrating superior intelligence.

Since the turn of the new century, the global cyber battlefield has been raging, not only infiltrating the wars in Afghanistan, Iraq, and Libya, and the "Color Revolutions" in the Middle East and North Africa, but also becoming inextricably linked to the Syrian war and the anti-ISIS Islamic organization. On the horizon is a new era of intelligent cyber warfare.

9.4 Cyber Attack

Cyber attacks can be classified as small-scale or large-scale. Small-scale attacks may cause minor inconvenience to a few users (e.g., thousands) because most viruses and spyware in this category share certain characteristics that make them easily detectable by anti-malware software. In comparison, large-scale attacks can affect millions of users by restricting the Internet's functionality, blocking it from specific countries and regions, or stealing large amounts of user account information for fraudulent purposes. Prior to 2002, most cyber threats came from amateur viruses and other experimental malicious code, as well as financial fraud. Since 2002, cyber threats have expanded to include organized crime and state-sponsored cyber reconnaissance and attacks. The following are seven major categories of common cyber attacks.

9.4.1 Spyware

Typically, spyware is installed without the user's knowledge or consent and is used to gather these information. Attackers install spyware on users' desktops and laptops by probing for software vulnerabilities on the target machine or bundling spyware with other software to deceive users. Spyware is frequently used to track and analyze how users interact with a series of websites to promote targeted products and advertisements. The simplest method of user-tracking is to store user information using cookies. A cookie is a small amount of data that a website sends to the user's browser. It can be stored on the user's computer, and whenever the user visits the site again or another site associated with a spyware company, the cookie is sent back to the relevant site, allowing for the tracking of the user's site history. When a user tries to access a mainstream search engine, the spyware may redirect the user's browser to a search engine supported by advertisements.

Additionally, some spyware is designed specifically to steal user information. They search the user's hard drive for private files containing sensitive data and send them to the attacker. One of the most intrusive spyware is keyboard recording software, which, for example, records and transmits to the attacker the user's account number and password entered on a financial website's login page [1].

9.4.2 Bot Software

Bot software is primarily used to seize control of an infected host. As a result, it can monitor users' web-browsing activity, steal data, and record keystrokes. Bot software can set up a network, i.e., botnet, and infect dozens, if not hundreds, of user hosts. The botnet can be considered a supercomputer due to the massive distribution of bots on the infected PCs. A botnet is tens of thousands of times faster than a single computer at cracking keys or passwords. Bots are also increasingly associated with Rootkits, a sort of malicious software that can alter the operating system to deceive users, disguise the presence of attackers on the machine, and even prevent anti-virus and personal firewall programs from functioning properly. Rootkits represent a type of stealthy bot software frequently employed by cyber criminals [1].

9.4.3 Phishing

Phishing is a prevalent form of Internet fraud. It typically involves sending a large number of emails, ostensibly from legitimate businesses such as online banks or online stores, claiming that there is a problem with the recipient's account and then tricking the user into a link that appears to lead to a legit-imate commercial website but redirects the user to an attacker-controlled bogus website that requests the user enter their username and password or other sensitive account information. Phishing attackers are increasingly impersonating tax officials, government institutions, and agencies via falsified emails. Senior citizens are more susceptible to these unethical behaviors. For instance, in 2011, one attack on emails received by US encryption software businesses included an attachment titled "2011 recruitment plan.xls" with attack codes targeting Adobe Flash Player's zero-day vulnerability. The codes were eventually codenamed "CVE-2011-0609." If an attack is successful, the attackers may grab control of the targeted internal server and subsequently establish persistent control of the system using a remote-access protocol called "Poison Ivy" [1].

9.4.4 "Puddle" Attack

A "puddle" attack is a term coined by researchers at RSA, a US cryptographic software company, to refer to the practice of luring animals to specific areas for hunting. This concept also applies to Internet users, where infection of specific networks results in a puddle forming. A puddle attack is a variant of a webpage-attached Trojan virus attack in which malware is downloaded on to an end user's system via exploitation of a specific browser vulnerability. In contrast to phishing, this type of attack is tailored to users' browsing habits and waits for them to visit infected legitimate websites on their own. The gen-eral procedure is as follows: First, the attacker gathers information about the target user's browsing habits via open-source intelligence to compile a list of frequently visited websites and conducts targeted user modeling. Second, the attacker exploits these websites' vulnerabilities to inject malicious code. Third, the attacker waits for the user to visit infected websites before installing mal-ware on the user's system via webpage-attached Trojan viruses. Fourth, once the user's system is infected with malware, a remote-access Trojan (RAT) is downloaded to the infected system; the attacker can use this to take control of the user's system, stealing sensitive data and transferring it to the system under his control [5].

9.4.5 USB Flash Drive Infection

When a user's computer is not connected to the Internet, the most effective and likely means of virus transmission is via USB flash drives or detachable hard drives. This method is especially effective against critical infrastructure, like industrial control systems. For instance, the US-developed "Stuxnet" virus was used to disable numerous centrifuges at Iran's nuclear enrichment facility. A USB supports two distinct modes of operation for malicious application execution. It could infect the USB drive by renaming the autorun file "autorun.inf" or generating a rogue shortcut. When the user clicks on the shortcut, the malicious code is running.

When a malware-infected USB device is linked to the Internet, network disruption is possible. In the United States, a similar event occurred in a related firm when a third-party vendor updated a turbine control system using an infected USB stick, infecting the control system and resulting in 3 weeks of system dysfunction and severe financial loss. Similarly, an attack on a control system at a business in New Jersey resulted in the attackers acquiring control of the heating and air-conditioning system, posing severe security concerns [5].

9.4.6 Packet Flooding as a Denial-of-Service (DoS) Attack

Cyber attacks can cause harm by flooding one or more target devices with packets, disconnecting them from regular network connectivity. This type of cyber assault is referred to as "packet flooding." By deploying a small botnet of a few hundred workstations, an attacker can flood packets and take down the website of a typical medium-sized business. For example, if a website, like an online shopping mall, is attacked, customers will be unable to access it, and the business will be unable to operate, which might be used as a negotiating chip by criminals attempting to extort money. The cyber attack against Estonia in 2007 was a massive operation involving numerous interconnected botnets. Each week, an attacker will normally sustain a packet flooding attack for about 3 days. If the flooding persists for an extended period, most ISPs will identify and prevent the flooding traffic. Flooding attacks on a large scale may be directed at specific businesses, a country's related systems and Internet infrastructure, or even the entire Internet, causing thousands of Internet users to experience disruption. Synchronize (SYN) flood attacks, HTTP flood attacks, and Domain Name System (DNS) amplification attacks are the most prevalent and typical DoS attacks [1].

SYN flooding attacks are most successful during the handshake phase of a TCP session. The TCP three-way handshake protocol is used to send data across the Internet. To establish a connection, the initiating host generates a SYN packet with a sequence number that serves as the starting sequence number for all subsequent packets broadcast from the initiating host to the receiving host (e.g., web server). If the recipient accepts the connection request, it sends a SYN-ACK (acknowledgment) packet to the initiating host, indicating that it has responded to the sequence number received and synchronizing the connection's new beginning sequence number. The initiating host then sends an ACK packet, including the sequence number of the recipient. Thus, the three sequence number exchanges between the two hosts are referred to as three-way handshakes. With each transmitted packet, the sequence number of the loaded data increases for subsequent transmissions over that connection.

The SYN flooding attack violates the three-way handshake protocol by initiating a TCP connection and purposefully blocking the entire connection procedure in the second stage. For instance, if the attacker sends a SYN packet and the recipient answers with a SYN-ACK packet, the attacker then ceases to send ACK packets, leaving the target computer with a half-open connection waiting for a response. When repeated hundreds of times per second, this insufficient exchange renders the target system incapable of responding to subsequent regular requests. Using flow sensors that identify traffic-like features, network managers can avoid this type of attack.

To circumvent the aforementioned features of flooding traffic, an HTTP flood attack impersonates legitimate traffic flow. It comprises a high volume of erroneous genuine requests, rendering detection systems incapable of distinguishing legal from malicious traffic.

Amplification attacks against the DNS include the attacker sending small query packets to hundreds of third-party DNS servers, each of which forces the server to generate a larger response packet, therefore amplifying the traffic burden. To route this traffic to the victim host, the attacker sends each query packet with a forged source address for the victim host, making it appear as though it is sent from the victim host, and the reply traffic sent from the DNS servers to the victim host will exhaust the victim host's bandwidth. While attackers do not directly target DNS servers, they are employed as traffic amplifiers to cause another victim host to become overburdened. Since 2005, DNS amplification assaults have resulted in a deluge of red traffic at up to 20 Gb/s, a rate comparable to the bandwidth of most enterprise backbone routers or huge e-commerce devices.

It is feasible to map the traffic flow and behavioral characteristics of a single flood attack over an extended observation time, enabling the installation of preventive measures. However, if an attacker uses a novel combination of these tactics over a protracted period – several weeks or even longer – defending against them may become incredibly difficult, and even more difficult when cross-border collaboration is required.

9.4.7 *Critical Infrastructure Vulnerabilities*

Another method of executing large-scale attacks is investigating critical infrastructure vulnerabilities, such as backbone routers and DNS name servers. Most key infrastructure components may contain vulnerabilities and weaknesses that an attacker may exploit to obtain access to the system and crash the target unit. For example, an attacker can even intercept Internet traffic destined for a bank in another country and steal critical data from the traffic [1].

Cyber attacks are deliberate human actions that fall under the heading of cyberspace operation and cognitive confrontation. In the future, when artificial intelligence technology is developed and applied, cyber attackers may employ diversified and intelligent assault tactics and behaviors, posing new and substantial challenges to cyber defense.

9.5 Cyber Defense

Cyber defenses are divided into two broad categories: web-based and host-based. Web-based defense may require the collaboration of network administrators, such as ISPs, large organizations, institutions, and network equipment and software vendors whose products may perform specialized security functions. Host-based defense requires specific software to be deployed on all networked systems to secure many terminals while retaining scalability.

9.5.1 Web-Based Defense

The focus of web-based defense is to prevent large-scale cyber attacks. To achieve this goal, security tools and means need to be established from distributed control systems.

9.5.1.1 Firewall

A firewall can filter network traffic according to configured rules, allowing certain data types to join the network while blocking others. For instance, it can permit Internet-based web traffic while restricting network management traffic flows. Advanced firewalls can limit the traffic that originates from or terminates at a specified source or destination address. The most powerful firewalls can detect application-specific data, critical characters, or words contained within packets. Typically, firewalls are established at the connection nodes between two networks, such as the business network's boundary with the Internet. These firewalls are usually configured to provide access to the external Internet, filtering outbound traffic liberally while limiting inbound traffic strictly. Naturally, some organizations, such as large enterprises, institutions, and military units, may substantially restrict external access. Certain countries limit all outbound traffic through firewalls or prohibit any online activities that promote harmful political rhetoric or religion. In late 2007, due to political unrest, the Myanmar government cut off Internet connectivity, restricting both the entry and outflow of information. Countries without firewalls, whose global access is guaranteed by many original design manufacturers (ODMs) and global Internet, would struggle to execute full network blocking [1].

9.5.1.2 Web-Based Intrusion Detection System

A web-based intrusion detection system monitors Internet traffic in order to identify potential threats. When an attack is discovered, the system notifies network managers, acting as an early-warning system for network burglars. Businesses and government agencies have installed detection sensors at their Internet gateways. These sensors are utilized to determine whether an assault has made it through their front entrance. Additionally, some firms have deployed network detection sensors within their internal networks, establishing several thresholds to detect all threats.

Currently, the majority of intrusion detection solutions are based on feature-based detection. For each known attack activity, the appropriate vendor might include a detailed description of the attack features in the packet. Once the traits are identified, it signals that the associated attack is underway. These features are typically made available to the public regularly, and thousands of them are available via both commercial and non-commercial means. Certain specialized methods are used to detect specific behaviors, determine when network activities differ from standard network behavior, and assess whether the activity falls within the category of a cyber attack. For instance, the occurrence of many pending TCP three-way handshake procedures may signal the execution of recurring, concurrent SYN flood attacks. Web-based intrusion detection systems can aid information security agencies and teams that evaluate data on cyber attacks and collaborate on preventing network breaches in certain industries, such as the financial and military sectors. The system protects the country from cyber attacks by installing equipment at ISPs and network interconnection nodes, monitoring all outbound and inbound data flow to and from the country and detecting coordinated attacks. However, a portion of traffic transported through a satellite will be challenging to the monitor. Due to the large volume of network traffic, the state-based approach to cyber defense focuses exclusively on the traffic flow features, ignoring the content of individual packets [1].

9.5.1.3 Web-Based Intrusion Prevention System

Web-based intrusion prevention systems are a hybrid of firewalls and intrusion detection systems. When the prevention system detects packets containing attack data, it may reject or re-link them in order to prevent the attack activity, thereby securing the network. Naturally, legitimate traffic may be misinterpreted as an attack, which is why some protection systems may issue warnings rather than block the data directly. Certain systems include defense capabilities that allow for the real-time examination and review of traffic. Others operate in the background, analyzing representative samples of traffic. The in-line work mode examines all packets, significantly slowing down traffic and making it ideal for systems requiring a high level of confidentiality and warning, whereas the out-of-line pattern is appropriate for defending massive normal systems [1].

9.5.1.4 Network Encryption

Initially aimed at end-user and computer application programs, network encryption expanded to incorporate a network device layer, resulting in multi-layered encryption security from both the network and the end system. Three types of end-user encryption exist: pretty good privacy (PGP) programs developed by encryption enthusiasts for email and file encryption; secure shell (SSH) for remote login and access protection; and secure sockets layer (SSL) for communications between web browsers and websites.

The Internet protocol security (IPsec) protocol combines authentication and encryption of network devices. It can be integrated into either the IPv4 or IPv6 protocols in order to enhance network-level encryption. While using predefined shared keys is somewhat convenient when only a few end-to-end systems are utilized, key preservation and distribution become more difficult when more than a dozen interconnected systems are employed [1].

9.5.2 Host-Based Defense

Host-based defense refers to a technique for guarding against cyber attacks targeted at hosts. While web-based protection provides defensive monitoring of multiple devices, determining the behavior of specific machines is difficult. Host-based defenses are capable of monitoring and defending against all aspects of a computer's functioning and activities. They comprise anti-malware technologies, a host-based defensive system, a personal firewall, and encryption of host data. A personal firewall is functionally identical to a network firewall but with a limited reach.

9.5.2.1 Anti-Malware Tools

Anti-malware tools include anti-virus and anti-spyware software. They once belonged to two different and autonomous markets but have become increasingly entwined. Up-to-date anti-malware programs often combine three detection techniques: signature-based, heuristic-based, and behavioral-based detection. Each has its own set of advantages and disadvantages. While signature-based detection is incredibly effective, it necessitates the continual and rapid updating of a signature database. However, it is almost impossible to predict the signatures of malware. Heuristic-based detection exploits the fact that attackers frequently repurpose functional code modules from previous malware to detect and initiate security systems. However, the attack will remain undiscovered if the attacker avoids repeating code sequences when creating new malware. The third anti-malware technique, behavioral-based detection, detects malicious programs based on their normal working behavior. For instance, malware frequently affects browser settings and opens, writes to, and closes hundreds of files on the system in a matter of seconds. Anti-malware software can detect and eliminate malware by monitoring certain actions [1].

9.5.2.2 Host-Based Intrusion Prevention System

Unlike web-based intrusion prevention systems, which examine network traffic to detect attacks, host-based intrusion prevention systems scan the programs running on each terminal. The principle is to study lawful program operations' features, detect abnormal program behavior, and operate within specified norms. Such operations, however, run the risk of producing false alarms. False detections may fail important applications, forcing some firms to suspend or deactivate protection measures to resume normal operations [1].

9.5.2.3 Host-Based Encrypted Files

On the host computer, encrypted files are generally used to protect data. Encryption can be used to secure individual files or directories and entire disks, operating systems, and applications. While encryption of files is substantially faster than encryption of disks, the latter is significantly safer because an attacker may bypass host-based encryption techniques in a variety of methods, including by generating hidden temporary files, gaining access to data via a valid account, or attempting to recover the decryption key or password required to execute an attack [1].

9.5.3 Security Tactics

9.5.3.1 Response Plans

When an organization's security vulnerabilities are found, it is critical to take prompt corrective actions. A more prudent strategy would be to educate yourself about the security of critical data storage, isolate and control critical data and servers within the network, formulate internal security management mechanisms, detect data breaches, and monitor communication traffic for anomalies inside and outside the network.

9.5.3.2 Terminal System Security Mechanisms

The system's virus database should be updated to the most recent version and be patched immediately upon the vendor's release of a patch or version upgrade. Additionally, the system should keep up to date with third-party programs, as obsolete plug-ins that are subject to vulnerability attacks may be incorporated in browsers. End users should be equipped with effective malware detection and protection tools.

9.5.3.3 User-Centered Security Tactics

When users are online, they should take caution and avoid clicking on tempting yet dubious links. They should create strong, complex passwords and regularly change them. They should refrain from using the same password across many websites. They should exercise caution while opening email attachments. Users may commit the virtual machine to untrusted websites to protect the host's security, prevent disclosing personal or sensitive information online, and avoid using personal USB or external storage devices on business intranets or networks.

9.5.3.4 Network Security

Additional security layers, email-filtering software, firewalls, and a robust DNS are all possible measures to improve network security. Other possible measures include monitoring network communication, identifying user browsing habits and associated domain names, collecting the network's various incoming and outgoing resources, establishing reliable intelligence feedback for early warning, encrypting sensitive data flowing in and out of the network, doing regular log analysis for indications of attack, enhancing the virtual LAN, splitting the main network into smaller networks for suitable isolation, and configuring dependable access.

9.5.3.5 Security Assessment and Patch Management

All devices and applications on the internal network should be examined regularly for known and unknown vulnerabilities, and system security should be verified through relevant testing and simulated assaults. It is critical to have a patch management strategy that regularly updates security software and virus databases via manual and automatic patching.

9.5.4 Next-Generation Defensive Measures

Since 2016, artificial intelligence technologies have been widely employed in cyber defense, manifested in automatic encryption, vulnerability detection and repair, malware behavior recognition, and IoT protection. Simultaneously, fresh cyber defense strategies and tactics have evolved due to new technologies such as virtualization, blockchain, and the Intranet. The convergence of "protection + intelligence" will become a prominent trend in the future, as indicated by the following:

1. Machine learning, deep learning, and feature vectors applied to develop effective behavior-based malware detection and cyber attack identification systems based on traffic and TCP handshake characteristics.
2. Cutting-edge web-based technologies, applied for the efficient building of virtual sandbox environments and the prevention of browser-based attacks.
3. Virtualization solutions, applied to separate jobs and schedules within the operating system, enabling malware detection and restriction.

4. Virtual private networks (VPNs), applied to construct an Intranet, where a data communication tunnel is encrypted on the public network with a hierarchical and incremental "isolated" transmission between the virtual gateway and the target address, forming a "private network" on the public network. The operational principle is that each source and destination address establish a virtual gateway connected to the public network's IP. The sender encrypts the source address and content and sends them across the public network using the IP address of the virtual gateway. When data are received, the virtual gateway on the receiving end deconstructs the header of the public network transmission and decomposes the data load before sending it to the destination address.

5. Whitelisting management, which allows whitelisted software to run while restricting the operation of other software.

6. Blockchain technology, applied to provide another layer of trust to enable end-to-end transactions and interactions.

9.6 Electromagnetic Attack and Defense

In a limited sense, electromagnetic attack and defense are related to electronic warfare. In a broad sense, it includes communication, radar, navigation, adversary identification, hydroacoustic, and photoelectric confrontation, as well as camouflage, stealth, wireless injection, interference replacement, and other electronic confrontation and deception involving multiple media, multiple networks, and physical space. Electronic warfare's necessary means and technologies include aircraft, airborne, shipborne, ground-based, aerospace electronic warfare equipment, and emerging electronic warfare technologies.

Between 2015 and 2018, Russia exhibited considerable electronic warfare capabilities on the battlefields of Ukraine and Syria, bringing electronic warfare into the global spotlight. According to the Russian Ministry of Defense's Information and Mass Media Department, in the early-morning hours of January 6, 2018, the Russian air defense system in Syria spotted 13 small aerial objects (i.e., drones) en route to hit Russian military units. Ten of the drones flew close to the Khmeimim airbase, while three flew close to the Tartus port depot. Russian soldiers then executed a counter-drone swarm operation flawlessly. The Russian army allegedly used direct fire damage for "hard killing" and ground jamming to interfere with drones for "soft killing." According to statistics, Russia's Pantsyr S1 air defense system destroyed seven drones. Anti-aircraft guns and close-range air defense missiles were also used to intercept targets, including drones, air-to-ground guided weapons, and helicopter gunships. Simultaneously, Russian radio-technical troops in Syria successfully commanded six drones, three of which were remotely controlled and landed outside the base, while the remaining three crashed during landing. "Leer-2", an electronic warfare system installed on a Tiger armored vehicle, could simultaneously detect and track several targets and employ jamming technologies to defeat drones. "Leer-3" was made up of a command-and-control center and three Orlan-10 drones that could monitor and disrupt communications networks, as well as transmit fake signals. The Orlan-10 drones carried an electronic warfare payload for detecting and suppressing sources of electro-magnetic radiation within a 6-kilometer radius. The "Repellent" system could simultaneously jam 12 global positioning systems (GPS), Galileo, and other satellite and wireless communication bands within a radius of 30 kilometers.

According to Russian Defense Ministry scientists, based on a technical study of the recovered drones, it was speculated that the terrorists most likely launched the long-range attack from roughly 100 kilometers distant. This was the first time in Syria's history that anti-government forces have deployed drones, operating them from up to 50 kilometers away. Russian military experts thoroughly studied the drone's structure, internal components, and explosives. After scanning the drone's data pad, the takeoff position was determined. According to the Russian military, this new sort of attack "could only be undertaken by countries equipped with modern satellite navigation and remote delivery of explosives to precise locations" [6]. The drone swarm strike was subsequently revealed to have originated 50 kilometers away in a zone held by moderate Syrian opposition forces. In 2015, the Russian Air and Space Forces began operations in Syria to target terrorists. In December 2017, Russian President Vladimir Putin ordered the withdrawal of Russian forces from Syria but retained the Russian airbase at Khmeimim and the navy base at Tartus.

The US emphasized electronic reconnaissance, electronic jamming, and electronic deception throughout World War II. The power has recently begun to reconsider its position on electronic warfare, exploring various projects to boost its electronic warfare capabilities. Terry Halvorsen, Chief Information Officer of the DoD, revealed that given the critical nature of electromagnetic attack and defense, the DoD was considering the electromagnetic spectrum as a separate battlespace, ranking sixth behind land, sea, air, space, and cyber-space, and would shift policy toward electronic warfare and increase invest-ment through institutional, strategy, regulatory, technology, testing, and procurement approaches, thus creating a new battlespace.

Electronic dispersed, swarmed, and decentralized devices will become more difficult to identify or directly and efficiently attack. Global jamming of suspected locations will consume a large amount of energy, in which oper-ational equipment would also be exposed, risking combatants' safety. Thus, it is vital to develop electronic equipment that identifies and destroys the enemy at the lowest resource cost possible, minimizes the adversary's chances of dis-covery, and minimizes the impact and collateral damage to your electronic systems.

The future development of emerging electronic assault and defense tech-nologies, particularly cognitive electronics, is speeding up the introduction of new capabilities for cognitive electronic warfare. These features include intelligent electromagnetic interference (EMI) detection via cognitive radio, cooperative jamming (CJ), automatic frequency hopping, self-adaptive compatibility, and identification with present or future communication systems. Additionally, it incorporates directed energy and information from the future, enabling cross-domain assault and defense via GPS or other navigation timing systems.

Electronic warfare has entered the cognitive era due to artificial intelligence advancement, with cognitive technology becoming the trend. The progress of self-adaptive capabilities in radar and communication systems and the proliferation of frequency-dependent devices have resulted in an increasingly congested spectrum. The US military has advanced the notion of cognitive electronic warfare since 2010, researching cognitive jammers, behavioral learning, self-adaptive electronic warfare, and self-adaptive radar confrontation. In 2016, flying trials and prototype flight validations were conducted in compliance with the research. The plans include equipping the F-35 fighter with cognitive electronic warfare capabilities and a next-generation jammer to improve autonomous observation, real-time response, efficient countermeasures, and assessment feedback capabilities. Simultaneously, the US military is constructing an electromagnetic management system that blends electronic spectrum and electromagnetic operations management to provide a single operational architecture for precise spectrum control.

Major powers are investing heavily in electronic warfare systems that include detection, tracking, early warning, destruction, interference, and camouflage deception in response to the growing threat posed by drones due to the systems' active interference and full energy destruction capabilities. There are currently several counter-drone electronic warfare systems available, including Boeing's "Silent Attack" counter-drone laser weapon, Raytheon's "Phaser" high-power microwave counter-drone system, the Dutch Airbus Group's electronic jamming counter-unmanned aircraft system (C-UAS), and UK Surveillance Systems' "Anti-Drone Defense System". They have all garnered tremendous attention.

Electronic warfare technology for networked swarms has also been vigorously developed. For example, "Gremlins" is a DARPA program to construct a reusable electronic warfare drone swarm capable of penetrating hostile airspace and attacking adversaries by weakening missile defenses, interfering with communications, undermining internal security systems, and even performing cyber attacks. In 2016, the US DoD began developing the Perdix Drone Swarm System for electronic warfare tasks. When detached from the Aerial Delivery Dispenser, this system transforms into a swarm capable of acting as air defense system baits or performing intelligence, surveillance, and reconnaissance tasks alone.

In the future, it will be necessary to study electromagnetic space holistically, to develop methods and models for cognition and control of electromagnetic space on a broad to granular scale, to delve deeply into the electromagnetic information intentionally and unintentionally radiated by adversary sensors or reflected by targets in order to achieve rapid cognition, identification, and localization of adversary electromagnetic threats. The armed forces should start with the electromagnetic field and wave, reconstructing electromagnetic space, developing new theories and technical solutions, and eventually constructing an integrated information network system of radar, communication, navigation, broadcasting, and remote sensing capable of mitigating strong interference and induced deception, detecting and identifying stealthy targets, and performing accurate and effective operations.

9.7 Public Opinion Control

In the digital age, psychological warfare on the Internet and the ensuing cognitive conflicts are critical ways to control public opinion. Public opinion has been a critical battleground for military forces charged with occupying and controlling target areas, eradicating terrorist groups, as well as participating in future military operations other than war (MOOTWs). Cyberspace is a virtual realm that is inextricably linked to individuals and society. It results in non-linear mutations such as universal perception, rapid propagation, temporary linkage, and emergent blowout. Expert manipulation of public opinion and targeted "publishing" of sensitive material can quickly result in a shift in public opinion, psychological disorientation, and societal upheaval during times of war or calamity. Thus, by utilizing big data and cyberspace to monitor and manage public opinion, as well as to conduct psychological warfare against terrorists, key adversary figures in the conflict, and other target groups, future military operations will gain a new operational capability.

When deployed effectively, cyber-psychological warfare capabilities at the strategic level can be used to influence and modify a country's mass public opinion and cause societal instability, eventually resulting in regime change and government extinction. Tunisia's Ben Ali, Libya's Gaddafi, and Iraq's Saddam Hussein are three well-known victims. They all suffered losses because of US cyber attacks and Wikileaks "disclosures," which directly or indirectly contributed to the regime's demise. Gaddafi and Saddam were removed by cyberspace and ground operations, whereas Ben Ali was deposed mostly due to poor online public opinion.

Tunisia's "Color Revolution" in early 2011 was the Arab world's first public uprising. It is sometimes referred to as the "Jasmine Revolution" and is widely considered the first revolution sparked by WikiLeaks. The US's disclosure of evidence of President Ben Ali's corruption and associated crimes via Wikileaks and the Central Intelligence Agency (CIA) essentially brought an end to President Ben Ali's reign in Tunisia, where he ruled with an iron fist for 23 years until being exiled abroad within a month.

According to international news reports, the people's revolution began with a fruit vendor's suicide. In mid-December 2010, a 26-year-old educated yet unemployed young man set himself on fire while putting up a roadside fruit kiosk after the police confiscated his vegetables and fruits for operating the stall without a license. This increased comparable suicides. The news of the fruit vendor's self-immolation spread swiftly via social media platforms such as Facebook, escalating public outrage over the country's already high unemployment rate and food costs. WikiLeaks played a critical role at this moment, disclosing a June 2009 US diplomatic message in which the Ben Ali family was portrayed as a mafia that controlled every sector of the country's economy and in which the first lady reaped enormous profits from the establishment of aristocratic schools. Another 2009 message portrayed the scene at a party held at Ben Ali's son-in-law's mansion: Roman-era relics were strewn about, visitors sipped yogurt carried in by private aircraft from a hamlet in southern France, and a pet tiger roamed the lawn. Another letter headed "Corruption in Tunisia: Yours is Mine" stated that in Tunisia, members of the president's family will eventually obtain any money, land, house, or yacht they wish. There was a statement describing how the populace despised the first family. US Ambassador Gerdyk in Tunisia wrote: "Corruption in limited circles of power continues to grow in severity. Even regular folks are aware of this. Everywhere, there was widespread unhappiness." These letters revealed the country's ruling elites' corruption. Within a few weeks, word of mouth and social networking sites disseminated the content of these letters. The public had long been aware of the presidential family's corruption. However, these messages revealed the details of high-level corruption, and it was these naked details of corruption, combined with the various social problems, that eventually brought people to the streets, and evolved into a popular revolution.

Facebook played a critical role in Tunisia's turmoil. Numerous websites are banned in Tunisia, yet Facebook is not, with one in every ten Tunisians having a Facebook account. The Internet functioned as a channel for protest messages, exacerbated official repression of demonstrators, and increased popular anger. Ben Hassan, the head of a citizen movement in Paris, France, believes that Facebook had a major role in disseminating people's discontent and enabling previously cautious and fearful ordinary people to begin breaking the silence. "A month ago, we had little hope for the revolution's success, but the people are finally rising," said Manai, a professor of geology at the University of Tunis. Despite Ben Ali's last-minute promises to reform, dissolve his cabinet, re-elect parliament in 6 months, and abstain from running for re-election, enraged crowds rioted in Tunis, Tunisia's capital. Ben Ali eventually quit abruptly, to the chagrin of long-standing ally France. Later, Saudi Arabia was convinced to welcome Ben Ali's family. According to *Foreign Policy* magazine [7], WikiLeaks revelations sparked Tunisia's revolt, and it was called the world's first "wiki revolution".

The US government and Western countries routinely engaged in public opinion and propaganda wars with the Gaddafi regime throughout the Libyan war. They continuously exposed the details of Gaddafi's eight sons and one daughter's business dealings in oil, gas, hotels, media, distribution, communications, and social infrastructure, claiming that "tens of billions of dollars flow into their pockets every year", revealing that Gaddafi's second son Saif spent $1 million on Mariah Carey, a female singer, to perform only four songs, and exposing a series of bizarre quotes from Gaddafi. As a result, the media asserted that the constant exposure of the Gaddafi family's corruption and other follies threw Libya into instability and finally resulted in the demise of the Gaddafi dictatorship and his family.

According to evidence, a week before the Iraq War, the US military Shute system seized control of the Iraqi military command-and-control system and the Iraqi Ministry of Defense's website and sent targeted emails to thousands of officers convincing them that "the game was over," making the US military effectively "winning without fighting." The following is the text of an email sent to hundreds of Iraqi officers via the Iraqi Ministry of Defense's email system on the eve of the conflict:

> This is a message from US Central Command. As you know, we may be instructed to invade Iraq in the near future. We will defeat any forces that oppose us, as we did several years ago, if we do so. We do not want to harm you or your troops. Our goal would be to displace Saddam and his two sons. If you wish to remain unharmed, place your tanks and other armored vehicles in formation, abandon them, and walk away. You and your troops should go home. You and other Iraqi forces will be reconstituted after the regime is changed in Baghdad [8].

The email resulted in rows of equipment being thrown aside along the road from Kuwait to Baghdad, and US troops drove in unopposed until they confronted a small force in the heart of Baghdad. The email demonstrates three points: First, Iraq's command and communication infrastructure was targeted and may have been taken over by cyber troops. Second, Saddam lost connection with his commanders and troops, rendering him unable to implement the operational command of the forces and limiting himself to a small-scale resistance in Baghdad's central business district via face-to-face leadership of the garrison. Third, the email acted as a psychological deterrent, causing a huge number of Iraqi commanders to have a mental breakdown, exposing them to the full wrath of a powerful army. Simultaneously, in the absence of higher leadership directives, they were forced to ditch their equipment and flee the battlefield, hiding in their homes. The email was leaked by the Special Advisor to three US Presidents on Cybersecurity and Counterterrorism in a retrospective account of the Iraq War published in 2014 in the freely available book *Cyber War* [8].

Through the use of big data and artificial intelligence technologies, information on social networks, instant messaging, and cultural backgrounds is gathered about key target figures from the government, military, religious groups, political parties, and non-governmental organizations, as well as leaders in terrorist organizations, in order to understand better their personalities, interests, hobbies, religious beliefs, social relationships, and perceptions of specific events. Microblogs, instant messaging software, SMS, television, and radio may be utilized for psychological interventions with target groups, military occupation control, deployment, and emergency management.

As a result, it is necessary to meet the requirements of diverse multi-level psychological warfare operations, to investigate the modes, methods, and procedures of peacetime and wartime, strategic and tactical, offensive and defensive, military and civilian, popular and elite psychological warfare in cyberspace. It is important to develop an intelligent search and application system that is civil–military integrated, safe, trustworthy, and target-oriented. Precise and efficient information source control and information delivery to combat areas can be achieved via mobile communication, digital television broadcasting, the Internet, and other widely available information media. This will facilitate psychological intervention and public opinion guidance against adversaries.

In the future, Internet psychological warfare will gradually give way to a new type of virtual space operations, resulting in a new operational form of virtual and physical space cross-fertilization. History and modern conflicts demonstrate that psychological warfare operations against an adversary can effectively support an attack on critical physical targets and critical information targets. It is also capable of directly attacking critical targets in the cognitive realm. Psychological warfare fills a void that cannot be filled by physical or information warfare activities and, in some cases, may even be vital to the triumph of a war.

9.8 The Future of Internet and IoT

With the advancement of technology, cyberspace is expanding horizontally to include an in-depth interconnection of people, machines, things, and diverse applications, and vertically to include the Deep Web, Dark Web, and even more secretive cyberspace, with issues such as complexity, diversity, obscurity, and security becoming increasingly prominent. Future network technology developments include virtualization, high-performance computing, IPv6, blockchain, embedded sensor networks, multi-source heterogeneous information fusion, and data correlation search.

9.8.1 Virtualization

Virtualization aims to increase the capacity and efficiency of diverse networks. Software-defined networking, which consists of computing, storage, and other resources distributed across multiple systems or distributed computing machines, is used to collect, store, and process data in a unified manner by creating a virtual machine or virtualized environment. It is gradually evolving into various public, private, centralized, and distributed cloud platforms. Virtualization technology ushered a new era of Internet and online application development.

9.8.2 High-Performance Computing

High-performance computing consists of three components: the processing capability of the terminal, the use of parallel processors, and the establishment of a large-scale commercial server complex. The development and application of quantum computers are of great significance. In March 2018, Google announced the launch of Bristlecone, a 72-qubit general-purpose quantum computer with a 1% error rate. Microsoft has spent more than a decade developing quantum computers. Although it has not yet launched mature and commercially accessible goods, it claims that its developing prototype will outperform Google's and IBM's prototypes. It is expected to release improved items in the near future. Scientists are investigating photonic computers based on optical signal storage and biocomputing technologies, with parallel processing capabilities on biochips. In the future, as quantum, photonic, and biological computing technologies mature and become more widely applied, they will usher a new era of human computing, bringing significant impetus and leaps to cyberspace and intelligent operations.

9.8.3 IPv6

IPv6 is an acronym for "Internet protocol version 6," which was developed by the Internet Engineering Task Force to replace the existing version, IPv4, and claims to be capable of generating a website address for every grain of sand on Earth.

IPv4 is a commonly used Internet protocol with a 32-bit address length. There are around 4.3 billion IPv4 addresses available worldwide. IPv6 addresses are 128 bits long and have the 128th power of 2, implying that the IP address space approaches infinity. The limitation of IPv4 is the scarcity of network address resources, which severely limits Internet application and development. IPv6 resolves the issue of limited network address resources and removes barriers to Internet connectivity for a range of access devices.

The International Internet Society celebrated World IPv6 Launch Day on June 6, 2012, marking the official launch of the global IPv6 network. At 00:00 global standard time on the same day, some well-known websites, including Google, Facebook, and Yahoo, began permanently supporting IPv6 access.

IPv6 addresses have a length of 128 B, four times that of IPv4 addresses, and are represented in hexadecimal format instead of IPv4's decimal format. IPv6 addresses can be expressed in three ways: in decimal notation, in 0-bit compression notation, or embedded IPv4 address notation.

The IPv6 packet header format is shown in Table 9.2.

An IPv6 packet comprises an IPv6 header, an extension header, and protocol data at the higher layer. The IPv6 header must be a set length of 40 B. It contains the message's basic information. The extension header is optional and may be zero, one, or more characters long. The IPv6 protocol

Table 9.2 IPv6 packet header format

Version	In the version of the protocol, the value is 6
Traffic class	It is mainly used for quality of service (QoS)
Flow label	It allows labeling packets belonging to the same flow
Payload length	It specifies the length of the payload, in bytes, that the packet is encapsulating. If there are extension headers, it indicates the type of the first extension header; if not, it indicates the type of the upper-layer protocol. It primarily realizes the various functions of IPv6
Next header	It specifies which header follows the IPv6 packet header
Hop limit	It corresponds to the IPv4 time to live (TTL) field. The number decreases for each forwarding. When the TTL decrements to 0, the packet is discarded
Source address	The address of the initial sender of the packet
Destination address	The address of the intended recipient of the packet

uses the extension header to perform a variety of purposes. The upper-layer protocol data is the information carried by the IPv6 packet, including ICMPv6 messages, TCP messages, user datagram protocol (UDP) messages, or other types of communication (Table 9.2).

IPv6 has the following advantages over IPv4:

1. IPv6 offers a more expansive address space. IPv4 provides a maximum address length of 32 and a maximum address count of 2^{32}. IPv6 addresses have a maximum length of 128 bytes. Therefore, the maximum number of IP addresses is 2^{128}. In comparison to the 32-bit address space, its address space is 2^{32}–2^{128} times larger.

2. IPv6 employs a more compact routing table. IPv6 address allocation begins with the swarm concept, which enables the router to represent a subnet with a record in the routing table, significantly lowering the length of the router's routing table and increasing the router's packet-forwarding performance.

3. IPv6 delivers greater multi-cast support and flow management, allowing for the large growth of multimedia applications on the network and establishing a solid network platform for quality of service (QoS) control.

4. IPv6 adds automatic configuration support. This enhancement and extension of the dynamic high configuration protocol (DHCP) protocol make network management (particularly LAN management) more convenient and faster.

5. IPv6 is more secure. Users on IPv6 networks can encrypt and validate IP messages at the network layer. IPv6's encryption and authentication mechanisms ensure the secrecy and integrity of packets, significantly boosting the network's security.

6. IPv6 is scalable. IPv6 enables the protocol to be expanded to accommodate new technologies or applications.

7. IPv6 features a more advanced header format. IPv6 introduces a new header format with distinct choices from the base header that may be put between the base header and the upper-level data if desired. This simplifies and accelerates the routing process, as the route does not need to select most of the possibilities.
8. IPv6 introduces new possibilities for additional functionality.

9.8.4 Blockchain

Blockchain technology, a novel distributed infrastructure and computing paradigm, verifies and stores data using blockchain data structures. It generates and updates data using distributed node consensus techniques, secures data transfer and access using encryption, and programs and manipulates data using smart contracts comprised of automated scripting code. In a limited sense, blockchain technology is referred to as distributed ledger technology, and it is primarily concerned with distributed data storage, database operations, and file activities over the Internet. According to some experts, blockchain technology designs products that enable open, transparent, and traceable data. We can design extremely rich services and product formats based on the properties of blockchain technology. It has enormous potential and application scenarios in banking, government, enterprise, and cross-industry sectors.

Decentralization, openness, and transparency are characteristics of blockchain technology, enabling everyone to contribute to database records. The fundamental principles and notions are as follows:

1. Transaction: an operation that modifies the ledger's state, such as adding a record.
2. Block: it contains the results of transactions and states that occur throughout time and serves as a consensus on the ledger's current state.
3. Chain: a log of the complete state change, constructed by chaining together blocks in the order they occur.

If we regard the blockchain as a state machine, each transaction represents a single effort to modify the state, and each consensus-generated block represents the participants' acknowledgment of the result of all the transactions contained in the block that resulted in the state change.

Blockchain is a decentralized database storage system that enables collaborative work. Unlike traditional databases, which are centralized and restricted to a single firm or a small number of concentrated persons, blockchain enables anybody with the ability to set up a server to participate.

As a protocol system, blockchain technology is composed of six layers: data, network, consensus, incentive, contract, and application. Each layer performs a critical function independently, cooperating to create a trust system in which all participants contribute.

The blockchain could be separated into three architectural layers: protocol, extension, and application. The protocol layer can be subdivided into the storage and network layers, which are distinct but interdependent.

The protocol layer is the lowest layer; it includes the storage layer and the network layer. This layer establishes the network environment, constructs transaction channels, develops node reward criteria, and implements distributed algorithms and cryptographic signatures, among other things.

The extension layer is analogous to a hard disk on a computer: to begin, to implement a variety of transaction kinds, and to continue, to extend the application in a specific direction. The "smart contract" is a common example of extension-level application development. The term "smart contract" refers to a "programmable contract" or "contract intelligence", which means that when certain criteria are satisfied, the contract is automatically implemented, such as automatic security transfer or payment.

The application layer is analogous to the numerous software programs on a computer; it is a product that people can interact with directly and may also be thought of as the browser of a B/S architecture product.

Blockchain technology performs five main core functions.

1. Decentralization

    Due to the distributed nature of accounting and storage, there is no centralized hardware or governing body. Each node has equal rights and obligations, and the data blocks in the system are collectively maintained by the nodes equipped with maintenance services. Blockchain technology evolves from a "public ledger" or "public database" to a "public computer."

2. Transparency

    Blockchain features a transaction confirmation technique that involves cooperative bookkeeping and cross-certification of the transaction's various parties. The blockchain platform enables the openness and transparency of all transaction behaviors and associated procedures. The system is now operational. Except for the private information of transaction parties, the blockchain data are available to all, and anybody can query the data and construct associated apps via the open interface, ensuring that the entire system's information is highly transparent.

3. Autonomy

    The self-running blockchain uses consensus-based specifications and protocols (e.g., a collection of open and transparent algorithms) to enable all nodes in the system to exchange data freely and securely in a de-trusted environment, effectively replacing trust in "people" with trust in machines.

4. Tamper resistance

    Blockchain is essentially a distributed bookkeeping technology that can be thought of as a database of decentralized nodes. Rather than a single party, all parties participate in the bookkeeping using some signature private keys and consensus mechanism algorithms. To ensure that the data are not tampered with or corrupted, the bookkeeping behaviors are all traceable. The blockchain's timestamp service and proof of existence ensure that the precise time and events surrounding the creation of the first blockchain are kept in perpetuity. Once data are confirmed and put into the blockchain, they are permanently kept. Changes to the database on a single node are invalid unless more than 51% of the system's nodes can be controlled concurrently. As a result, the blockchain's data stability and reliability are exceedingly high.

5. Anonymity

   Since the trade between nodes is governed by a fixed algorithm, the data interaction is free of verification. The blockchain's program rules will assess whether the activity is genuine. Thus, counterparties are not required to acquire one another's trust by disclosing their identities, which is extremely beneficial for credit accumulation [9].

Blockchain technology is essentially a decentralized and distributed data storage, transport, and verification framework that replaces the Internet's present reliance on central servers with data blocks, allowing for recording all data changes or transaction items on a cloud system. This end-to-end verification results in creating a "foundation protocol," a new type of distributed artificial intelligence that establishes a new connection and common interface between the human brain and machine intelligence. The convergence of blockchain technology with next-generation Internet, IoT, and artificial intelligence, in particular, will establish the technical groundwork for the deep integration of the digital and real economies. Blockchain technology is very secure and capable of recreating many upgraded versions of the Dark Web using encrypted communication channels that do not require a centralized network.

The technological qualities of blockchain technology can be stated as "TRUE" and "DAO," where TRUE stands for trustable, reliable, usable, efficient, and effective, and DAO represents distributed and decentralized, autonomous and automated, organized and ordered. Blockchain technology's numerous benefits and features can be applied to many military applications, including distributed networks, information transmission, data security, intelligent identification, autonomous decision-making, swarm attack and defense, comprehensive logistics capability, and acquisition management.

Each tier of a typical blockchain system's six-layer architecture contains key technical elements for military intelligence control. Among these, the data layer assures the security, dependability, and trustworthiness of military systems, data, and intelligence. The network layer contributes to the realization of an autonomous and decentralized military network system. The consensus layer encapsulates several consensus algorithms that enable military management to attain autonomy and credible decision-making. Through programmable digital currency and incentive systems, the incentive layer prevents all forms of misconduct and incentivizes positive behavior. The contract layer enables the automation and intelligence of military administration, as well as the reduction of unpredictability, diversity, and complexity introduced by human and social elements. Blockchain technology has a transformational effect on military management [10].

9.8.5 Embedded Sensing and IoT

Embedded sensing and the IoT encompass all facets of military and civilian applications. Civilian applications include traffic control, smart homes, industrial automation, data collecting, monitoring, and supply chain management. Sensor networks and combat environment detection, target identification, autonomous swarm attack and defense, wartime mobilization, and complete logistical capability are just a few military applications. Currently, automobiles and other motor vehicles contain inbuilt sensors, which operate in a closed system. Inter-vehicle communication, or connection between the car and other information systems, enables unmanned driving in a networked environment. While piloting and controlling unmanned vehicles in military contexts are very different from civilian environments, there are numerous similarities.

Mobilization and security of operations are critical components of embedded sensing and IoT applications. For instance, a distributed management system for container logistics may be constructed using a backbone storage depot. The system provides users with fundamental applications such as administration, real-time material status information, order status, scheduling, inventory strategy, off-route alarms, transportation schedules, and route planning. Meanwhile, based on big data mining, it is possible to calculate the application demand for certain equipment, sending feedback to the production site and the building and layout of the warehouse. Based on past data, early-warning models may be developed to help limit the occurrence of accidents. A destruction-resistant transport network can be constructed to ensure that the system continues to operate during a conflict.

The system's design, construction, and operation must take network security into account, including data encryption, anti-virus protection, firewall technology, intrusion detection, and security vulnerability screening, as well as system online and offline data backup and safe operation.

The system's physical connectivity comprises three components: warehouse storage, container transportation, and a logistics management system. The first refers to the location where materials are stored; the second refers to the mode of transportation used to transport materials between warehouses, typically in containers. The container can be thought of as a temporary mobile storehouse. In addition, radiofrequency identification (RFID) technology can be used to create records of inbound and outbound data automatically. IoT can be used in warehouses and container carriers to convey real-time information about items to the management information system.

Satellite communication, 3G/4G/5G mobile communication, train information networks, terminal information networks, highway toll station information networks, and special radios are all examples of back-transmission methods. The management system maintains the data and makes it available to the user via a web service.

9.8.6 Multi-Source Information Fusion

The multi-source information network is prevalent in civilian IoT domains and will play an increasingly vital role in the military IoT domain.

Since 2010, the growth of Internet+ industrial resources has accelerated the civil sector, enabling any visible object, such as refrigerators, televisions, and washing machines, to be connected in a network. Multi-source network information has permeated numerous spheres of human productivity and life.

In the military field, the future complex, intelligent battlefield, and adversarial environment will necessitate the development of heterogeneous networks capable of autonomous planning, dynamic connectivity, functional reconfigurability, and self-healing. The technical focuses are as follows.

First, the fusion, offense, and defense of virtual and physical network information, which include open-source big data correlation and authentication for sky–space–ground intelligence, surveillance, and reconnaissance (ISR) systems; reconnaissance, command, communication, combat, and protection; cyber warfare, psychological warfare, public opinion warfare, legal warfare, we-media, and other cross-media public opinion control and cognitive confrontation; and electric poaching.

Second, intelligent networks for complicated battlefields. It is based on a self-organizing sky–space–ground network that adapts to the mission area's geography, automatically detects interference spectrum, hops frequencies autonomously, and reconfigures flexible linkages to provide long-duration coverage of typical missions and battles. For example, it is necessary to undertake building studies in metropolitan regions and conduct indoor and outdoor navigation, communication, and detection using integrated network systems to support varied operation missions.

Third, a unified defense against complicated networks. Based on artificial intelligence, IPv6, virtualization, blockchain, and mimetic defense, research on active network immune protection, intelligent random encryption of information data, addressing methods free of root servers, and end-to-end communication free of central servers enables a fundamental shift in the passive defense model against the network attack chain's hidden attributes and resources, establishing new integrated defense means and capabilities on chips and software systems.

Fourth, precise searching and positioning. A megatrend accelerated by the emergence of embedded sensors is localization through the converged analysis of people (cell phones), things (RFID), and information (data tags) as they move around. Collecting such information makes it possible to track and infer the behavior of a human being, regardless of that person's willingness to be found.

Fifth, a novel cross-domain attack and defense mechanism. It places a premium on emerging technologies for jamming GPS. Foreign organizations have investigated novel methods for launching remote wireless injection attacks on GPS, including exploiting GPS receiver software and hardware vulnerabilities, forging the content of navigation messages (ephemeris), launching remote wireless injection attacks via radiofrequency, forging location and time information, attacking the operating system, modifying the GPS-based systems' synchronization time base, and paralyzing the receiver (such as financial systems).

9.9 New Models and Trends

In November 2018, Israel's Homeland Security and Cyberspace Defense Exhibition took place at Tel Aviv's Israel Convention Center, with 172 companies exhibiting, 80 of which focused on cyber products, 76 on homeland security products, and 16 on fintech products, as well as ten related lectures and interviews (Figure 9.7). Israel's three major military industry groups (Israel Aircraft Industries (IAI), Rafael, and Elbit) demonstrated their homeland security, land, sea, air, and space coordination, regional offense and defense, and other combat systems. Numerous innovative Israeli small to medium-sized enterprises (SMEs) demonstrated their

Figure 9.7 2018 Israel Homeland Security and Cyberspace Defense Exhibition in Tel Aviv.

technologies and product scenarios in counter-drone, smart cities, quick face recognition, fifth-generation network security, and network ranges. The exhibition planned a study visit for conference delegates to Israel's largest science and technology innovation center, Beer Sheva Network Park, where the Israeli military demonstrated how network data centers were constructed. The exhibition, in general, represented emerging patterns and trends in the development of cyberspace operations and associated technology products.

To begin, a comprehensive system of cyber operations and security products is under establishment. From the terminal to the network, software to hardware, hacker attack to protection, information deception to awareness countermeasures, network communication to electronic confrontation, end defense to source defense, traditional cyber to intelligent cyber, simulation training to practical products, a systematic series of operational research products has been developed. It is a treasure trove of information. The depth of study, the variety of items, and the breadth of scope are all beyond conception. Israel has increasingly placed a premium on cyber security and technologies. The Israeli government has announced its intention to significantly advance cyber security through a series of tax breaks, talent incentives, and other policies that integrate regional and global government and business resources to create an open, free, and cooperative cyber environment. For example, scientists at the Weizmann Institute of Science have demonstrated that hackers can launch a full-scale attack on national infrastructure using a light bulb and other common household devices by hacking a Philips smart light bulb connected to a wireless network and triggering a chain reaction of the rapid spread of a large-area "worm" virus. The findings have gained traction in the world of cyber security.

Second, a cyber security discipline and training system is being built. This encompasses everything from fundamental theory and simulation training to hands-on operations, product creation, and instructional demonstrations. In the booth's center was a special cyber academy, which offers a comprehensive series of cyber teaching materials, conducts online and offline education, and is open to international students. Additionally, there were numerous specialized teaching and training systems on display at the event. Cyberbit's Cyber Range, for example, is an ultra-realistic network simulation platform and network training venue that immerses customers in a environment that accurately simulates their network topology, security tools, and normal and malicious traffic, significantly improving on traditional training methods and displacing the previous generation of desktop push network range training systems. It can train network security professionals similarly to how fighter pilots are trained by providing off-the-shelf training packages and additional application cases, incident response team training, penetration tester training, capture-the-flag drills, and critical infrastructure security training among other things. With 20 cyber training facilities and thousands of students enrolled each year, Cyber Range has grown to be the most widely deployed cyber range platform, offering customizable training scenarios, training programs, and simulated cyberinfrastructure to fit the unique demands of each user.

Third, cyber security products and operational concepts of the fifth generation are approaching. Check Point demonstrated its fifth generation of network security devices during the exhibition. The company believes that since 2017, there has been an unprecedented shift in cyber attacks, with the majority of them being large-scale, multi-vector mega-attacks that severely damage businesses and their interests, with an unprecedented level of sophistication and impact, exhibiting fifth-generation properties and characteristics. However, most businesses have established just second- or third-generation security systems. There is a generational divide that renders such defenses ineffective. Each age has a different level of cyber security:

First generation: in the late 1980s, virus attacks on standalone PCs afflicted all organizations, precipitating the creation of anti-virus software.

Second generation: in the mid-1990s, Internet-based attacks impacted all companies, resulting in the emergence of firewalls.

Third generation: at the turn of the twenty-first century, application vulnerability exploitation afflicted most businesses, resulting in the proliferation of intrusion prevention systems (IPS) solutions.

Fourth generation: around 2010, the introduction of targeted, unknown, evasive, polymorphic attacks plagued most businesses, resulting in an increase in anti-bot and sandbox technologies.

Fifth generation: around 2017, large-scale, multi-vector attacks utilizing new attack tactics appear. Advanced defense mechanisms are required to protect against such fast-moving attacks. Detection-only methods are insufficient.

At the moment, fourth-generation security technologies are unable to defend against fifth-generation threats against information technology settings such as networks, terminals, cloud installations, and mobile devices. Bridging the dangerous divide between fifth-generation cyber attacks and previous generations of cybersecurity needs a paradigm change away from second- and third-generation patching and best-in-class deployment methodologies toward a unified security foundation, i.e., a full security architecture.

The Infinity architecture from CheckPoint is a comprehensive network security architecture that protects against fifth-generation large-scale network assaults on all networks, terminals, clouds, and mobile devices (Table 9.3). The architecture incorporates best-in-class, verified threat prevention solutions across the organization's network, cloud, and mobile information technology infrastructures, real-time shared threat intelligence across and inside the enterprise, and a unified, comprehensive security management framework. To address the complexity and inefficiency of security that comes with increased connectivity, the architecture uses a single threat intelligence and open application program interfaces (APIs) to protect all environments from targeted attacks.

Table 9.3 CheckPoint's Infinity complete network security architecture

Clouds	
Infrastructure (IaaS)	Software (SaaS)
Advanced threat prevention	Zero-day threat protection
Adaptive security	Sensitive data protection
Automation and orchestration	End-to-end SaaS security
Cross-environmental dynamic policies	Identity protection
	Multi- and hybrid cloud
Network	
Headquarter	Segments
Access control	Access control
Data protection	Multi-layered security
Multi-layered security	Advanced threat prevention
Advanced threat prevention	Wi-Fi, DSL, PPPoE
Mobile	
Remote access	App protection
Secure business protection	Network protection
Protect docs everywhere	Device protection
Terminals	
Threat prevention	Access/data security
Anti-ransomware	Access control
Forensics	Secure media
	Secure documents

IaaS, infrastructure as a service; SaaS, software as a service; DSL, digital subscriber line; PPPoE, point-to-point protocol over ethernet.

Fourth, an effective system for civic–military collaboration on innovation and a model of "leading the civil with the military" are on the horizon. The Beer Sheva Cyber Campus is home to Israel's National Center for Cybersecurity Research and is rumored to be the Israeli Defense Forces' Cyber Command location. The zone was established with military research institutions as its nucleus, resulting in gathering innovative SMEs involved in cyber security (Figure 9.8). Military and civilian personnel are located just a few meters apart and share the same structure, forming a true trinity of the military, universities, science, and technological industries. The alignment of needs, talent exchange, transformation of results, and daily management between the military and local, school, and enterprise communication is seamless and efficient, enabling the deep integration of science and technology innovation on a civil–military level. Numerous entrepreneurs in the zone are war veterans. The Israeli Defense Forces itself is a conventional school, offering classes in everything from leadership and collaboration to mathematics, science, chemistry, biology, and technologies such as communications, electronics, and network information. Soldiers acquire new talents throughout their military career and then apply them to innovation and business following their retirement. The army is also quite receptive and supportive of retired personnel. Many of the company's founders visit their previously serving army weekly

Figure 9.8 Advanced Network Technology Zone in Beer Sheva, Israel.

for exchanges, allowing them to produce numerous creative goods directly related to the necessities of real battle. For instance, the army squad's weapon synergy system for preventing accidental injury is a new technical product designed by retired people in response to the demands of real battle. Thus, Israel's civil–military integration is distinguished by integrating talents and military personnel with front-line real-world combat experience who can truly participate in civil–military integration, ensuring that technological innovation is tightly coupled with military requirements and truly contributes to the development of the combat force.

Fifth, intelligent applications and trends are emerging. At the exhibition, artificial intelligence technology was applied to the majority of fields and systems within cyber security, with particular emphases on automatic detection of multiple viruses and system vulnerabilities, identification and learning of multiple attacks and malware behaviors, autonomous encryption of information transmission, and data sending and receiving, automatic defense of network systems and terminals, attack traceability, and intelligent counterattack; and also in areas such as face recognition, detection, and perception, unmanned systems, and simulation training.

In the future, the creation of fifth-generation cyber attacks and defensive products and behaviors will undoubtedly extend to a diverse array of intelligent operation systems. Distributed, cross-domain, and heterogeneous network information systems, IoT systems, cloud systems, human–computer interaction systems, and even pure artificial intelligence systems may become cyber and cross-domain attacks and defense targets. They should be the focus of future attention, research, and prevention.

References

[1] Kramer, F. D., Starr, S. H., & Wentz, L. K. 弗兰金·D.克拉默, 斯图尔特·H.斯塔尔,拉里·K.温茨. (2017). *Saibo liliang yu guojia anquan* 赛博力量与国家安全 (G. Zhao 赵刚, X. H. Kuang 况晓辉, L. Fang 方兰, D. X. Wang王东霞, F. Xu 许飞, & J. Tang 唐剑, Trans.). Beijing: National Defense Industry Press.

[2] *Xinxixitonglingyukejifazhanbaogao*信息系统领域科技发展报告. (2016). *Shijie guofang keji niandu fazhan baogao* 世界国防科技年度发展报告. Beijing: National Defense Industry Press.

[3] National Academy of Sciences. (1991). *Computer at risk: Safe computing in the information age*. Washington, DC: National Academy Press.

[4] Office of the Secretary of Defense. (2014) *Quadrennial defense review 2014*. www.defense.gov/

[5] Sood, A., & Enbody, R. 阿迪蒂亚·苏德, 理查德·尹鲍. (2016). *Dingxiang wangluo gongji: you loudong liyong yu eyi ruanjian qudong de duojieduan gongji* 定向网络攻击: 由漏洞利用与恶意软件驱动的多阶段攻击 (Y. J. Sun 孙宇军, G. T. Geng 耿国桐, & X. Z. Fan 范笑峥, Trans.). Beijing: National Defense Industry Press.

[6] Qi, Y 齐莹 (2018). Kongbu fenzi dongyong wurenji xiji e zhuxujidi weiguo Meiguo bei yi tigong zhichi 恐怖分子动用无人机袭击俄驻叙基地未果 美国被疑提供支持. Sina 新浪网. www.huanqiu.com/

[7] Dickinson, E. (2011). *The first WikiLeaks revolution*? https://foreignpolicy.com/2011/01/13/the-first-wikileaks-revolution/

[8] Clarke, R. A., & Knake, R. K. (2014). *Cyber war: The next threat to national security and what to do about it*. New York, NY: Ecco.

[9] Zhu, Z. W. 朱志文. (2016, October 12). *Yiwen kandong yukuailian jiagou sheji* 一文看懂区块链架构设计. Babite wang 巴比特网. www.8btc.com/article/106022.

[10] Wang, F. Y., Yuan, Y., Wang, S., Li, J. J., & Qin, R. 王飞跃, 袁勇, 王帅, 李娟娟, 秦蕊. (2018). Junshi qukuailian: cong buduicheng de zhanzheng dao duicheng de heping 军事区块链: 从不对称的战争到对称的和平. *Journal of Command and Control* 指挥与控制学报, *4*(3), 175–182.

10 Hypersonic Confrontation

Since the 1990s, weaponry technologies such as hypersonic aircraft and missiles, kinetic energy interceptors, directed energy weapons, electromagnetic cannons, and high-energy destructive warfare components have fundamentally altered traditional strike and destruction patterns. Combining hypersonic and intelligent technologies effectively will usher in a new era of hypersonic intelligent conflict. Hypersonic penetration and counter-penetration will become a primary pattern of operation in intelligent warfare.

Usually, weapons with a speed exceeding Mach 1 are called supersonic weapons, and those with a speed exceeding Mach 5 are called hypersonic weapons. Hypersonic weapons in a narrow sense include conventional weapons, vehicles, and ammunition. In a broad sense, they also include high-powered microwave weapons, laser weapons, electromagnetic cannons, electromagnetic pulse (EMP) bombs, and high-firing rate guns.

10.1 Hypersonic Operations

Hypersonic operations will be conducted in the future as highly dynamic, highly responsive offensive and defensive confrontation based on network data and artificial intelligence (AI). The operations will be strategic and tactical, offensive and defensive, single-equipped and swarmed, and will involve both fixed and moving targets.

Strategically, conventional long-range strike weapons, such as hypersonic cruise missiles, hypersonic gliding missiles, and unmanned aerial vehicles (UAVs) in air and space operations, can effectively penetrate existing operational defense systems, significantly enhancing strike capability against time-sensitive and deeply buried targets and exerting strategic strike effects comparable to or exceeding those of nuclear weapons. These weapons will enable revolutionary strategic striking methods and operational patterns for strategic deterrence. With the advancement and development of the US X-51A, X-37B, SR-72, HTV-2, SR-91, and B-3, as well as other hypersonic weapons and aircraft, the goal of an "1-hour" global strike is becoming a turns into real.

DOI: 10.4324/b22974-12

Tactically, as medium- and close-range hypersonic weapons and munitions become increasingly networked, swarmed, and intelligent, hypersonic intelligent confrontation will become an integral part of the entire operations process and at all strategic and tactic levels.

From an offensive standpoint, hypersonic weapons, vehicles, and munitions capable of precise strikes at variable mobility at extremely high speeds will present significant challenges to the adversary's defense systems. Hypersonic gliding technology enables aircraft to fly non-ballistically at speeds of up to Mach 20. Once hypersonic gliding technology is developed, "1-hour" global strikes will become possible; at the moment these are almost indefensible. This technology, however, is complicated. The US conducted two flight tests of the hypersonic vehicle HTV-2 in April 2010 and August 2011, both of which failed. The X-51A cruise hypersonic weapon developed by the US military also failed two flight tests in March 2011 and August 2012.

From a defensive standpoint, it is extremely difficult to quickly detect, track, and locate hypersonic weapons due to their variable trajectory. Interceptor weapons must operate at higher speeds in order to counter hypersonic weapons, and it will be nearly impossible to intercept the weapons using conventional defense systems or interception methods. Point-to-point defense with faster weapons is ineffective and costly. As a result, new technological capabilities must be developed. There are only a few mature and reliable anti-hypersonic solutions available at the moment. The majority of them are still research and development (R&D) concepts.

At the turn of the twenty-first century, the US military proposed the Conventional Prompt Global Strike (CPGS) program, which would allow conventional weapons to strike any target on Earth within an hour. The CPGS program can be implemented in stages. In the initial stage, a conventional modification of the Navy's Trident missile is used. The US Navy's Submarine Global Strike Missile Program and the US Air Force's Boost-glide Missile Program represent the next stage. The subsequent stage involves the development of hypersonic cruise missiles.

Hypersonic weapon development at the moment is concentrated on single-vehicle solutions for strategic fixed targets. Hypersonic weapons will evolve in the future toward formation, network, swarm, and systematization. They will become offensive and defensive weapons capable of operating at long and short ranges for deterrence or strategic and tactical operations.

10.2 Hypersonic Weapons and Munitions

Long-range hypersonic weapons are being developed in two configurations: guided boost-jump-slide weapons and guided ram-cruise weapons. Tactical hypersonic guided weapons, high-speed kinetic energy anti-tank missiles, electromagnetic cannons, and high-speed artillery are all examples of medium- and short-range hypersonic weapons. A hypersonic cruise weapon is a system that uses ramjets and other propulsion methods to achieve orbital transformation and defense penetration. A boost-slide hypersonic weapon is a weapon system propelled into near space or beyond the atmosphere by a first-stage solid rocket, which increases speed and range through a series of glide jumps utilizing gravity or a second rocket.

Since the mid-1990s, Defense Advanced Research Projects Agency (DARPA) and the US Navy have collaborated on several hypersonic missile development projects, the most notable of which are the Fast Hawk cruise missiles, ultra-high-speed strike missiles, and joint ultra-high-speed cruise missiles. Fast Hawk is a low-cost ground attack missile powered by a ramjet engine, cruising at speeds between Mach 4 and Mach 6. The US Navy first proposed the Hypervelocity Strike Missile Program in 1998. A cruise missile with an average speed of Mach 7, a maximum range of 1,100 kilometers, and the capability to penetrate 11 meters of concrete was to be launched on this mission. The missile's development began in 2004. In 2001, DARPA, the US Navy, and the US Air Force launched the Joint Hypervelocity Cruise Missile Program to develop a common hypervelocity missile with a flight speed of Mach 8 and a maximum range of 1,400 kilometers. The first ground test of this missile's hydrocarbon fuel scramjet occurred in 2003, and the full-scale test of the demonstrator aircraft occurred in 2004.

Since 2002, the US has also operated the CPGS program, which was renamed the Conventional Rapid Strike Program in 2017, with the specific objective of developing weaponry and operational capabilities of striking time-sensitive targets. In some ways, conventional rapid strikes are equivalent to hypersonic strikes. Before 2035, the US military may be equipped with two types of hypersonic weapons, according to R&D progress: hypersonic cruise missiles and hypersonic boost-glide vehicles. These include the US Air Force's hypersonic strike weapon, DARPA's hypersonic aspirated weapon and tactical boost glider, and the US Army's advanced hypersonic weapon.

The US has concentrated its efforts within the Conventional Rapid Strike Program framework on developing a tactical-range High-Speed Strike Weapon Program (with a range of approximately 1,000 kilometers) and a strategic-range Advanced Hypersonic Weapon Program (with a range of more than 6,000 kilometers). Two subprojects of the High-Speed Strike Weapon Program have completed their initial design reviews: the Aspirated Hypersonic Weapon Program and the Tactical Booster Glide Weapon Program. The US Army is investigating a control method for hypersonic vehicles with a flight altitude of 30–50 kilometers and a top speed of 5,800 meters per second.

In the 1980s, Russia began developing hypervelocity missiles. Due to external factors, the program was suspended in the post-Cold War era, i.e., the 1990s. Russia restarted research on hypersonic weapons in the new century. Since 2010, Russia has accelerated the development of hypersonic weapons, focusing on three weapons. The tactical concept is "aspirated cruise + boosted glide". Vanguard, a silo-launched hypersonic missile system capable of exceeding Mach 20 and with a range of more than 10,000 kilometers, was officially delivered to combat duty in December 2019. Dagger, the operational-level air-launched hypersonic missile system, began operational duty at the end of 2017. It has a maximum speed of Mach 10 and a maximum range of 2,000 kilometers. Zircon, a tactical hypersonic anti-ship missile, is scheduled to enter service in 2022 with a top speed of Mach 9 and a range of 1,000 kilometers. In October 2019, China's 70th National Day parade featured the Dongfeng-17 hypersonic missile, with a top speed of Mach 10 and a range of 1,800–2,500 kilometers. France and Germany are also working on hypervelocity missile development. France has developed an ultra-high-speed missile development program (Flying Fish) with the goal of increasing anti-ship missile flight speeds to between Mach 6 and Mach 7.

The key technologies underlying hypersonic weapons and munitions are ballistic surprise defense for out-of-area launch, integrated design technology for ramjet/intake/ballistic, aerodynamic and thermal protection, highly dynamic and rapid response guidance control in large airspace, submunition/end-sensitive bomb technology, composite guidance technology, high-speed automatic sensing and decision-making technology, swarm network strike technology, and efficient hypersonic weapons and munitions.

10.3 Hypersonic Aircraft

Speed, stealth, range, and mobility are all critical characteristics of future aircraft designed for long-range reconnaissance, detection, and attack strategies.

10.3.1 X-51A Hypersonic Aircraft

Rapidity is critical to the US CPGS Program. Specifically, the aircraft developed under this program must exceed five times the speed of sound, with the X-51A hypersonic aircraft serving as the most representative example. Many years ago, Pentagon policymakers drew a lesson from the incident on August 20, 1998, in which the US Abraham Lincoln carrier battle group launched several Tomahawk cruise missiles against Taliban training camps in eastern Afghanistan in an attempt to eliminate Osama bin Laden. The Tomahawk cruise missile's top speed was 885 kilometers per hour, and it took the missiles up to 2 hours to complete the 1,770 kilometers range. Bin Laden, however, departed the training camp an hour before the missile arrived. The failure of this operation caused irreparable damage to the US Department of Defense and accelerated the development of hypersonic weapons. The X-51A is a continuation of the National Aero-Space Plane (NASP) Program and the X-43 program. The NASP Program's objective was to develop and validate a scramjet engine-powered X-30 demonstrator. The aircraft was to take off from a conventional runway, reach an entry speed of at least Mach 25 into space, enter Earth's low orbit in a single stage, fly into space, re-enter the atmosphere, and finally land on the runway. Although the NASP Program was enticing, it was overly ambitious and technically demanding. Finally, it was canceled in 1992.

Since then, NASA successfully flew the X-43A demonstrator in 2004, demonstrating that the scramjet engine could generate enough thrust to accelerate the vehicle. Following that, NASA shifted its investments in various aeronautical research programs to the space sector, and the X-43 program was halted. The US Air Force then continued research on hypersonic technologies and invested in the HyTech Program, which later became the HySet Program. The two new technical studies paved the way for the introduction of the X-51A.

The US Air Force invited Boeing and Pratt & Whitney to collaborate on the Scramjet Engine Demonstrator – Waverider (SED-WR) demonstrator aircraft in January 2004, with Boeing building the airframe and Pratt & Whitney producing the engine. The US Air Force officially designated the program as the X-51A in September 2005.

The X-51A program's primary objective was to conduct flight tests of the US Air Force's HyTech scramjet engine. This engine runs on heat-absorbing hydrocarbon fuel and can increase the vehicle's Mach number from 4.5 to 6.5. There are additional purposes for the program:

1. To collect the scramjet engine's ground and flight test data for developing computational tools for engineering design.
2. To validate the heat-absorbing scramjet engine's survivability in-flight conditions.
3. To determine whether the scramjet engine can produce sufficient thrust during free flight tests.

The X-51A program, which aims to complete propulsion tests, is not merely a demonstration of HyTech's scramjet technology. To integrate engine and vehicle technology, two research processes must be coordinated: hypersonic propulsion flight tests and aircraft R&D. The aircraft's R&D process was equally important, featuring the following.

10.3.1.1 Pneumatics

The X-51A "Waverider" featured a flat head, four deflectable winglets in the body's center, and an abdominal air intake. The aircraft is designed using an analytical or numerical solution of a three-dimensional supersonic flow with a surge system. Along the flow line, a cone with a pointed triangular shape is cut to form the shape of the aircraft. The sonic or hypersonic aircraft configuration has attached excitation waves at all of its leading edges. This design contributes to the X-51 engine's combustion improvement by providing the following benefits:

1. The upper surface of the wave multiplier configuration is parallel to free flow. It has low pressure, whereas the lower surface has high pressure, similar to a conventional profile with the same designed Mach number.

2. Following excitation, the high pressure on the lower surface does not leak up the leading edge to the upper surface. There is no pressure communication between the high pressure after the wave and the low pressure on the upper surface, resulting in a high lift-to-drag ratio for the waveriding configuration in comparison to the common shape.

3. After being compressed by the surge, the incoming flow along the compression surface is confined within the leading edge of the surge, resulting in a more uniform flow field along the lower surface, which can eliminate lateral flow at the engine inlet, thus also increasing the aspirated engine's intake efficiency, and making this configuration convenient for integrated cartridge/engine/intake design.

4. Because there is no pressure communication between the upper and lower surfaces of the vehicle, there is no interference between the flow fields on the upper and lower surfaces, and the upper and lower surfaces can be treated independently, significantly simplifying the aircraft's preliminary design and calculation process. Combined with the scramjet engine, the integrated wave multiplier configuration can result in a hypersonic wave multiplier with superior performance.

10.3.1.2 Heat Resistance

To support hypersonic flight and direct re-entry from space into the atmosphere, the aircraft's surface must withstand temperatures of up to 4,500°C. To accomplish this, the demonstrator's body was coated with a layer of heat-resistant ablative material, and its belly was lined with the same thermal insulation tiles used on space shuttles.

10.3.1.3 Supercombustion Stamping

The X-51A's scramjet engine is an efficient aspirated one. It has a greater range and payload capacity than rocket engines. Since this engine uses oxygen from the surrounding air to maintain propulsion, it does not require the same amount of fuel and oxidizer as rocket engines, which have a massive launch mass. Due to the low number of moving parts in scramjet engines, they operate as reliably as turbine engines even in the most demanding operational environments.

10.3.1.4 Ignition Technology

The ignition of the X-51A scramjet is complicated during flight. To begin, compressed air entering the inlet is isolated, and the airflow is adjusted to a stable pressure suitable for the combustion chamber's functional requirements. The airflow is then mixed with atomized JP-7 jet fuel for ignition and combustion. Because JP-7 fuel does not self-ignite at Mach 4 or higher, it must be blended with ethylene liquid. The ignition process begins with the ignition of a small amount of easily combustible ethylene contained on board, and the fuel is burned by injecting the ethylene into the combustion chamber and mixing it with JP-7 fuel.

10.3.1.5   Fuel Technology

The X-51A demonstrator runs on hydrocarbon fuel, the same JP-7 aviation fuel used in the SR-71 Blackbird's J58 turbojet engine. This hydrocarbon fuel is a ready-to-use fuel that is non-flammable, non-volatile, and easily stored. According to the X-51A program manager, hydrocarbon fuel must absorb a certain amount of heat from the engine structure before it can flow through a heat exchanger to cool the structure supplying the combustion chamber. This heat exchanger is a groove machined directly into the engine case wall and can cool the combustion chamber to temperatures exceeding 1,650°C. The heat exchanger converts the pre-treated fuel into a hot gas, containing more than 10% more energy than the liquid fuel.

The X-51A's waverider technology is a novel flight mechanism that is quite unlike that of conventional aircraft. It is particularly well suited for hypersonic flight at the edge of the atmosphere and possesses unfathomable military utility. A suborbital hypersonic vehicle's flight path is as unpredictable as that of an airplane, and it will not circle the Earth in a fixed orbit repeatedly. Interception time is limited, and interceptions are difficult. However, suborbital hypersonic vehicles such as the X-51A have several drawbacks: they cannot be detached from the parent aircraft, the operating window of a scramjet engine is extremely narrow, and any error could render supercombustion unsustainable.

In May 2010, during flight tests, the US X-51A set a record for the longest duration of scramjet engine operation at 143 seconds and a maximum flight speed of Mach 4.87. In May 2013, the US military completed a 240-second Mach 5.1 flight test, drawing lessons from two previous failures in March 2011 and August 2012. According to RAND, this technological advancement is on a par with the aviation industry's transition from propeller to jet. The *Christian Monitor* compared the X-51A's flight to that of "Superman" and the technical difficulty of its scramjet to lighting a match in a hurricane and keeping the flame alive. "We equate this leap in engine technology as equivalent to the post-World War II jump from propeller-driven aircraft to jet engines," said Air Force program manager Charlie Brink [1].

The X-51A aircraft has three major advantages over cruise missiles.

To begin, the X-51A is highly responsive. While it takes an hour for a subsonic cruise missile to reach a 1,000 kilometer target, the X-51A does so in less than 10 minutes.

Second, it possesses strong penetration capabilities against defense systems. An existing cruise missile relies heavily on ultra-low-altitude flight and stealth technology to penetrate an adversary's defenses. Because of its slow speed, the missile is easily intercepted following exposure. However, existing air defense weapons are incapable of accommodating the X-51A flying at a high altitude.

Third, it possesses incredible destructive capability. With incredible kinetic energy, the X-51A has a penetration range of more than 10 meters against steel and concrete strike targets. It is particularly well suited for striking deep-underground command centers and other solid targets.

10.3.2 SR-72 Hypersonic Reconnaissance and Strike (R/S) UAVs

On May 19, 2015, *Popular Science* magazine revealed Lockheed Martin's SR-72 hypersonic R/S UAV (Figure 10.1) [2]. According to the report, the SR-72 will be an unmanned aircraft capable of performing reconnaissance missions at a top speed of 4,000 mph (6,436 km/h), with strike capabilities. The aircraft, conducting R/S from an altitude of 80,000 feet (about 24.4 kilometers), will avoid attacks by flying at speeds of up to Mach 6. This is twice the speed of the retired SR-71 high-speed strategic reconnaissance aircraft. According to Lockheed Martin Space Systems Company, the aircraft will be deployed in 2030 and will be capable of reaching any region of the world in less than an hour.

For several years, aeronautical engineers from Lockheed Martin Space Systems Company and Aerojet Rocketdyne Company have been designing the SR-72 at the Skunk Works. First, the aircraft would require a hybrid propulsion system: a conventional spot turbojet engine would propel the aircraft to Mach 3, while a ramjet and scramjet engine (a dual-mode engine) would provide the remaining acceleration. To bridge the gap between the maximum flight speed of turbojet engines and the lower speed limit for dual-mode engine start-up, engineers developed the dual-mode or hybrid engine operating in three modes: after accelerating the aircraft to Mach 3 using (1) turbojet power, the dual-mode engine continues to accelerate the aircraft to Mach 5 using (2) ramjet power before switching to (3) scramjet power.

Figure 10.1 SR-72 Hypersonic reconnaissance and strike unmanned aerial vehicle (R/S UAV).

During the hypersonic flight, the aircraft's airframe structure will be exposed to extremely high levels of aerodynamic heating, which conventional steel cannot withstand. When flying faster than Mach 5, aerodynamic frictional heating causes the aircraft's surface temperature to rise to 2,000°F (1,093°C). Structure steel bodies will melt at this temperature. As a result, engineers considered composite materials resembling the high-performance carbon fiber, ceramic, and metal composites used on the front ends of intercontinental ballistic missiles and space shuttles. Additionally, any connections must be sealed to prevent air leaks at hypersonic speeds, which could cause the aircraft to disintegrate and result in accidents similar to the Columbia space shuttle disaster.

Aerodynamic properties are an issue. The amount of stress placed on an aircraft varies according to flight speed. When an aircraft accelerates in the subsonic segment, the aircraft's center of lift shifts back. Once the aircraft reaches the hypersonic segment, the center of the lift shifts forward again due to the leading edge's drag. If the lift's center is too close to the center of gravity, instability will result. The vehicle's shape must be optimized to account for these changes and avoid disruptions.

Payload adds another layer of complexity. The SR-72's specific mission load is still classified, and, likely, the associated load has not been designed yet. It takes extraordinary engineering to conduct image reconnaissance or bomb-dropping missions at Mach 6. To complete the turn, the aircraft must sail hundreds of miles and rely on a powerful guidance computer to establish a line of sight from the flight altitude to the target. Opening the weapon bay at Mach 6 is a significant technical challenge. As a result, the SR-72 will require upgraded sensors and weapons operating at this high speed.

10.3.3 Bomber B-3

The US Air Force is developing the Bomber B-3, a hypersonic, long-range strategic bomber. According to *Popular Science* magazine [3], the Air Combat Command, which is in charge of the long-range bomber, has proposed a new set of stealthy strategic bomber concepts, including the F-117, B-2, and B-3, to meet the US Air Force's operational requirements in the year 2030.

According to reports, the Bomber B-3 is powered by a ramjet or scramjet engine that enables it to fly at five times the speed of sound, or even faster, across the Atlantic Ocean in less than an hour and bomb any location on Earth in a matter of hours.

In terms of the Bomber B-3's historical context [4], *American Aviation Week* reported in early 2009 that the B-2 stealth bomber had a long range and a large weapon-carrying capacity, but it could only operate at night, and its unit price of up to US$2 billion was prohibitively high. The B-52 bomber had a large payload and a long range, but it was extremely vulnerable to air defense systems, limiting it to long-range attacks. Despite being a supersonic bomber, the B-1B lacks stealth and long-range weapons. While F-22s can approach and destroy heavily protected ground targets, they are limited to carrying two medium-sized bombs and cannot conduct long-range attacks without aerial refueling.

US Air Force officials announced the lack of a bomber for long-range attacks, carrying a large payload, surviving in dangerous environments, and operating effectively around the clock regardless of weather conditions. As a result of this warfare gap, the Bomber B-3 was developed.

The idea for the US military to develop the Bomber B-3 came around 1995, just before the US Air Force began secretly designing the aircraft at Wright-Patterson Air Force Base in Ohio.

On September 16, 2009, while attending the annual meeting of the US Air Force Association in Maryland, US Secretary of Defense Robert Gates expressed support for the US Air Force developing new long-range (strategic) bombers to improve long-range strike capability. The US military's top brass had been indecisive about developing the B-3. During the Air Force Association's annual meeting, Gates noted that China had made significant investments in cyber, anti-satellite warfare, naval defense, air defense, and even missile manufacturing. The US must accelerate its pace of strategic consolidation in the Pacific.

Although Gates did not reveal the new bomber's model, it is widely believed to be called the B-3. According to the US Air Force's plan, the bomber was built as a prototype in 2018 and deployed shortly after extensive testing. After 2040, the B-3 will thus become the US Air Force's "strategic bastion."

US military aircraft such as the F-117, B-1, and B-2 are already stealthy. The Bomber B-3's shape and structural design take stealth into account completely. The B-3, as the US Air Force's future primary bomber, will feature extensive stealth capabilities for radar, infrared, and visual detection.

Although the Bomber B-3's external surface area is reduced, its integrated fuselage provides additional internal space, allowing it to carry a heavier bomb load. The B-52 heavy bomber is capable of carrying a maximum of 27 tons of bombs. The B-2's maximum bomb load is also 22 tons. The B-3's internal space is greater than that of the B-2, and with a practical design, the B-3's bomb load can equal or exceed that of the B-52, making it the best of the heavy bombers. The Bomber B-3 can still carpet bomb like the B-52 thanks to the Rapid Bomb Delivery System technology, despite dropping bombs from its tail.

Following the Cold War, the US Air Force deployed strategic bombers aggressively overseas, with Guam being the preferred location. The US Air Force believed that strategic bombers launched from Guam could fly over the Taiwan Strait in less than 3 hours, sufficient time to respond to emergencies throughout the Asia-Pacific region.

The US already possessed the world's most advanced and largest fleet of bombers. According to a 2013 analysis of the US Air Force's bomber fleet [5], the US military owned enough B-2 stealth bombers, B-1Bs, and B-52 strategic bombers to launch lethal strikes against targets in any country. However, in the fiscal year 2013 military budget request, the Pentagon included a special fund to develop cutting-edge stealth long-range bombers, demonstrating that the US Air Force must prioritize developing stealth bombers capable of launching long-range strikes. According to internal US military analysis, the majority of current bombers were built in the 1970s and 1980s. As a result, adapting these bombers to future battlefield requirements will be difficult in the face of China, Russia, and other countries with a robust air defense system. The current US bombers will almost certainly be ineffective. As a result, developing next-generation bombers has become a critical task for the US to maintain military superiority in the future.

10.4 Directed Energy Weapons and Electromagnetic Cannons

Lasers, microwaves, and particle beams are the most common directed energy weapons. After years of R&D, laser weapons, high-power microwave weapons, EMP bombs, and other weapons technologies have entered the sprint stage. Relevant technological advances that result in an operational capability will significantly enhance attack and defense capabilities and serve as a better alternative to in-service equipment and operational means; this is critical for future intelligent and new types of operation.

10.4.1 High-Power Microwave Weapons

High-power microwave weapons, also known as radiofrequency (RF) weapons, use pulsed or continuous waves of high-energy microwave beams to kill or temporarily disable personnel or electronic equipment. Microwave weapons with high power offer the advantages of rapid engagement, high precision, controllable killing effect, and low cost per launch, all without regard for magazine capacity. With near-all-weather capability, its beam is wider than laser weapons, which can simultaneously kill multiple targets and benefit from anti-electronic equipment, earning it recognition. Microwave weapons with a high-power output include vehicle antenna microwave weapons, artillery microwave bombs, and cruise microwave missiles.

Microwave weapon research began in the 1970s, and the microwave weapons that have been developed are primarily soft-killing high-power microwave weapons. The US has developed the weapons in successive generations, including the microwave bomb active rejection system, RF vehicle/ship brake, and the Vigilant Eagle anti-missile defense system. Simultaneously, microwave sources with high power and conversion efficiency should be developed to enhance microwave weapons' killing ability. The US Army is concentrating its efforts on developing EMP artillery shells and multi-purpose EMP operation units that can be used to short-circuit the electronics in cell phones, radios, global positioning system (GPS) jammers, computers, and even vehicle electrical ignition systems, effectively disabling all types of tactical information systems. The US began research on anti-electronic high-powered microwave missiles in 2009. According to US media reports, Boeing completed the first 1-hour flight test of an "anti-electronic device high-power microwave advanced missile", or EMP cruise missile, in December 2012 in Utah. During the test, the missile attacked seven predetermined targets and had a noticeable effect. Even the camera used to record the process was rendered immobile, indicating significant technological advancements in device miniaturization, beam precision control, pulse power efficiency, and other areas. According to some experts, this technology has ushered in a "revolutionary era of modern warfare". Completed in 2015, the electronic pulse cruise missiles were delivered in small numbers to the US Global Strike Command for electronic confrontation and cyber warfare. In 2016, reports indicated that these cruise missiles would soon be fitted to US Air Force bombers. Russia has been developing microwave anti-aircraft and anti-missile weaponry. European countries such as Germany, France, and Israel concentrate their efforts on counterterrorism technology based on miniaturized microwave weapons. The UK is using UAVs to investigate the development of airborne high-power microwave weapons technology.

10.4.2 Solid-State Laser Tactical Weapons

Tactical high-energy laser weaponry is a type of weaponry that employs high-energy/high-power laser beams to make direct attacks on targets. As a medium/near-range offensive and defensive weapon technology, it employs thermal, impact, radiation, and other damage mechanisms to inflict soft or hard kills on targets. It applies to a wide variety of vehicles on land, sea, and air. Laser weapons with a maximum output of 300 kW will be used in area defense or battle group operations. High-energy tactical laser weapons offer the advantages of rapid engagement, high accuracy, controllable killing effect, and low cost per launch and are not limited by magazine capacity, allowing them to better meet the requirements of a variety of future combat missions.

Laser weapons research began in the 1960s on an international level. Around 2010, a breakthrough in slatted solid-state laser and fiber laser technology accelerated the application of laser weapons, with various countries actively testing the weapons on various vehicles. Currently, the power output of high-energy tactical laser weapons ranges between 30 and 60 kW, with a range of several to tens of kilometers, and the output is increasing to more than 100 kW as technology advances. The US Army, Navy, and Air Force have been conducting field trials of a general-purpose 150 kW laser weapon, while shipboard, vehicle-mounted, airborne, and portable variants are also being actively tested. Germany, Israel, and other countries are developing high-energy tactical laser weapons mounted on vehicles and ships. Figure 10.2 presents the laser weapon system at the 2021 Zhuhai Airshow.

The key technologies used in solid-state tactical laser weapons include a high-power pulse power supply, high-precision target tracking, a pump laser,

Figure 10.2 Laser weapon system (2021 Zhuhai Airshow).

a combined long- and short-pulse laser, a long-/short-pulse laser damage mechanism, and compensation for atmospheric optical effects.

The US Joint Directed Energy Conversion Office began operations in 2018. The office's predecessor, the High Energy Laser Joint Technology Office, was primarily responsible for coordinating and managing the US Department of Defense's research and development of high-energy laser technologies. After reorganizing as a joint directed energy conversion office, high-power microwave technology was given official oversight, with a greater emphasis on transformation and application, as well as efforts to advance directed energy technology from laboratories and test sites to operational deployment.

For decades, the US has invested in directed energy weapons, developing various prototypes that are currently deployed under tests for counter-unmanned aircraft systems (C-UAS), rocket-propelled grenades, and mortar shells.

10.4.3 Electromagnetic Cannons

An electromagnetic cannon is defined by a high muzzle velocity, a high muzzle kinetic energy, a long range, and a short flight time for the projectile, with a muzzle velocity of more than 3 km/s. In 2010, the US Navy conducted a test firing of an electromagnetic cannon equipped with a 10.43 kg projectile, 33 MJ of muzzle kinetic energy, 2,500 m/s muzzle velocity, and a range of nearly 200 kilometers. After 2012, a series of reduced-power level life tests was conducted to verify the pulsed power supplies, rail materials, and weapon systems' reliability. This futuristic electromagnetic cannon will be used for air defense, anti-missile defense, and fire suppression, as well as long-range precision strikes against ground forces.

An electromagnetic rail cannon accelerates projectile launch by utilizing strong current induction created by the rapid discharge of a large-capacity, high-power pulse power supply. This technology will enable projectile launches to enter the era of electric propulsion from chemical energy, which is critical for leapfrogging conventional weapons and equipment development.

The development of electromagnetic rail cannons began in the 1980s. The US Navy launched the electromagnetic rail cannon Innovative Naval Prototype Project in 2005, with the first phase beginning in 2006. Two 32 MJ laboratory single-shot electromagnetic rail cannon prototypes were successfully fired in 2012. It entered the second phase of development in 2013. The second development phase was completed in 2017, resulting in a technical maturity level of 5–6 for the engineering prototype gun.

In 2016, the US Army successfully demonstrated General Atomics' 10-MJ Lightning electromagnetic rail cannon system. The US Army also researched the XM1 electromagnetic rail cannon and the full-cantilever electromagnetic rail cannon (i.e., 120 mm electromagnetic mortar).

After years of theoretical and technological accumulation, China has achieved results in electromagnetic rail cannon research and is currently developing with varying muzzle kinetic energy. The electromagnetic cannon's critical technologies include miniaturization of the power supply, guided munition adaptation and high overload, launcher design, armature design, and launch control.

10.5 Intelligent Hypersonic Attacks

In the age of AI, hypersonic attacks can be classified into two types: strategic and tactical. The majority of strategic hypersonic attacks are directed at discrete strategic targets behind adversaries to carry out large-depth attacks. It is necessary to establish a global network information support system, incorporate intelligent elements throughout the operation process, and significantly improve the overall level of single-vehicle and system operations.

First, it is necessary to conduct intelligent detection and perception of battlefield situations and targets, comprehend the nature of the strike territory and targets, and master precise geographic location, distribution, and environment. Second, it is necessary to foster intelligent cooperation between single vehicles and swarm operations in all-weather and whole-process conditions, as well as to develop protection measures and contingency capabilities such as stealth, anti-interference, orbit change maneuvering, and bait spreading to deal with various detection, interference, and interception means. Finally, it is critical to accurately assess the effect of hypersonic strategic strikes in order to create a closed loop around the operational process and strike link, thereby facilitating the execution of secondary strikes. Hypersonic strategic strikes are typically directed at fixed and swarm targets that are clearly marked and easily detected by defense layouts and reconnaissance. Thus, pre-war simulation training can be conducted using machine learning to accumulate data, continuously optimize models, and even validate them using live fire, all of which will contribute to the increasingly intelligent hypersonic vehicle AI or swarm AI.

Tactical hypersonic attacks must deal with complex and diverse targets and environments and cooperate closely with other tactical strike means while defending against each other's interception means. As uncertainty and unexpected situations exist, and it is impossible to fully grasp all information before a war through machine learning and simulation training, wartime learning should be conducted to accumulate wartime data and experience. Wartime uncertainty can be minimized by comprehending the adversary's information, mastering the adversary's equipment details and tactical applications, and conducting extensive simulation training and machine learning.

Technically, strategic and tactical hypersonic attacks impose more stringent requirements on hypersonic weapons' power, control, guidance, and target recognition systems. Hypersonic attacks will face technical challenges such as thermal vehicle protection, target detection response, precise control of rapid orbit change maneuvers, and multi-bomb network cooperation. It is necessary to address heat resistance, response time, overload characteristics, material and device reliability and stability, communication, command and control, and data processing under extreme conditions.

10.6 Intelligent Hypersonic Defense

At the moment, the development and employment of hypersonic weapons have become a priority for the world's leading military powers. Its counterpart, the capability to intercept hypersonic weapons, will also become a strategic emphasis for military counterbalance. While some armed forces' existing air defense and anti-missile capabilities initially achieved information technology detection and automated interception, they cannot respond to future global interception missions requiring rapid strikes within 1 hour. With the advancement and competition of hypersonic attack and defense systems, the speed of hypersonic weapons will increase; maneuverability, bait technology, and defense penetration capabilities will improve; and the requirements for defense systems' detection, rapid response, coordination, interception, and damage capability will rise. Among them, resolving the problems of hypersonic target detection, tracking, rapid response, intersection, and destruction, as well as their mutual adaptability, is the central issue for improving intelligent defense.

Based on future hypersonic missiles and aircraft characteristics, the fundamental theory and system framework for anti-hypersonic targets should be developed, including target detection, command and control, interception means, and support measures. New technologies, methods, approaches, and measures for anti-hypersonic targets should be studied systematically and comprehensively.

The military should be concerned with possible types, trends, and characteristics of hypersonic weapons in the future to clarify the limitations of conventional interception means, investigate novel means of weapon detection, seek out innovations in interception and verification methods, and enhance comprehensive interception capability. Research into the target and damage characteristics of hypersonic weapons should be bolstered. Additionally, it is necessary to research novel interception mechanisms, optimal interception capability, adaptability of detection equipment, and adaptability of interception equipment technology.

At the moment, there are virtually no defense systems or means for intercepting hypersonic orbital mobile weapons, i.e., the defense capabilities and means are severely limited. The following are possible future interception methods. To begin, technology capable of electronic interference and EMP damage can be adopted. However, it is also necessary to consider the plasma effect created by the collision of high-speed plasma between the air and operational vehicles or ammunition at an extremely fast speed (more than ten times the speed of sound), which may reduce the penetration efficiency of EMPs and thus weaken the attack effect. Second, defense systems should include a high-velocity barrage and an intelligent swarm. High relative speeds will cause devastating effects, similar to the effects caused by birds crashing civilian planes. Third, energetic materials and special chemicals should be rapidly dispersed in the air where high-speed weapons fly. These materials can be transformed into nanoparticles or dust, which combine with aerosols in the air. If even a trace of the materials enters the ramjet combustion chamber, an instantaneous explosion occurs.

To summarize, critical technologies for intelligent defense against hypersonic weapons include hypersonic weapon gridded detection network systems, long-range photoelectric air–space early warning, high-sensitivity infrared detection, high-power microwave weapons, EMP bombs, electromagnetic cannons and dense submunitions, swarm autonomous attack munitions, new-generation kinetic energy interceptors, and anti-hypersonic destruction testing and evaluation.

References

[1]   Waverider X-51A X-51A飞行器. (2021). Baidu Baike百度百科. https://baike. baidu.com/item/X-51A%E9%A3%9E%E8%A1%8C%E5%99%A8

[2]   Dillow, C. (2015). Inside America's next spyplane. *Popular Science*. www.popsci. com/inside-americas-next-spyplane/

[3] January 2012: The new science of stealth. (2012). *Popular Science*. www.popsci. com/announcements/article/2011-12/jaunary-2012-new-science-stealth/

[4] B-3 HongZhaJi B-3轰炸机. (2020). Baidu Baike 百度百科. https://baike.baidu. com/item/B-3%E8%BD%B0%E7%82%B8%E6%9C%BA

[5] Theohary, C. A., Kapp, L., Burrelli, D. F., & Jansen, D. J. (2013, January). *FY2013 National Defense Authorization Act: Selected military personnel policy issues*. Washington, DC: Congressional Research Service Careers.

11 Joint All-Domain and Cross-Domain Operations

With advancements in advanced network information systems, multidisciplinary integration, cross-media attack and defense, and other critical technology groups, cross-services operations, cross-geographic mobile operations, cross-functional operations, and joint all-domain operations will become inevitable. Joint all-domain and cross-domain operations encompass physical domains such as land, sea, air, space, and virtual spaces, as well as cognitive domains such as cyber attack and defense, electronic confrontations, intelligence warfare, psychological warfare, and opinion and propaganda warfare. A critical component of intelligent warfare is manifested in the use of cloud computing, big data, networking communications, advanced geographic information, spatio-temporal networks, and cross-domain interoperability to develop cross-functional domain combat capabilities in the physical, information, cognitive, and social domains, as well as joint all-domain operation capabilities for the land, sea, air, space, electrical, and network military services.

11.1 Battlefield Expansion and Force Synergy

The advancements of science and technology, the practice of war, and the revolutions in military affairs (RMA) have resulted in a continuous expansion of the operational space and field. Armed forces have become increasingly complex in composition. Significant changes have occurred in the way operations are coordinated. Operational space is constrained in primitive societies to small areas such as crowds and tribes. In this context, hand-to-hand confrontations are the main operational pattern. With the advancement of smelting and the availability of iron and bronze, warfare in the Cold War era increasingly manifested itself as wars between neighboring countries. The combination of men, horses, and cold weapons, as well as planar group synergy and large-scale contact operations, reflected its operational pattern. The development of gunpowder and firearms significantly increased the technological content of warfare during the hot-weapon era. Linearity, firepower synergy, and close non-contact were all characteristics of the operational pattern. This accelerated the Western powers' conquest and plunder of the rest of the

DOI: 10.4324/b22974-13

world. Land, sea, and mechanized aerial vehicles have emerged since the rapid development of steam engines, internal combustion engines, and electrical technology in the nineteenth century. The operational space expands beyond neighboring countries to include intercontinental warfare across seas, oceans, and borders. Operational patterns are becoming more three-dimensional: more mobile, cooperative, and contractual in nature. Since the mid-twentieth century, as computers, communication, and network technologies advanced, the operational domain expanded from the physical space to information and cyberspace, enabling system confrontation, information attack and defense, and integrated joint and precision operation synergy. With the advancement of intelligent technology and its widespread and in-depth application in the military field, operational space will expand beyond the physical and information domains and into the social, cognitive, and biological domains in the future. Thus, the integration of multiple domains, cross-domain attack and defense, virtual–real interaction, manned–unmanned cooperation, human–machine integration, swarm operations, and other aspects of force synergy is demonstrated. It is more evident at the strategic and campaign levels in the synergy of the military services' forces in joint all-domain operations, multi-domain operations, and cross-domain attack and defense.

Based on cross-media technology and equipment in the future, various land and sea, air and sea, submarine, anti-submarine, drilling, cross-atmosphere, and other means of cross-geographic domain operations will become increasingly enriched. As the tactical level of cross-service multi-domain cross-operations, integrated operations, and joint operations becomes increasingly enriched, they will gradually become the norm (Figure 11.1).

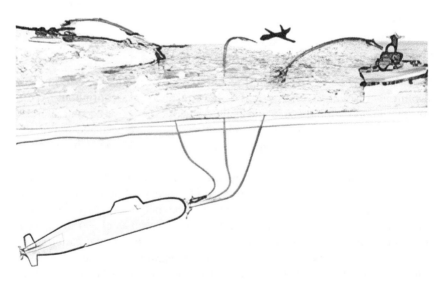

Figure 11.1 Submarine-launched air defense, anti-ship, and land strike weapon concept.

The find-and-destroy capability will be normalized without the need for commanders. Cross-functional domain operations, i.e., cross-physical, information, cognitive, social, and biological domain operations, will become a new growth point for future military capabilities. For example, a military power has conducted new research to strike targets using global positioning system (GPS) data, exploiting the software and hardware vulnerabilities of GPS receivers, forging ephemeris, time, location, and other information, and launching remote attacks via radiofrequency, posing significant hidden danger and harm to financial and electric power information systems that use GPS time reference.

The US Department of Defense has concentrated on joint operations, cross-domain operations, and force projection, sustaining military superiority and freedom of action on a global scale. The US Army and Air Force developed the concept of "air–land operation" in the late 1970s and began implementing the concept in the early 1980s to enable the North Atlantic Treaty Organization (NATO) to respond to Soviet and Warsaw Pact threats – a strategic, effective response and resistance on the European Central Front. In practice, the US military adopted the "air–land operation" doctrine in the Battle of Panama and the Gulf War.

In February 2010, US Secretary of Defense Robert Gates issued a revised version of the Quadrennial Defense Assessment Report, formally defining "air–land operations" as a joint operational concept and authorizing the US Air Force and Navy to conduct additional research in this area. In the fall of 2012, the deputy secretaries of the four primary US military services reached an agreement outlining a framework for conducting air–land operations. Under the vital impetus of the US government and military, air–land operations have evolved from an idea to a systematic theory in a matter of years and have been gradually integrated into US defense policy and Army operation regulations.

The Washington Center for Strategic and Budgetary Assessments and several other think tanks have publicly released several scenarios for air–sea warfare [1]. Here are some interpretations of the publications, which include the following key points:

1. The US Air Force's anti-space operations are intended to blind countries' space-based maritime surveillance systems and protect its operations from anti-ship ballistic missile targeting, but the reason for developing the X-37B air-to-space test aircraft remains unknown.
2. In January 2011, the Joint Star aircraft completed verification of the network weapon framework that would allow the US to track sailing ships via the Joint Star aircraft and attack them with AGM-154C glide bombs launched by F/A-18E/F fighters.
3. The US Navy's Aegis ships can destroy incoming missiles as part of the ballistic missile defense system, providing forward protection for US Air Force forward bases.

4. Long-range strikes can destroy ground-based long-range surveillance systems (e.g., over-the-horizon radar) and missiles that pose a threat to US ships and overseas bases, while US submarine-launched missiles can destroy an adversary's integrated air defense system, paving the way for an Air Force strike.

The air–sea operation is a joint operational pattern developed by the United States for high-end military operations in the Western Pacific Theater. It aims to maintain the US military's ability to successfully project military power into the Western Pacific, maintain military superiority in the region, and seize and maintain the initiative in the region's air, sea, space, and cyberspace domains, with the explicit goal of monitoring China.

The concept of "distributed killing" was initially proposed to address the US Navy's lack of anti-ship capabilities. With the widespread use of disruptive technologies such as big data, cloud computing, and artificial intelligence (AI) in the military, its connotation has gradually expanded to include the concept of multi-dimensional integrated warfare based on operation clouds and derived from the Third Offset Strategy's critical concept. In 2014, in response to a shortage of anti-ship capabilities, the US Naval War College proposed distributed killing operations based on military chess exercises pitting littoral combat ship formations against land and sea targets. In January 2015, a US Navy ship unit senior commander published a thesis titled "Distributed lethality" in the *United States Naval Institute Proceedings* [2]. Since then, Vice Admiral Rowden, the commander of the US Navy's surface forces, has continued to promote this concept through public speeches and online media, steadily expanding its influence and connotation. Lieutenant General David A. Deptula, a former US Air Force First Deputy Chief of Staff, pioneered the "operation cloud" concept in 2012. Through the use of a highly decentralized, self-evolving, and self-compensating shared-information network system, multiple servers can rapidly exchange data from sensors and shooters in various fields via command, control, intelligence, surveillance, and reconnaissance networks, as well as integrate the forces of various operation systems, thus increasing efficiency and achieving economies of scale. In a nutshell, an operation cloud is a cloud-based network that enables extensive data sharing among multi-dimensional operational vehicles on land, sea, air, and space. The objective is to improve the effectiveness of the operation system of systems (OSoS). The concept of an operation cloud has gradually become an integral part of the US Army's operational design, equipment development, and operational verification processes, with the support of the US Department of Defense.

Aviation Week published the operation cloud concept map in August 2014, describing "the development of an air–space cloud of air superiority for the operational side" [3] with multi-domain operation units that include in-orbit space reconnaissance/communications/navigation satellites, air early-warning aircraft, F-15/16 and other fourth-generation aircraft, surface carrier battle groups, F-22/35 stealth fighters that penetrate enemy air–space, and RQ-180 unmanned reconnaissance aircraft. On September 23, 2014, the US Air Force's F-22 Raptor fighter jets led the first joint airstrike group in conducting airstrikes against Islamic State extremist targets in Syria, kicking off the US Air Force's "cloud operation" verification process.

In summary, the US military's proposed concepts of "air–sea integrated warfare," "distributed killing," and "operation cloud" are all essentially multi-domain and cross-domain operations.

11.2 Multi-Domain Warfare

On November 11, 2016, the US Army formally incorporated multi-domain warfare into a revised version of their operational combat order, with the goal of encouraging the Army's expansion beyond land to other domains such as sea, air, space, cyber, and electromagnetic space, thus transforming the US Army's traditional "assisted" role. The US Army proposed the concept of multi-domain warfare as part of its Third Offset Strategy to address potential "anti-access/area denial" challenges and multi-domain threats and investigate ways to coordinate ground operations and deal with "advanced and equivalent enemies". From a military doctrine development standpoint, the US Army's multi-domain warfare concept is rational and necessary, and the innovation is significant, which is why it has garnered widespread support from the US Navy, Air Force, Department of Defense, and theater commanders. This is a significant shift in the US Army's and the US military's approach to joint operations, and it will have a profound effect on the development of the US military's operational capabilities.

11.2.1 Motives

The US military is confronted with threats across multiple domains, and its technological advantages are eroding. To maintain its power, the US military must develop the concept of cross-service and cross-domain joint operations. The US military believes that future operations will be markedly different than previous and current operations. In the future, US forces may face well-matched adversaries with superior situational awareness and lethal precision-guided weapons capable of limiting US joint forces' mobility and movement. Adversaries could undermine the US military's air and sea dominance by limiting the use of space, cyberspace, and the electromagnetic spectrum. They could breach the US military's potentially crippling defenses, rendering US joint forces incapable of maintaining a sustained advantage in any area.

From a joint operations perspective, the US Army proposed the concept of multi-domain warfare as a means of elevating the Army's status and role in future joint OSoS and developing new operational and equipment systems. Since 2009, the US Air Force and Navy have promoted the concept of "air–sea operation", but the Army's role is limited in this concept, and Army equipment lags behind that of the Air Force and Navy. To avoid marginalization, the US Army introduced the concept of "strategic ground power" in collaboration with the Marine Corps and Special Operations Command. This concept emphasized the Army's critical role but excluded other services. Following that, the US Army stressed that multi-domain warfare had created a framework within which all military services should collaborate to solve common problems. As a result, the concept of multi-domain warfare has piqued the interest of all branches of the armed forces.

11.2.2 Concept

Although multi-domain warfare was formally introduced in October 2016 as a novel concept, it has a long history. The RAND Corporation conducted a study, "Employing land-based anti-ship missiles in the western Pacific", in 2013 with funding from the US Army [4]. The study concluded that deploying land-based anti-ship missiles in the Western Pacific could effectively constrain adversaries' mobility and military operations. The US Army incorporated RAND's recommendations into its 2014 Army Operation Concept.

The US Army and Marine Corps jointly released a white paper titled "Multi-domain battle: Combined arms for the 21st century" on February 24, 2017 [5]. It explains the context, necessity, and concrete strategy for multi-domain warfare. The US Army and Marine Corps must defeat an adversary physically and cognitively by expanding synthetic arms in all fields. The future US Army and Marine Corps will integrate and coordinate to provide an advantage in various fields, including seizing and maintaining superiority, defeating adversaries, and achieving military objectives.

Since 2017, the US Army in the Pacific Theater has incorporated multi-domain warfare into military exercises. The US Army in the European Theater has done so in the fiscal year 2018.

11.2.3 Mechanisms

The US Army's primary demand for integration into the multi-domain warfare concept is to develop a flexible and responsive ground formation capable of delivering combat forces to diverse battlefields, seizing positions with comparative advantages, controlling critical terrain to consolidate results, ensuring joint forces' freedom of movement, and physically and cognitively defeating high-end adversaries.

The air–land operation focuses on the use of air power to augment the capability of ground forces, while ground forces are supported by air forces. Ground forces can also assist air and naval operations in multi-domain warfare. Air power is used to support ground forces in US military operations, and air support is available when necessary. The ground forces will be equipped with new armored vehicles, long-range missiles, networks, and electronic warfare equipment to perform Air Force and Navy functions. The Army's self-protection capability will be enhanced, allowing it to attack the enemy on the ground and support sea and air operations. Mark Milley, the US Army's chief of staff, stated that ground forces must breach the exclusion zone and facilitate air and naval operations. This contrasts with the previous 70 years' practice. Historically, the Air Force and Navy assisted the ground forces. "The army will sink warships, defend against enemy air and missile attacks, and exercise superiority over our troops in the air and space," he added [6].

In terms of operational methods, air–sea operations refer to operations conducted "from the outside, gradually infiltrating the inside". The US military deploys as far away from an adversary as possible. The US military begins by destroying an adversary's most remote system, then advances closer and destroys an adversary's relatively remote system, and so on. The weakness of this method is that it advances too slowly, causing significant losses to allies trapped in an opponent's "anti-access / area denial" circle. US forces can "advance from the inside out" in multi-domain warfare to cripple adversaries' strategic facilities. Anti-access / area denial systems are complex systems. The US military can exploit the system's inherent flaws by first creating a hole, then cracking it wider to paralyze an adversary from the inside out.

11.2.4 Characteristics

Multi-domain warfare is a concept that begins with expanding operational domains beyond the land to the air, sea, space, cyberspace, and the electromagnetic spectrum. It focuses on new domains and intangible countermeasures, such as the information environment and cognitive domain. The term "multi-domain" encompasses all domains, such as physical, informational, and cognitive. Future conflicts will erupt in each of these domains. To survive and maintain superiority, US forces must maximize their use of all domains and employ a variety of weapons, including firepower, electromagnetic, cyber, and psychological strikes.

The multi-domain warfare concept is innovative in that it emphasizes the breakdown of barriers between military services and domains, the integration of disparate force elements, and the consideration of all operational spaces, from submarines to satellites, tanks to aircraft, destroyers to unmanned aerial vehicles (UAVs), and combatants to cyber hackers. All of these should be connected seamlessly to achieve synchronized cross-domain firepower and full-domain mobility to seize the physical, cognitive, and temporal initiative.

Multi-domain warfare embodies several joint operating principles. The first principle is simultaneous operations, which entail attacking simultaneously in multiple locations and areas of the land, sea, air, space, and network, inflicting multiple blows on the adversary and physically and psychologically overwhelming them. Second, in-depth action, i.e., attacking enemy reserves, striking command-and-control nodes, and crippling logistical supplies to make it difficult for the enemy to recover. Third, sustained action, i.e., conducting continuous operations and denying the enemy breathing space.

Fourth, concerted action, i.e., the performance of multiple related and mutually supportive tasks in multiple locations, maximizing operational effectiveness at critical times and locations. Fifth, adaptability, i.e., the ability to utilize a variety of capabilities, formations, and equipment. By operating in concert across multiple domains and depths, one can create multiple dilemmas for adversaries, constrain their mobility, reduce their flexibility and endurance, and disrupt their planning and coordination, thus ensuring the joint force's mobility and effectiveness across domains.

11.2.5 Measures

Following the proposal of a multi-domain warfare concept, the US Army has been actively developing new capabilities, such as organizational restructuring, ordinance development, combatant training, and equipment system development, to complement the concept and enhance the US Army's overall combat effectiveness.

In March 2017, during the Association of the United States Army's Annual Multi-Domain Warfare Global Military Power Symposium, Commander Perkins of the US Army Training and Ordinance Command proposed that the US Army prioritize eight key capabilities for future operations: cross-domain firepower, combat vehicles, expeditionary mission command, advanced defense, cyber and electromagnetic spectrum, and a future vertical takeoff and landing vehicle. These eight capabilities are critical in the development of the concept of multi-domain warfare.

Long-range firepower and precision ammunition are the primary means by which US Army executives counter their adversaries' "anti-access/area denial" capabilities. The long-range precision fire missile will replace the US Army's current army tactical missile system. In 2018, the US Army proposed developing a strategic artillery piece with a range of 1,600 kilometers and a hypersonic strategic firepower missile with a range of 2,250 kilometers. The US Army revealed to the world in May 2021 that the range of the land-based long-range hypersonic weapon under development would exceed 2,775 kilometers.

The US Pacific Command views multi-domain warfare as a joint operational concept and actively promotes it. Admiral Harry Harris, the Pacific Command commander, requested in a May 2017 seminar that the Navy, Air Force, Marine Corps, and other services incorporate this concept into the exercise and prepare for the Army sinking warships during the "Pacific Rim 2018" exercise.

"I'd like to see examples of when joint and coalition forces operate in other areas in a complex environment. Army ground forces sinking a ship, destroying a missile, and destroying the aircraft that launched the missile almost simultaneously," Harris explained [7].

Harris stated he would not allow the Navy to concentrate exclusively on deep-water operations, the Marines on beachheads, and the Army on inland operations, claiming, "We need to be highly collaborative, with no single service dominating and no single domain having fixed boundaries. The theater commander must be capable of striking the target from any domain to fight and win the war tonight."

To carry out this concept, the US Pacific Army established a "Multi-Domain Task Force," an operation force that specializes in air, sea, space, and cyberspace operations in addition to ground operations, and made them available to Harris as needed. The task force, and the US Army in general, must overcome five obstacles: hardware, software, connectivity, procedures, and training.

11.3 Joint All-Domain Operations

Since the US Army proposed multi-domain warfare, the US Air Force has proposed joint all-domain operations, backed by the US Department of Defense and the US Army, Navy, and Air Force. The US military has begun to transition away from multi-domain towards all-domain operations. Late in 2019, the US Joint Chiefs of Staff established a joint committee with the four military services to demonstrate a concept known as "joint all-domain operations". The Department of Defense elaborated on the concept in detail in the "Defense Highlights" on February 18, 2020, that joint all-domain operations will cover all operational domains, including land, sea, air, space, cyber, electromagnetic, cognitive, and "gray areas," and will serve as a new operational pattern for the US military in the future as it contends with evenly matched strategic competitors [8].

The US military's transition from multi-domain warfare conducted by the Army to joint all-domain operations conducted by the entire military has had little effect on the concept's connotation. The new concept has a broader vision and is more inclusive of services. As a result, it has received widespread recognition and support from the Ministry of Defense, the Federation of Representatives, and theater commands. This concept promotes the development of the US military's capability in the direction of "integration, concentration, speed, and precision" in order to increase the global military advantage over "balanced adversaries" and accomplish the strategic goal of deterring and defeating the adversary's "anti-access/area denial" system.

The US Air Force/Space Force, along with the Army, Navy, Special Operations Forces, and industry, launched the Joint All-Domain Command and Control (JADC2) system in December 2019. The US military can connect all troops and battle vehicles via JADC2 that serves as the foundation for "cross-domain operations". All military applications can be connected to the JADC2 military Internet platform. It can use big data, AI, and machine learning to track friendly and adversarial forces at any time and to enable real-time information sharing across a multi-domain operation space that includes land, sea, air, and air networks. The program leverages the Air Force

Automated Battlefield Management System (ABMS) to develop software and algorithms that enable AI and machine learning to network and compute massive amounts of data from multiple sources at speeds and accuracies far exceeding current levels, seamlessly connecting disparate battle vehicles to forces engaged in real-world combat scenarios.

The US military conducted four JADC2 exercises from December 2019 to February 2021 to validate the refinement of the joint all-domain command-and-control systems. The US Army proposed a "project fusion" program in the first half of 2020 to integrate a series of projects into joint forces for testing and exercises. The Congressional Research Service officially released the report on the JADC2 system in August 2020. The US Army and Air Force signed an enhanced interoperability cooperation agreement in September 2020 that will last until 2022. The agreement will integrate the Army's Project Fusion Program with the Air Force's ABMS program, enabling commanders to make faster-informed battlefield decisions.

The US Army conducted the "Project Fusion 2020" exercise at Yuma Proving Range, Arizona, in August 2020, utilizing the Pacific Air Force's low-orbit reconnaissance satellite and Greyhawk UAVs to conduct wide-area reconnaissance of the target area and transmit reconnaissance data via transmission layer satellites to the Titan Earth Station at Joint Base Lewis-McChord, Washington. The "Prometheus" AI system identifies targets by converting reconnaissance data to target data. The data are transmitted to the Joint Operations Center at Yuma Proving Ground via transport layer satellites. The FireStorm AI-assisted decision-making system matches fire-strike vehicles to target threats automatically and transmits target data to 155 mm howitzers and Greyhawk UAVs to enable strikes on targets. A hovering Reaper UAV transmits video of enemy tanks on fire following an attack to the FireStorm AI system for assessment of strike effectiveness. From discovery to strike, the kill chain takes only 20 seconds. After 1 minute in the air, the shell strikes the target, and the evaluation takes 10 seconds, for a total of 90 seconds (Figure 11.2).

From August 3 to August 16, 2021, the US Navy's largest maritime military exercise in 40 years involved over 25,000 sailors and Marines from the Second, Third, Sixth, Seventh, and Tenth Fleets, as well as three expeditionary units of the US Marine Corps. It was the first time the concepts of distributed maritime operations (DMO), expeditionary advance base operations (EABO), and littoral operations in a confrontation environment (LOCE) were put into practice. The US forces that participated in the exercise also utilized UAVs, unmanned surface vehicles (USVs), unmanned underwater vehicles (UUVs), and other tools. The *Sea Hunter*, the US Navy's first anti-submarine USV, worked alongside the littoral combat ships on the West Coast. The *Overlord*, another large unmanned USV, joined the Atlantic Fleet's East Coast training. The exercise's objective was to demonstrate and assess the interoperability of unmanned weapons and equipment, as well as their coordination with manned equipment [9].

Figure 11.2 The US Army's Project Fusion 2020 exercise flow chart.

According to the US Department of Defense website (news reported on March 17, 2022), the US Deputy Secretary of Defense signed the JADC2 Implementation Plan and released a public version of the JADC2 Strategic Summary on March 15, 2022 [10].

11.4 Cross-Geographic Domain Operations

Cross-geographic domain operations will become the norm for all-domain, multi-domain, and cross-domain operations in the future. Land operations are classified as land–air joint to-ground operations, sea-to-air joint to-ground operations, sea-to-air cooperative submersible to-ground operations, and space-to-ground cross-domain operations. Under the overall planning of sea operations, there are primarily sea, air, space, submarine, and shore cross-domain attack and defense, as well as air–submarine cross-domain collaboration for ship-to-submarine operations and joint land–air sea-to-submarine operations. Air operations cover a variety of activities, including air–ground coordinated air-to-air sensing, ground–air integrated air defense, anti-missile, air–air integrated anti-missile, formed long-range three-dimensional delivery, close air-to-ground fire support, and others. There are primarily ground, air, sea, and space-based anti-defense and anti-missile systems. Space-based information enables ground, sea, and air operations, while suborbital/near-space data enable high-speed ground, sea, and air detection and attack (Figure 11.3). Cross-media weapons supporting cross-geographic domain operations include submarine-launched missiles, anti-submarine weapons, drilling bombs, amphibious and multi-dwelling vehicles, trans-atmospheric maneuvering

Figure 11.3 Imagery of future space-based laser weapons for ground strikes.

vehicles, trans-atmospheric anti-defense and anti-missile systems, high-speed ground-to-sea strikes in near space and space, and ground/sea/air-based anti-defense and anti-missile weapon systems.

11.4.1 X-37B Reusable Orbiter

The US reusable orbiter X-37B is a typical ground–air–space cross-domain operational vehicle. In 2010, the US X-37B completed a 224-day first orbital flight technology demonstration flight, followed by a 469-day second flight, a 674-day third flight in late 2012, and a 718-day fourth flight on May 21, 2015. On September 7, 2017, the fifth classified mission flight began with the launch of Space Exploration Technologies Corporation's "Falcon 9" rocket from the Kennedy Space Center. On May 17, 2020, the Atlas-5 carrier rocket launched the sixth space flight, which may evolve into a space confrontation weapon platform in the future.

11.4.2 Submersible Aircraft and Air–Water Strike Vehicles

The air–water strike vehicle serves as an innovative and highly effective method of penetration and strike. It exploits the near-surface/submerged defense weakness of surface ships and the gap in submarines' long-range air defense capability to achieve strong penetrations and efficient strikes against enemy surface ships, submarines, and other key targets in key areas by leveraging the advantages of flight, defense penetration, and strikes in air and water. Defense Advanced Research Projects Agency (DARPA) proposed the Submersible Aircraft project in 2010, which resulted in an aircraft capable of both flying and diving underwater. It combines the speed and range of airborne vehicles with the stealth capabilities of underwater vehicles.

Air–water vehicles are capable of maneuvering in both air and water. Due to the 800-second density difference between air and water, the air–water environment requires structural variability, dynamic adaptability, guidance control switching, and cross-media load adaptability for both media. The vehicles must adopt new technologies such as intelligent materials, complex adaptive deformation control, cross-media flight intelligent control, and supercavity navigation to achieve cross-air/water media flight. Among its core technologies are an integrated design of air–water vehicle structures, high-speed out/in water, near-surface gliding flight, high-speed underwater navigation, underwater high-energy-density power, and efficient underwater damage.

11.4.3 Long-Range Gliding Air–Sea Cooperative Operational Aircraft

Due to noise wave interference and complex conditions at sea, aircraft carrier formations and large fleets have difficulties defending against near-surface flight weapons. Simultaneously, electromagnetic waves decay rapidly in seawater, and radar is incapable of detecting areas deeper than the surface. These conditions act as a natural deterrent and safeguard for cross-media weapon systems. A stealthy cross-media weapon system allows it to avoid early-warning radar detection, giving it an unmatched advantage in completing stealth attacks on aircraft carrier groups.

A long-range gliding cross-media air–sea cooperative operational aircraft is one or more long-range air vehicles transported to the battle area by a carrier aircraft to detect and track enemy aircraft carrier groups or large fleets using ground surveillance radar, air early warning, and satellites. The air vehicles determine an aircraft carrier's or fleet's geographic location and movement, as well as its anti-aircraft and anti-submarine fire capabilities. It can then transmit these data via the carrier aircraft to the cross-media air vehicles. Then, as the aircraft approaches the battle area, it releases a cross-media vehicle formation. A cross-media aircraft group flies to a predetermined sea area near the fleet, forming an air cooperative operation unit and an underwater cross-media cooperative operation unit.

1. The airborne cooperative operation unit conducts formation penetration and effective strikes against aircraft carriers or the surface portion of the fleet using cooperative guidance and control.
2. The underwater cooperative operation unit, via its integrated control system, determines the carrier's or fleet's defense blind spot and automatically solves the media crossing point, as well as the corresponding movement time and distance. It then sinks beneath the water's surface, forms an underwater group, and selects the appropriate strike mode for underwater attacks on the carrier based on the battlefield situation.
3. Individual air and underwater system operation units can attack aircraft carriers and large fleets. When the technology matures, they will also collaborate in a network to launch coordinated attacks on aircraft carriers, much like a wolf pack does.
4. When the underwater cooperative unit enters the water, it will close the wings to minimize the overload shock. After entering the water, the underwater system operation unit can communicate and collaborate via sonar in order to conduct simultaneous, continuous, and direct attacks on aircraft carriers and large fleets.
5. When an air operation unit and an underwater operation unit attack concurrently, they can be launched at different times depending on the situation. After entering the water, the operation units synchronize the attack times of the air and underwater units by interacting the air information between the lead bomb and the slave bomb.

11.5 Cross-Functional Domain Operations

At strategic levels, cross-functional domain operations entail cross-virtual and physical space operations, as well as cross-physical, information, social, and biological domain operations. At tactical levels, the operations include information, mobility, firepower, air defense, security, and other cross-domain military operations.

11.5.1 Virtual and Physical Fusion Operations

Cross-domain operations in virtual and physical space entail collecting and fusing intelligence from multiple sources, situational analysis, detection and strike integration, front and back integration and support, public opinion shaping and cognitive confrontation, cyber attack and defense, and infrastructure control. After correlation and validation, the battlefield information acquired in virtual space can directly provide information to the firepower system in physical space. Operational equipment and personnel can upload frontier perception data to the back-end cloud, where they can access high-value calculation results and battlefield-related data. This creates a positive cycle that improves operational effectiveness significantly. The virtual space system engages in targeted confrontation based on opinion, information warfare, and psychological warfare, deceives and interferes with adversaries, shapes favorable environments for one's side, and promotes the development of operational advantages in physical space. It uses cyberspace to conduct cyber attacks against social infrastructure, influence operational processes, and accomplish military objectives in physical space.

11.5.2 Cross-Domain Intelligence Support

Complementary are the battlefield environment information and other data gathered via open-source channels such as the Internet in virtual space. Correlations can be made between the open-source data in virtual space and the target information on the battlefield gathered by reconnaissance systems on land, sea, air, and space in physical space. They can be used in combination to determine the effectiveness of strikes. All types of command systems at all levels can be supported and integrated based on big data, intelligent image, voice, and spectrum recognition, and cross-domain intelligence information transmission. Subdomain and thematic situations can be formed, enabling the standard view, consensus, and sharing of battlefield situations at all levels. Simultaneously, all levels and types of command institutions, task forces, and manned/unmanned vehicles can achieve information connectivity, particularly at the tactical command and control and fire control levels, opening up the information links for joint operation detection, control, and evaluation between military services.

11.5.3 Cross-Domain Command and Control with Interoperability

Future joint all-domain and cross-domain operations will be similar to typical cross-service operations. A pervasive network of air–land operations should facilitate this mode of operation. Information links and data interaction across domains and vehicles are enabled at the tactical level using unified data and information transmission standards and protocols. It is possible to integrate and coordinate the command-and-control functions of various services and arms, as well as the interoperability of vehicles such as UAVs, unmanned surface vehicles (USVs), precision-guided weapons, and even satellites. It will be possible to achieve battlefield monitoring, information fusion, overall planning and management, air/space/sea/land conflict detection, and armed services interconnectivity. This will enable cross-domain command and control at all levels, resulting in synchronous mastery of operation processes and task states across multiple domains, real-time assessment of combat situations, and rapid operational coordination.

11.5.4 Cross-Domain Fire Support

Cross-domain fire support is based on a collaborative and interoperable networked environment and operational planning software. It aims to achieve joint mission fire targeting, the deep hinge of the command-and-control system and various fire strike vehicles, as well as cross-domain fire support among ground, underground, air, space, sea, and underwater. It is guided by operational plans supported by a unified data model and a synchronized collaborative operational model. Cross-domain fire support can be classified into three distinct patterns. The first is fire-guided strike and assessment of damage for armed services based on frontier information. Second, direct operational control of fire vehicles based on standard information interfaces and data exchange conditions. Third, cross-domain assistance in response to requests from other services.

11.5.5 Cross-Domain Defense

Long-range precision firepower, air defense, and anti-missile weapons of the Army can provide coordinated ground defense for the Navy and Air Force. Meanwhile, the Navy and Air Force's mobile firepower can provide timely support defense for the Army in its preferred operational domain, striking incoming targets while simultaneously striking the threat's source, thus forming a defense system based on cross-domain interactions and integrated strikes. The same applies to air and naval defense. In cyberspace and cross-domain defense, a military must defend against cyber attacks from multiple domains while minimizing negative public opinion, avoiding infrastructure damage, and coordinating domains for post-war governance and social order restoration via integrated operations in the physical, information, and cognitive domains.

11.5.6 Cross-Domain Support

The Army, Navy, and Air Force support resources can be integrated with the assistance of a unified network information system and cloud platform.

The platform overcomes service and field limitations by information sharing, standardized design, distributed reserve, and flexible and rapid configuration of support resources. It also enables operational, equipment, and logistics support. Simultaneously, civil support resources will be integrated into the military support system to coordinate military and civilian security forces and resources in both peacetime and wartime, thus achieving civil–military fusion.

11.5.7 Land-Based Multi-Domain Joint Operations

Land-based multi-domain joint operations are relatively complex compared to maritime, air, space, and underwater operations.

While modern science and technology have enhanced operational capabilities ranging from perception to action, the land battlefield remains more complex and foggy than space, air, sea, and even cyberspace. To begin, the land environment is the most complex. Land battles can take place on plains, hills, watery rice fields, mountainous jungles, plateaus, deserts, near-shore islands, large and small towns, and megacities, among other places where targets are difficult to identify due to constraints imposed by multiple factors and technical means. Second, operational tasks and patterns on land vary significantly. Urban, island, mountain, border and naval defense, special, and counterterrorism operations, stability maintenance, and overseas military operations are all examples of land battles. Third, land battles span the most operational domains, including land, sea, air, space, network, electromagnetic, psychological, and social fields, involving open-source information resources, the Internet of Things, infrastructure control, key targets, and crowd tracking, as well as network and new media-based confrontations based on public opinion, psychological warfare, and legal issue confrontation, among others. Fourth, it requires participation throughout the process, from pre-war governance to post-war governance. Other military services may fight and withdraw, but the Army or ground forces must restore order and governance following the war. Thus, in the age of AI, a land-based multi-domain joint operation will serve as the focal point for joint operations and cross-domain actions.

11.6 Intelligent Focus

While joint operations are at the heart of multi-domain warfare, military service synergy and mutual support are at the heart of cross-domain operations. Compared to general intelligence in terms of detection, control, combat, evaluation, and protection, the intelligence of joint all-domain and cross-domain operations has distinct characteristics and focuses, encompassing five aspects.

11.6.1 Joint Mission Planning

Since joint all-domain and cross-domain operations involve land, sea, air, space, and cyberspace, it is critical to plan and design joint operations holistically, emphasizing the theater of operations. From strategic to operational to tactical levels, between joint land and joint air, naval, space, and cyberspace operations, while there are some commonalities in operational patterns, objectives, environments, and mission planning, significant differences persist. Joint operational mission planning for different battlefields is critical for intelligence because it focuses on joint all-domain and cross-domain operations. The higher the operational level, the more complex the operation systems are, and the more these systems rely on people and commanders for decision-making, particularly during the pre-war planning stage, while AI is primarily used in auxiliary decision-making and supporting capacities. The lower the level, the clearer the task, the simpler the system, and the more autonomous decision-making and AI can be used to resolve it. At the intermediate level, the majority of issues are resolved through a human–AI hybrid decision-making model. The more precise the pre-war data collection, the more refined the model algorithm will become. The more thoroughly the adversaries are studied, the greater the likelihood of victory, regardless of whether the decision is made by a human or an AI.

11.6.2 Cross-Domain Intelligence

Cross-domain networked detection, communication, navigation, and other information support are critical in all-domain and cross-domain joint operations. It is necessary to concentrate on over-the-horizon network detection, communication, and navigation technology covering maritime and underwater, indoor and outdoor, above-ground and underground, air–space, air–ground domains, and a space–time reference platform based on satellite navigation, inertial navigation, and other advanced navigation technologies. Simultaneously, network information should possess a cognitive function capable of automatically detecting electronic interference and network viruses, initiating automatic frequency hopping and security protections, and ensuring smooth networking communications and timely, secure, and reliable data exchange.

11.6.3 Military Effectiveness Expansion

To facilitate joint all-domain and cross-domain operations in the age of AI, each military service should strengthen its intelligence and cultivate relevant capabilities in OSoS, networking communications, vehicle mobility, firepower, defense, and support. For example, during land joint multi-domain operations, in addition to conducting traditional air-to-ground operations, it is necessary to develop sea-to-ground mobility and strike capabilities, as well as underwater and submarine-to-ground, near-space-to-ground, and space-based-to-ground mobility and strike capabilities. Another example is that during joint maritime operations and integrated air–sea operations, land-to-sea, near-space-to-sea, and space-to-sea capabilities are required to disrupt maritime supply lines and conduct strikes against enemy coasts and naval bases.

11.6.4 Intelligent Collaboration of Operation Forces

In joint all-domain or cross-domain operations, various military services should develop intelligent and cooperative command, action, evaluation, and support under a unified operational mission-planning framework supported by network information. A unified interface and data exchange format should be promoted to achieve strikes upon detection or coordinated strikes upon detection at strategic, operational, and tactical levels.

11.6.5 Unified Standard Specifications

The unified standard specification entails the development of two distinct types of specification systems: those for operation command management and those for information data exchange. First, standardization and unification of command-and-control systems, operational patterns, processes, terminologies, operational descriptions, conceptual connotations, operational rules, and processes are required. Second, it is necessary to standardize space–time benchmarks, information processes, interfaces, data exchange formats, and protocols.

References

[1] US Congress. (2012). *National Defense Authorization Act for fiscal year 2012.* www.congress.gov/112/plaws/publ81/PLAW-112publ81.pdf.

[2] Rowden, T., Gumataotao, P., & Fanta, P. (2015). Distributed lethality. *United States Naval Institute Proceedings.* www.usni.org/magazines/proceedings/2015/january/distributed-lethality.

[3] Pentagon's 'combat cloud' concept taking shape. (2014, September 29). *Aviation Week Intelligence Network.* https://aviationweek.com/aerospace/pentagons-combat-cloud-concept-taking-shape.

[4] Kelly, T., Atler, A., Nichols, T., & Thrall, L. (2013). Employing land-based anti-ship missiles in the Western Pacific. www.rand.org/pubs/technical_reports/TR1321.html.

[5] United States Army. (2016). *United States Army white paper*: Muti-domain battle: combined arms for the 21st century. https://community.apan.org/cfs-file/__key/docpreview-s/00-00-00-97-20/161013-Army-MDB-White-paper-v53a.pdf.

[6] Myers, M. (2017). *Milley: Future conflicts will require smaller Army units, more mature soldiers.* www.armytimes.com/news/your-army/2017/03/21/milley-future-conflicts-will-require-smaller-army-units-more-mature-soldiers/.

[7] Lin, Z. 林治远. (2017,September 19). *"Duo yu zhan": Meiguo lujun zuozhan xingainian* "多域战": 美国陆军作战新概念. Sohu 搜狐. www.sohu.com/.

[8] US Department of Defense. (2020). *Report on joint all-domain command and control.* www.defense.gov/.

[9] Ress, D. (2021). *'It's called being agile': Fixing a pier, setting up fueling station all part of what Navy's global exercise demands.* www.stripes.com/.

[10] US Department of Defense. (2022). DoD announces release of JADC2 implementation plan. www.defense.gov/News/Releases/Release/Article/2970094/dod-announces-release-of-jadc2-implementation-plan.

12 Cognitive Confrontation

For thousands of years, cognitive clashes have primarily manifested themselves through intellectual and knowledge competitions between humans. Humans, as planners and participants in warfare, as well as creators of machine intelligence, play an irreplaceable role and continue to be the decisive factor in warfare in the age of artificial intelligence (AI). However, with the advent of machine AI and its ongoing optimization, upgrading, and improvement, machine AI will play an increasingly significant role in warfare, perhaps even surpassing the role of humans. Cognitive confrontation will change dramatically in the future, as evidenced by the clashes between humans, between humans and AI, and between AIs. Both humans and AI will be critical.

In broad terms, AI-based confrontation in the age of AI falls under the category of cognitive confrontation, which includes three distinct domains: physical space, virtual space, and human mentality. Due to the broad scopes and contents of cognitive confrontation, this chapter will focus on a few key points.

12.1 Perceptual Confrontation

Combatants will need to conduct operations on land, sea, aeronautic space, astronautic space, network, and other complex battlefields in the future, confronting not only multiple adversaries, targets, and modes of attack and defense, but also multi-dimensional space perception challenges posed by unfamiliar domains, territories, cities, society, media, and public opinion. To defeat adversaries in a perceptual conflict, make battlefield information more transparent and detailed, and comprehend the battlefield more deeply than adversaries, the military must concentrate on five aspects of capacity building.

To begin, a networked multi-source perception system that integrates military and civilian technologies must be established. It is necessary to employ military space-based, air-based, ground-based, and sea-based detection means; to exploit image, infrared, video, synthetic aperture radar (SAR), electronic reconnaissance, multi-spectral, magnetic detection, gravity gradient, hydroacoustic, and other reconnaissance means for detecting, tracking, locating, and analyzing fixed, mobile, high-speed, underwater, and underground targets and complex combat environments; and to meet the requirements of

DOI: 10.4324/b22974-14

various weapons and equipment. At the same time, possessing only military information data is insufficient to prevail in a perception conflict. The combatant must leverage information resources such as the civilian Internet, the Internet of Things (IoT), civilian satellites, and social media, as well as web crawlers and big data, to conduct multi-source searches and build multi-dimensional correlation models of various geographical environments and targets. The majority of military detection methods are limited to detecting the shape, size, electromagnetic signature, and other physical characteristics of the operation environment and targets. The questions of what institutions, personnel, and facilities surround an operation environment, whether it is a military target or a civilian target, a government building or a cultural relic building, and what kind of structure, distribution, and units are contained within the building, etc., must be correlated and verified through the use of the above resources. It is also necessary to expand research on remote search and discovery, full tracking and surveillance, all-domain detection coverage, all-weather operation, and in-domain information fusion and utilization, in order to address the detection, discovery, identification, and intent discrimination of diverse targets (e.g., stealthy and moving targets) and human behaviors, and to develop a full-spectrum perception capability covering land, sea, aeronautic space, and astronautic space.

Second, advanced perception and detection technology must be developed and applied. Novel detection technologies such as terahertz, magnetic field, quantum correlation imaging, and hyperspectral imaging may assist in addressing the issue of insufficient information for low-observable or undetectable targets under complex conditions. Hyperspectral imaging is a common application technology among them. It employs specialized spectral acquisition techniques to image a target in narrow spectral bands within the visible and infrared spectrums, followed by complete super-resolution or hyper-fine observation using the spatial and spectral information obtained from the image. This technique enables both observation of a target's external shape and analysis of its internal properties, as well as the most subtle changes in key parts of the target, allowing for direct qualitative and quantitative analysis and identification of the target.

In the late 1980s, hyperspectral imaging technology was developed based on multi-spectral scanning imaging remote sensing technology. In the 1990s, the US Navy developed a hyperspectral airborne prototype. The US Army and Air Force invested heavily in hyperspectral imaging technology in the twenty-first century, hoping to develop all-day hyperspectral imaging systems for brigade and divisional unmanned aerial vehicles (UAVs). They have successfully demonstrated improvised explosive device (IED) detection on the US Army's Shadow 200 tactical UAVs using BAE Systems' hyperspectral sensor system.

The advent of quantum dot spectroscopy recently paved the way for the miniaturization and integration of hyperspectral sensor systems with conventional optical imaging systems. Quantum infrared sensors can boost battlefield sensors' detection sensitivity to a single photon level for detecting targets with extremely low visibility, providing soldiers with superior vision and observation capabilities. Quantum sensors with single-photon sensitivity can improve temporal and spatial resolution by utilizing entangled photons and higher-order coherence, resist interference from specific noise, reduce the size, power consumption, and cost of infrared sensors, and effectively maintain troops' situational awareness advantage on the battlefield.

In the future, it will be necessary to study the theory and method of detection imaging processing under the condition of incomplete data, as well as technologies for information autonomous perception, sensor self-organization and cooperation, and front-end intelligent information processing, in order to reduce transmitted data flow and improve sensing information acquisition efficiency and accuracy.

Third, a standardized system for the fusion, transmission, and processing of information data must be established. The collection, storage, processing, distribution, transmission, and utilization of multi-source and heterogeneous information data, as well as the use of standard systems with data that computers and combatants can quickly identify, are required to facilitate the command and control of weapons, equipment, and troops. Given the wide variety of data formats used by different systems, sources, uses, purposes, and user objects, a unified data classification and transmission protocol must be established at the underlying layer to enable computers to identify, store, compute, and process data, enabling end-to-end transmission and cross-domain information sharing and awareness for different users. When operating in a multi-domain or cross-domain environment, the target information detected, discovered, and tracked via space-based information and the Internet should be transmitted to the military–civilian general-purpose data center and front-line command posts battle vehicle terminals or individual soldier portable systems. The processed results should then be returned to the data center for correlation and validation with space-based and secondary Internet information. This process necessitates multiple data links in order to ensure the efficient transmission and interaction of information under conditions of joint all-domain, multi-domain, and cross-domain operations.

Fourth, there is a need to accumulate various operational data, both directly and indirectly. Continuous data collection on target characteristics, enemy equipment, human identity, exercises and training status, strategies and tactics, personnel quality, and combat experience is necessary to update and maintain information "freshness". In peacetime, it is necessary to accumulate information on critical operational areas and the adversaries' operational means, habits, and strategies. Additionally, the data collected should be classified according to the image, electromagnetic, and voice characteristics, as well as operation indicators, personnel training level, and equipment effects to ensure timely access to the information during operations.

Fifth, commercial satellites should be applied in perceptual confrontation. They are available for civilian as well as military purposes. After acquiring the Iridium satellite system, the US military invested \$2.9 billion in 2015 to build a new generation of broadband information networks, launching 72 satellites with communications, meteorological, and multi-spectral payloads distributed over 780 kilometers and six orbital planes with interstellar links and Internet protocol (IP) capabilities, 1.5 Mbit/s downlink in the L-band and 30 Mbit/s in the KA-band. Global satellite communication and space-based battlefield reconnaissance, target recognition, and operational evaluation are all possible with the improved Iridium satellite system. It has high capabilities for information transmission and processing, potentially enhancing the US military's capabilities in IP-based networked reconnaissance, command and control, and communication. Integrating intelligent image, spectral, and electromagnetic spectrum identification functions, its onboard data-processing capabilities will be further enhanced, enabling it to interact with land-, sea-, and air-based data. The observe–orient–decide–act (OODA) loop duration will be significantly reduced as a result of increased user participation and faster data transmission and processing.

Several prominent Internet companies and investors, including SpaceX, O3b, and Japan's Softbank Group, have recently proposed constellation development plans that include launching hundreds to tens of thousands of small satellites. Assume that the world's military superpowers and commercial titans gradually fill the Earth's low- and medium-orbit space with satellites, reducing costs using intelligent technology and batch advantages. In that case, it will have a significant impact on the construction of traditional mobile communications and their associated base stations, as well as on data transmission, information services, terminal systems, and the global mobile communication and sensor industry. Space-based information systems are also connected to the Internet's big data, increasing the level of intelligent identification and interaction, soon rendering the global battlefield "transparent" and revolutionizing information perception during a conflict. The development of disruptive technologies such as holography is noteworthy in the field of perceptual confrontation. In April 2017, the US Army Research Laboratory's weapons and materials research division released a research report stating that holography has significant potential in military visual camouflage [1]. Holography is a technique for projecting virtual objects into three-dimensional images in physical space. It combines the advantages of three-dimensional imaging and stereoscopic simulation.

First, holography can be potentially applied in special operations. Traditional military camouflage is a relatively straightforward passive defense technique, while visual camouflage via holography is a practical method of active camouflage that features being "stealthy" to the naked eye. Holographic visual camouflage with a high degree of simulation, a strong spatial sense of visual deception, and chameleon-like environmental adaptability will significantly increase the effectiveness and survivability of combatants and operational vehicles. It is capable of concealing aircraft, ships, vehicles, and temporary command posts. Although this technology is still in its infancy, it is possible to achieve holographic camouflage with the development of holographic image acquisition, display, and new material breakthroughs.

Second, the use of holograms to deceive adversaries is a tactical possibility. The US Army Research Laboratory believes that by projecting images and slogans on to the cloud or a specific space via holography, a confusing holographic force could be created on the battlefield. The US successfully conducted military experiments on holography in the mid-1990s, projecting visions of planes, tanks, warships, or entire operation units on to the battlefield to confuse adversaries. In the future, if the holographic system is capable of storing and displaying large images at high resolution, the new tactics of using holograms to confuse adversaries may have unexpected operational effectiveness.

12.2 Data Mining

Previously, battlefield data and information were only accessible via closed systems and internal channels. In the future, open-source information from the Internet, mobile communications, radio, and television will be integrated using big data and AI, while civil satellites will supplant traditional channels and sources of battlefield data. In peacetime, it is necessary to collect historical data on adversaries and deduce their strategies and tactics so that during wartime, combatants can quickly obtain information via calculations based on established models, which effectively compensates for the lack of battlefield information and the limitations of timeliness and objectivity. For example, Google Maps can be used to learn the geographical locations and image information for the majority of the world's countries, regions, cities, and streets. Civil aviation flight inquiry enables an accurate understanding of major airports' takeoff and landing patterns worldwide, subject to weather changes, military exercises, and unexpected events. A weather-forecasting service enables easy access to real-time weather conditions worldwide. The distribution of famous places, landmarks, hotels, and shopping malls in major countries and cities around the world, for example, can be learned through travel networking sites or apps.

At the strategic level, it is possible to excavate comprehensive data on major countries and regions' political, economic, military, scientific and technological, and cultural conditions throughout the world. This also includes the topography of landscapes and natural features (e.g., vegetation, rivers, and lakes) and ethnic composition, religious customs, social opinion, and ideology. With the assistance of big data, not only is it possible to analyze changes in the adversaries' defense spending, equipment research and procurement, military strength, war mobilization potential, and traces of war preparation, but it is also possible to forecast and warn about the movements of extremist forces or terrorists.

At operational and tactical levels, it is possible to precisely determine the geography, weather, city, population, and distribution of military and civilian facilities within a predefined battlefield or target area, as well as the shape and structure of buildings, incoming enterprises, and personnel distribution from the exterior and interior. It is necessary to conduct web crawls and data modeling around critical targets such as government agencies, military installations, airports and terminals, hotels and restaurants, cultural relics, network communications, radio and television, as well as water, electricity, oil, and gas. Large building characteristics and operational requirements enable the accumulation of historical and dynamic real-time data, the acquisition and correlation of multi-dimensional information, and the provision of adequate services for operational and tactical situational awareness. Big data technology combined with intelligence, surveillance, and reconnaissance (ISR) systems and human intelligence can track and locate the targeted behavior trajectories of key individuals and groups; analyze, judge, and warn the public; and identify, distinguish, and classify military and civilian targets. Critical fixed targets such as airports, ports, military installations, ammunition warehouses, military industries, and moving targets, underground targets, and targets inside buildings can be identified using cutting-edge sensor systems in conjunction with space-based information and network data.

At the technical level, models and algorithms must be developed to support rapid response, decision-making, and operations in complex battlefield environments. In contrast to the civilian environments, the battlefields are primarily concerned with perceiving and identifying adversary information in highly complex geographical environments, with a high level of confrontation and interference, requiring rapid response. Simulations and calculations must be performed in advance, as well as physical and semi-physical verification via exercise training. Various battlefield environments and target recognition models should be established through continuous optimization and iteration to forecast and predict battlefield trends. Machine learning and other technologies for acquiring, analyzing, and modeling military targets' image, electromagnetic, and spectral characteristics can improve target recognition probabilities and capabilities.

12.3 Decision-Making Competition

Competitive decision-making is at the heart of cognitive confrontation. Human commanders and virtual commanders, i.e., robotic agents, define the future of operational decision-making. The more tactically oriented operations are, the more AI virtual commanders are required. At the tactical level, the operational environment, adversaries, objectives, means, and methods are relatively certain, and the adversaries' circumstances and response strategies are straightforward to comprehend. As a result, the information sought by both parties is relatively complete and can be solved gradually via AI and machine learning. Additionally, tactical operations decisions could be made primarily by machines and supplemented by humans.

However, strategic operations involve a variety of operational elements, environments, and factors. Therefore, combatants often struggle to comprehend their adversaries' overall strength, operational capability, mobilization capability, war potential, and military-related diplomacy, economy, science and technology, society, and public reaction. As a result, decisions must be made by machines. The final strategic decision can be made only after collective research by senior commanders and the general staff. The higher the strategic operations level, the more integrated humans and machines are in making decisions, as high-level operational intent, objectives, goals, actions, and plans are highly classified. During a fierce conflict, both sides will use strategic deception to deny the other side accurate political, economic, and military strategy information, mislead their adversaries, influence their decisions, and ultimately gain the strategic initiative in the war.

In terms of strategic deception, it is possible to model and analyze the rules and laws of victory deception. Multiple simulation calculations can be used to verify, evaluate, and analyze objects and means in order to develop a set of systematic deception strategies capable of effectively weakening and interfering with opponents' decision-making abilities. Anti-deception requires the use of big data and machine learning to determine the authenticity of the information. In the future, deception behaviors will be numerous, and deception data will be massive, outpacing human analysis and discrimination capabilities and posing time constraints, necessitating the use of machine AI. Machine AI can now outperform humans in information deception and recognition in the era of big data and AI. In April 2017, Libratus, an AI robot, competed against four top human players in a game of Texas Hold'em. The robot learned a "bluff" skill that the humans could not perform due to psychological constraints, demonstrating that robots can also employ deception techniques.

The advantages of machine AI decision-making are reflected in the high dynamics, significant interference, and rapid response. For instance, in the fields of air defense and missile defense, hypersonic confrontation, electronic confrontation, swarm attack and defense, and even cyber attack and defense, algorithms and models generated through intelligent training and machine learning can be used to aid in decision-making and execution. Human decision-making and participation are still required for overall war planning, strategic analysis, judgment, phased operation evaluation, and analysis but are inextricably linked to machine-assisted decision-making. As a result, decision-making competitions will increasingly rely on models and algorithms generated by adversarial networks and parallel OSoS.

It is necessary to concentrate on task-oriented situational deep-mining technology, to simulate human learning and reasoning, problem-solving, judgment, and decision-making, and to develop technical means for battlefield data collection, rapid analysis, and cognition understanding, thereby assisting commanders, officers, and combatants in making scientific analyses and decisions about the battlefield situation.

12.4 Key Target Monitoring

Operational officers, commanders, senior generals, presidents, and key members of Congress are all influential groups with strategic influence over war and operations. They will be closely monitored in the future by both sides' intelligence and command organizations, and they will almost certainly be the target groups for decapitation operations. With the continued advancement and integration of the Internet, the IoT, big data, brain–machine interfaces, the Internet+ initiative, and AI, it is possible to conduct all-encompassing monitoring of strategic targets and, under certain conditions, intervene and control them.

Critical personnel's social relationship networks can be understood by generating diagrams involving their families, colleagues, superiors, subordinates, and friends using the Internet, social media, and ISR systems. This may help to understand the figures' investments, business, assets, and funds, as well as analyze and judge their values and worldview. The association model and database of key personnel's personal information can be established and instantly updated based on a multi-source information network.

Decapitation strikes and targeted killing require two critical components: detection and action. Everyone has a circle of influence, which includes their work, life, and social relationships. Once the information of this circle is mastered, the image of the individual can be depicted. Due to the high frequency of exposure, political and military leaders, among other public figures, are more likely to be detected, tracked, and located. Even if they wish to conceal their location, various pieces of information about their whereabouts can still leak their location. Once the locations and precise positions of critical targets under specific space–time conditions are determined, precise attacks can be launched. The US Department of Defense proposed the Data to Decision Technology Priority Development Plan in 2011. The US government released its Big Data Initiative in March 2012, elevating data engineering to a national strategic priority. The US military has undertaken several research projects, including the Data Extension Program. It is also worth noting that the US military used Palantir's big data correlation analysis technology and human behavior model algorithm to track Osama bin Laden and his sole messenger Abu Ahmed al-Kuwaiti, eventually locating bin Laden's apartment in the town of Abbottabad near Islamabad (Figure 12.1). Among the critical processes involved in locating bin Laden were the following:

1. In 2007, suspects in the "9/11" attacks made reference to bin Laden's contacts.
2. The Pakistani military provided the names and cell phone numbers of over 2,000 al-Qaeda operatives, allowing investigators to probe al-Qaeda leaders' communication networks and contacts in Pakistan, as well as create a map of bin Laden's messaging connections.

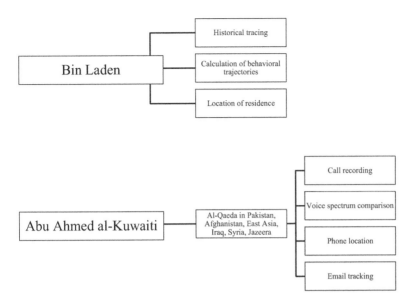

Figure 12.1 Behavior-tracking graph based on Palantir's big data correlation technology.

3. In August 2010, Ahmed was confirmed as the sole messenger between Bin Laden and al-Qaeda following a search of communications links with various al-Qaeda leaders throughout the world and the corroboration of incident tracking.
4. The US monitored Ahmed's contacts and behavior and obtained bin Laden's hiding place in the Pakistani town of Abbottabad through surveillance.
5. Bin Laden was assassinated in May 2011.

Throughout nearly a decade of tracking bin Laden, geologists examined the composition of rocks in the video, ornithologists examined the birdsong in the video, 12 senior behavioral psychologists examined bin Laden's thought patterns, over 100 local agents familiar with Afghan civilian conditions searched for clues to bin Laden's whereabouts, and over 1,100 US agents conducted intelligence research on bin Laden. However, the primary contributors to successfully tracking bin Laden's movements were identified as a Central Intelligence Agency (CIA) intelligence officer named "John" and Palantir, a mysterious big data company in California's Silicon Valley. Palantir, a data integration, information management, and quantitative analysis company headquartered in Silicon Valley, is a top-ranked big data company and Silicon Valley unicorn focused on identifying trends, correlations,

and anomalous behaviors of businesses, groups, and individuals using commercial, proprietary, and public datasets.

Computational models of human behavior are one of six disruptive technologies identified in the US Department of Defense's 5-year Science and
Technology Development plan for the period 2013 to 2017. Iran's military
leader Qassem Soleimani was bombed on January 3, 2020, demonstrating
how the US military's deep integration of human behavior computing models
based on big data association algorithms and unmanned operation systems
has gradually matured and developed into a practical combat capability. From
apprehending bin Laden to hunting Baghdadi (the leader of ISIS) and assas-
sinating Qasim al-Raymi (the leader of al-Qaeda in the Arabian Peninsula),
the US military has skillfully employed this technology, which will eventually
become a common method of intelligent warfare.

Soleimani was most likely a long-standing priority target of the US
military's disruptive technology program involving computational models
of human behavior. For several years, the US military had been modeling
his identity and social relationships, amassing a large amount of information and data about his activities in the overseas Al-Quds Brigade and his
relationships with his country's government and military, and conducting
several simulations of the decapitation strike. Soleimani was attacked in the
following manner:

1. The US military gained advance knowledge of Suleimani's itinerary
   through Iraqi government and military internal intelligence personnel/
   informants and developed an operation plan with numerous pre-plans.
2. Based on mobile phones, image surveillance, and license plate numbers,
   the US military tracked and positioned Mohandis, the Iraqi People's
   Mobilization Organization leader, and his assistants, drivers, and vehicles
   as they traveled to Baghdad Airport to pick up Suleimani.
3. After Soleimani's face, voice, fingerprints, and even DNA were obtained,
   once he appeared at airports, hotels, and planes, he would be quickly
   identified by US intelligence personnel/informants.
4. The MQ-9 UAV that assassinated Soleimani took off from Qatar's
   Uday Air Force Base. Its command and control were located thousands
   of miles away at the Joint Special Command at Kreiger Air Force
   Base in Nevada. The supply task was performed using the Air Force
   C2 command-and-control constellation network, airborne communication network, and over-the-horizon terminal based on the Global
   Information Grid (GIG).
5. Soleimani was assassinated (Figures 12.2–12.4).

On October 27, 2020, the Iranian Defense Ministry confirmed that
Mohsen Fakhrizadeh, a senior Iranian nuclear physicist and head of the
Defense Ministry's nuclear program, was attacked by "armed terrorists" near

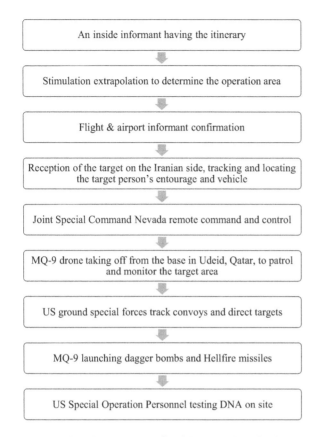

Figure 12.2 The whole process of Soleimani being killed by precise tracking.

the capital, Tehran, and died after an unsuccessful rescue attempt. Ayatollah Khamenei, Iran's Supreme Leader, stated on October 28 that the perpetrators would face severe punishment. Iran's President Rouhani described the attack as an act of terrorism committed by Israel. Israel stated that it "does not know who was responsible for the assassination." Although there were two conflicting accounts of the assassination, officials later claimed it was a precision attack carried out by an AI unmanned system.

On November 30, Shamkhani, the head of Iran's Supreme National Security Council, confirmed that the attack was carried out remotely using "special methods" and that "no one was present at the scene" [2]. Iranian Defense Minister Khatami provided additional details about the incident: "Initially, the car he was riding in came under fire. About 15 seconds later, a Nissan pickup truck loaded with explosives exploded about 15 to 20 meters from his car. The shooting and explosion caused his injuries and ultimately killed him."

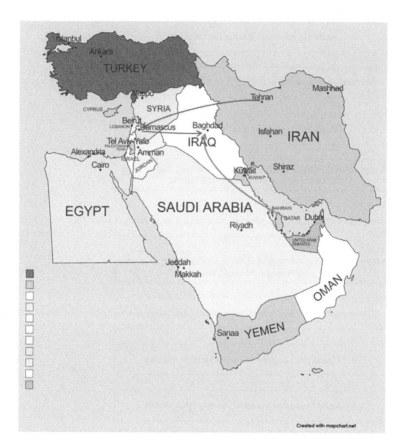

Figure 12.3 The MQ-9 unmanned aerial vehicle (UAV) that assassinated Soleimani took off from Uday air base in Qatar and intersected near Baghdad Airport to execute a precision strike.

On December 6, Fadavi, deputy commander of Iran's Islamic Revolutionary Guards Corps, stated that all 13 enemy bullets were fired from a machine gun in the Nissan, with the remainder being fired by guards on the scene (Figure 12.5). The Nissan's interior was equipped with an intelligent satellite system that was capable of tracking Fakhrizadeh. According to Iranian media, the scientist was assassinated using a "remote-controlled machine gun" or "satellite-controlled" weapon.

The assassination of Iranian nuclear scientist Fakhrizadeh is also an example of a successful attempt to integrate human behavior calculation models based on big data association algorithms with unmanned operation systems based on satellite or mobile communications. Fakhrizadeh may have

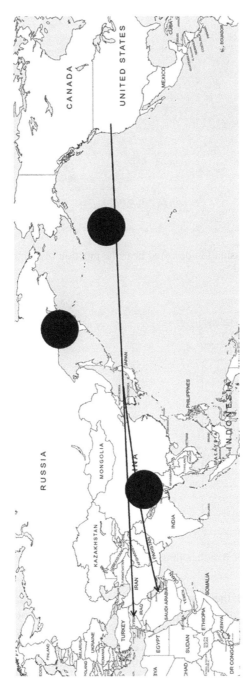

Figure 12.4 Command and control of the assassination operation at a Nevada military base 10,000 miles away in the US.

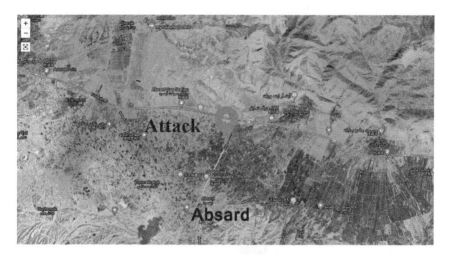

Figure 12.5 The assassination occurred near the junction of the bypass road leading to Absard.

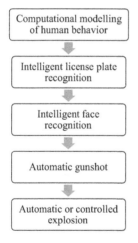

Figure 12.6 Process of remote killing of Iranian nuclear scientist.

been watched, and the assassination organization may have mastered the patterns and characteristics of Fakhrizadeh's daily behavior and developed multiple assassination plans (Figure 12.6). Since the attack occurred within Iranian territory, the countries involved may choose not to publicly acknow-ledge it in order to avoid international law conflicts and moral crises. Although no one was present, someone must be in remote command and control in order to carry out the precise assassination.

12.5 Social Media Warfare

On July 15, 2016, Turkey experienced a military coup in which rebels attempted to depose President Recep Tayyip Erdoğan. However, events unfolded differently than anticipated, and the coup against Erdoğan morphed into a coup for Erdoğan. Erdoğan's success was aided not by the army, aircraft, artillery, tanks, and armored vehicles, but by netizens who support the president on microblogs. The following is a summary of the events.

On July 15, 2016, officers of the Turkish Armed Forces' General Staff attempted to stage a military coup. On July 16, 2016, at approximately 00:00, the Turkish Radio and Television Association television station in the capital Ankara, which was controlled by coup soldiers, directed anchorwoman Tijan Karas to read aloud to television viewers a statement drafted by coup soldiers. The Fatherland Peace Committee, a military group, claimed in a statement that their army had seized power and imposed a nationwide curfew and military control laws. The coup soldiers invaded the offices of Doan News Agency, CNN Turk, and KanalD. After the military moved in, CNN's Turkish-language channel cut off live news coverage. The anchor in the studio first informed viewers that soldiers had entered the news station building and stated, "We don't know how much longer we can continue broadcasting." The anchor then reported that soldiers had entered the central control room and declared, "This is it. We have to pull out now." After that, the television station continued to broadcast live, but with no one on the live desk.

Turkish President Recep Tayyip Erdoğan has long been opposed to social media, claiming that these platforms are frequently used to spread anti-government sentiment, and has publicly chastised Twitter and Facebook on multiple occasions. Erdoğan, who was on vacation at the time of the coup, used social media to make a timely statement and garner support. Erdoğan was interviewed on CNN Turk via FaceTime, an iPhone video chat app, on July 16, 2016, at 00:24. Speaking into a mobile phone camera, he urged people to take to the streets in protest against the coup. The anchor raised his cell phone in the studio, faced the camera lens, and broadcast the entire process of Erdoğan's video chat. Erdoğan also addressed the public via Twitter, urging people to flock to airports and public squares in order to "reclaim ownership of democracy and national sovereignty". Erdoğan accused Fethullah Gülen, a former ally and religious figure living in exile in Pennsylvania, of masterminding the coup, claiming that his members were also involved. Erdoğan claimed that "some members" of the Turkish military had been "following orders from Pennsylvania". On the same day, Turkish Justice Minister Bekir Bozda stated on television that members of the Gülen movement were implicated in the coup. In a brief statement, Gülen categorically denied it and condemned the coup "in the strongest possible terms".

The Turkish presidential website stated in the early hours of July 16, 2016, that President Erdoğan was safe and sound and that a coup attempt by a "handful of soldiers" had failed. Erdoğan warned the public on the same day, via Twitter, that "regardless of the stage [the coup] has reached, we should continue to occupy the streets tonight ... because a new conflict could erupt at any moment."

On July 16, 2016, Turkish authorities declared the coup over. In comparison, the Air Force Commander-in-Chief and several senior generals remained in a hijacked state. As of noon on July 16, 2016, approximately 265 people had been killed in the coup, including 161 civilians and police officers and 104 renegade soldiers, and 1,440 others were injured.

The world was stunned by news of a military coup in Ankara, Turkey's capital. Although the military coup was a once-in-a-decade occurrence in a country where "the military protects secularization", the basic storyline was similar: the military seized power and then relinquished it to prevent Turkey from drifting too far toward extremist Islam. At the start of the confrontation, with a Leopard 2A4 and a Leopard 1A5 main battle tank on the streets of Ankara, it appeared as though Turkey's secular regime would undergo another correction.

Nonetheless, what ensued was shocking. It was established that Turkish President Recep Tayyip Erdoğan had not been assassinated or subdued during the coup. Second, it was discovered that the Turkish military was not as united as previously believed, and the Turkish First Army, based in the Istanbul area, declared its opposition to the coup. The coup forces seized several strategic locations in Ankara but were unable to control the media, allowing videos of Erdoğan summoning residents to the streets to go viral. When police and pro-government citizens reclaimed these strategic locations, the coup forces surrendered without a fight. The coup army was defeated, and the tanks proved ineffective. Numerous naive coup soldiers were physically assaulted by Erdoğan's supporters after they surrendered, a heavenly irony. The coup's failure to control the Internet and the media was a critical blunder. Ultimately, the coup against Erdoğan became a coup for Erdoğan.

12.6 Psychological Warfare and Mind Confrontation

In response to the ambiguity of the boundary between psychological warfare and cognitive confrontation, the proliferation of information threats, and the persistent and stage-specific characteristics of confrontation, it is necessary to analyze and study the general rules of psychology and behavior of the population, officers, and soldiers, elite groups, and key figures at both strategic and tactical levels. It is necessary to integrate the new media information science achievements and make extensive use of big data, social and human sciences, and technologies of sound, light, electricity, chemistry, and biology to develop strategies of "psychological warfare and cognitive confrontation" for developing data models, overpowering the cognition, emotion, and consciousness intervention of specific groups, and finally deterring and coercing adversaries. At the same time, it is necessary to accurately monitor and intervene in the mental states of officers and soldiers, overcome physiological and psychological obstacles on the battlefield, and maintain a good mental state.

At the moment, important people's mental, physical, and spiritual states can be monitored using brain–machine technology, and their health and mental state can be accurately assessed through analysis of their expressions, voices, and behaviors. In the future, important personnel may be subjected to indirect or direct consciousness intervention and cognitive influence via psychological, physical, chemical, and biological means ranging from visual, auditory, tactile, taste, and smell to language, emotion, thinking, subconscious, and dreaming.

In October 2018, the Association of the US Army published a report entitled *The Influence Machine: Automated Information Operations as a Strategic Defeat Mechanism*, stating that the use of algorithm-generated content, personalized targeting, and an intensive mix of information dissemination, aided by AI, can create "influence machines" to conduct information operations that will have an exponential impact [3]. The report argues that the strategic value of "influence machine" information operations far outweighs the benefits of other forms of AI. It can analyze emotions and prejudices, and identify psychologically vulnerable groups using machine learning. It can then rapidly and intensively "shoot" tailored "mental ammunition" at targeted groups in order to achieve psychological and cognitive manipulation.

Research and development efforts should be directed toward developing and applying chemical and pharmaceutical brain control techniques. On December 1, 2018, the evening following the conclusion of the G20 summit, a historic dinner meeting between the US and Chinese heads of state ignited a single word: fentanyl. Fentanyl is a potent analgesic that has been abused extensively in the United States, resulting in severe social problems. Following this meeting, the White House stated that it was imperative that Chinese leaders agree to designate fentanyl as a scheduled substance, which would mean that those who sell fentanyl to the United States would face the harshest penalties under Chinese law. The hallucinogenic drug used in the Russian Moscow theater hostage incident contained a fentanyl-like substance that put the terrorists into a coma, but damaged the nervous systems of several hostages, and ultimately resulted in their deaths from overdosed inhalation.

References

[1] Sano, T. (2017). *Holography: The next disruptive technology*. US Army Research Laboratory. https://apps.dtic.mil/sti/pdfs/AD1033176.pdf

[2] Rasmussen, S. E. (2020). Iran nuclear scientist was killed with 'new method,' Tehran says. *Wall Street Journal*. www.wsj.com/articles/iran-nuclear-scientist-was-killed-with-new-method-tehran-says-11606764574

[3] Association of the United States Army. (2018). *The influence machine: Automated information operations as a strategic defeat mechanism*. www.ausa.org/publications/influence-machine-automated-information-operations-strategic-defeat-mechanism

13 Global Military Operations

Global military operations are a critical starting point in competitions for strategic power. The dream of large countries is to develop capabilities for global operations and military operations other than war (MOOTWs). This is the standard configuration of world-class armed forces and a critical field for building military intelligence. The key to global military operations is to strengthen long-range intelligent operation capabilities, particularly emphasizing global networked information system construction, space-based information, small satellite constellations, cross-domain multi-source intelligent perception, long-range command and control, strategic delivery, rapid mobility, adaptive equipment, and comprehensive logistics capability.

13.1 World Powers' Strategic Demands

While local wars and conflicts frequently occur in Africa and Latin America, Eurasia and the Asia-Pacific region have always been the focal point of geopolitical and economic competition, as well as the epicenter of strategic military competitions between world powers.

Eurasia has historically been a center of population, civilization, culture, politics, economy, and military powers, and it continues to be so today. Eurasia comprises 92 countries and accounts for 36.2% of the world's land area and 70% of the world's population [1]. According to Mackinder's Land Power Theory, the world is centered on the Eurasian and African continents, with the United States and Australia serving as outlying islands. The Eurasian continental margin is home to seven of the world's eight major sea lanes.

Following World War II, the Asian-Pacific region's emerging economies grew rapidly and gradually established their strategic positions. By 2020, the world's top three economies had all been located in the Asia-Pacific region. The US gross domestic product (GDP), for example, is 21.8 trillion US dollars. China's GDP is 15.5 trillion US dollars. And Japan's GDP is 5.2 trillion US dollars. The above three countries together account for 68.5% of the world's top ten economies, or more than two-thirds of the total.

DOI: 10.4324/b22974-15

The world's top ten populous countries or economies with the most military expenditures are all concentrated in Eurasia and Asia-Pacific, as are the world's major economic cooperation corridors.

Following World War II, the US military's presence was nearly global, with over 5,000 military bases worldwide, nearly half of which were overseas. Following the Cold War's end, the number of US military bases was significantly reduced due to changes in international relations and US military strategy and opposition from the populations of the countries where they were stationed. Currently, the US maintains 374 military bases overseas, spread across more than 140 countries and regions, housing 300,000 stationed troops in 871 bases, including 242 naval bases and 384 airbases [2]. The US military's overseas deployment is divided into three major theaters, each with its base group: Europe, the Middle East, and Asia-Pacific.

The US must consider various factors when establishing a military base, including geographic locations, natural conditions, infrastructure, and political factors. Local bases serve as the core, while overseas bases serve as the front. The US military places a premium on front bases as well as intermediate and rear bases along strategic transportation routes. The US military bases are organized to maintain control of strategic points and safeguard maritime centers, with small troops deployed in advance and large troops on the mainland. The US military currently controls maritime centers such as the Gulf of Alaska, the Strait of Korea, the Makassar Strait, the Sunda Strait, the Strait of Malacca, the southern tip of the Red Sea, the Strait of Mande, the northern tip of the Suez Canal, the Strait of Gibraltar connecting the Mediterranean and Atlantic Oceans, the Strait of Hormuz in the Persian Gulf, the Strait of Florida north of Cuba, and the shipping lanes connecting the southern tip of Africa to North America, and the Greenland–Iceland–British shipping lanes, etc. [2].

The three base groups located in Central, Southern, and Western Europe continue to be the most significant overseas military bases for the US military. Germany is the North Atlantic Treaty Organization's (NATO's) primary military deployment and weapons depot, led by the US, with up to 188 US bases and over 210,000 conventional troops at its peak. At least 60,000 US soldiers remain stationed in Germany, the majority of them Army troops. These forces are believed to be preparing for a possible war with Russia and participating in regular NATO exercises.

The Asia-Pacific and Indian Ocean regions are critical strategic locations for the US, and the US military has announced that it will devote 60% of its overseas military presence to the Asia-Pacific region. For many years, the US military maintained numerous bases in this area, and the total number of overseas bases is second only to that of Europe, with the majority located in Japan, South Korea, and other countries and regions.

The US military maintains 14 bases in Japan, including three Army bases, four Marine Corps bases, four Navy bases, and three Air Force bases.

The US maintains two sizable permanent military bases in South Korea, one army and one naval. The US maintains military bases in Singapore, Kyrgyzstan, the Marshall Islands, Australia, and Antigua and Barbuda.

US military bases in the Asia-Pacific and Indian Ocean regions, with eight base groups, account for approximately 42.7% of US bases overseas [2]. These bases are broadly classified into three lines: the first line consists of four base groups located in Alaska, Northeast Asia, the Southwest Pacific, and the Indian Ocean, which control strategically vital shipping routes, straits, and seas. The second line is composed of Guam and base groups in Australia and New Zealand, which serve as the backbone of the first line's bases, acting as critical air and sea transportation hubs as well as surveillance and reconnaissance bases. The third line comprises base groups in the Hawaiian Islands, which provide rear support for operations in the Asia-Pacific region and outposts for the defense of the US mainland; Elmendorf Air Force Base in Alaska, also known as the Alaskan Air Force Command base; naval bases in Yokosuka and Okinawa, Japan; air force bases in Wushan and Seoul, South Korea; a naval base on Diego Garcia Island in the Indian Ocean; and Anderson Air Force Base and Apala Naval Base on Guam.

The US military uses overseas bases during peacetime to familiarize itself with the local climate and geography and conduct exercises and training. In times of war, overseas bases can provide immediate military support to the US military. They are ideal for aircraft takeoff and landing, replenishment of ships, transportation of military materials, and comprehensive logistics. Global military operations are difficult to conduct without global military bases and layouts.

13.2 Global Information Network System

Global military operations necessitate establishing a globally networked information system capable of global sensing, command, and control. With bases located worldwide, the US military is also investing heavily in developing a global network operation system of systems (OSoS). The US military, guided by the network-centric warfare doctrine, adopted an open architecture in building the Global Information Grid (GIG) that enabled command, control, communications, computers; intelligence, surveillance, and reconnaissance (C⁴ISR) to transition from information support to decision support. This demonstrates the US military's capability to achieve "cross-domain collaboration", "mission command", and "globally integrated operations", as well as to rapidly network and integrate globally deployed forces.

The US military is committed to civil–military fusion and places a premium on space-based resources in terms of command and communication. Satellite communications account for more than 70% of the US military's communications. The US military satellite communication system (MilSatCom) consists of: (1) a broadband satellite communication system, represented by the Defense Satellite Communication System-3 (DSCS-3) satellite; (2) a narrowband satellite communication system, represented by the UFO satellite; and (3) a military satellite communication system, represented by the Military Star satellite (Milstar). The satellite communication system is capable of providing real-time, secure, anti-jamming communication services to users at strategic, operational, and tactical levels, as well as covering all levels of US military users and weapon terminals worldwide. In terms of information services, the US military has built the Joint Information Environment (JIE) based on cloud computing, mobile computing, and big data technologies, launched the Army's Global Network Enterprise Construct (GNEC) program, and extended cloud services to soldiers at the edge of the tactical spectrum through programs such as the Content-Based Mobile Edge Network (CBMEN). Logistics and equipment support systems are critical components of the US military's operation system. It introduces the concepts of "complete asset visibility" and "precision support" to dispel logistics' "resource and demand fog", reducing support material backlogs and increasing support efficiency. Precision support requires visibility of assets in storage, transport materials and equipment status, troop security, and the sharing of material information between operation troops and security forces. Additionally, the US military established the Global Transport Network (GTN) to facilitate the global strategic delivery of operation forces and materials.

Both traditional and emerging military powers must develop global network information systems with space–ground connectivity, civil–military fusion, general and specialized purposes, self-assembling, anti-jamming, and anti-destruction features. In addition, global, networked, distributed, and intelligent military cloud platforms and service systems should be built based on civil intranets, encrypted networks, and confidential communication networks, as well as dedicated communication networks in specific areas of the battlefield, in order to ensure communication and data transmission on demand at any time and in any location and to improve global interconnection, cross-domain perception, mobile command, and joint operations.

13.3 Space-Based Resources Application and Control

Space-based information resources are critical for supporting global military operations. Currently, space powers are developing military satellites, with the US maintaining a global lead and accelerating the intelligence process. Russia, Europe, and other countries are all expanding their capabilities in this area as well. As of November 2020, the Union of Concerned Scientists' Statistical Database of Orbiting Satellites indicates that 3,300 satellites are in orbit in various countries worldwide. The US leads the way with 1,850 satellites, or 56% of all artificial satellites, including some military satellites. China is second with 400 satellites, or 12.1% of the total. Russia comes in third place with 200 satellites, or 6% of the total. There are 850 other countries. China, the United States, and Russia control 74.2% of the world's satellites collectively.

According to the *Blue Book Report on China's Space Science and Technology Activities*, 114 space missions were launched globally in 2020, exceeding the previous record of 1,277 launched spacecraft [3]. The United States launched 44, China launched 39, and Russia launched 15, accounting for more than 87% of all launches worldwide.

The US maintains a global leadership position in military satellites, and was the first country to employ military satellites in an operational capacity. US military satellites account for nearly 40% of all military satellites in the world. With allied satellite resources applied, the number of US military satellites available for application exceeds 50% of the total available worldwide.

In terms of satellite reconnaissance, the US has developed military-oriented, commercially supported military satellite reconnaissance and surveillance systems at least one generation ahead of other countries. The United States has developed systems for information fusion, such as the Eagle Vision ground station and the Digital Topographic Support System. In the US, more than 35 military reconnaissance satellites are in orbit, with high resolution (0.1 m), global coverage, and a strong capability for operational response. The US also obtained commercial imagery equipment with a maximum resolution of 0.31 m. Electronic reconnaissance satellites in high orbit can maintain surveillance 24 hours a day, while ocean surveillance satellites have global coverage and positioning accuracy of 2 kilometers. Additionally, the US has developed "quick-response" satellites and other tactical satellites to extend services to the command chain's end, enabling ground systems and equipment to perform faster operations.

In terms of environmental satellite detection, the US has established a network of detection satellites covering meteorology, oceanography, magnetic fields, gravity fields, and other environments, and has achieved global environmental monitoring with high spatial and temporal resolution through civil–military fusion and international collaboration. The US has developed integrated weather systems, meteorological information-processing systems, and other satellite platforms to ensure safe land, sea, and air operations. The US currently has over ten military environmental perception satellites in orbit, collecting timely data such as meteorological and oceanic conditions in militarily sensitive areas. Meanwhile, the US used dozens of civil environmental exploration satellites and gravity field measurement satellites from its allies to complete the global geography model with a measurement accuracy of 1 cm, a detection accuracy of 1 mGal for gravity field anomalies, and an 80-kilometer spatial resolution.

In terms of satellite communications, the US has developed a military-oriented, commercially supported military satellite communications system that encompasses four series: broadband, narrowband, protection, and relay. The system is designed to provide continuous and global coverage in critical regions. The US currently has more than 60 military communications satellites in orbit, providing global coverage, multi-band, multi-functional, and highly survivable capabilities that are at least a generation ahead of other countries. The US has established nine major teleports throughout the world to facilitate the integration of satellite communication systems into the GIG. The US Army, Navy, and Air Force each has tens of thousands of broadband, narrowband, and anti-jamming terminals. At the moment, the US Distributed Common Ground System (DCGS) is critical in integrating space reconnaissance intelligence with air, ground, and sea information. DCGS's standard interface, framework, and specifications have facilitated the integration of spatial data and equipment into military arms and services.

Due to the lengthy design, manufacturing, launch, and test cycles of traditional satellite systems, there are numerous limitations in responding to tactical battlefield applications with real-time response requirements. Since the late 1980s, small military satellites have grown in popularity, extending the capabilities of traditional military satellites to tactical applications. While small satellites have not yet been equipped with large-scale equipment and operational military applications, the US has conducted technology verification and application exploration in the fields of imaging reconnaissance, environmental surveillance, data relay, early-warning surveillance, and space countermeasures. The military has been demonstrating the feasibility of integrating small satellites into the global positioning system (GPS) to achieve enhanced navigation signals. In imaging reconnaissance, the US small satellites have had a maximum resolution of less than 1 meter. In data communications, the US Army has completed a test of voice communication using a 3 U CubeSat. In space countermeasures, the US has mastered low-orbit small satellite space attack and defense capabilities and is currently testing the feasibility of high- and low-orbit space target surveillance using a constellation of micro-nanosatellites.

At the moment, the US leads the world in small satellite reconnaissance and surveillance capabilities, with the Operational Response Space program serving as a pioneer. The US has revealed a series of imaging reconnaissance and environmental surveillance small satellite projects to validate the "theater command and control – mission response – imaging reconnaissance – satellite processing – information downstream" model.

Small satellites, the US military believes, are valuable because of their low cost, launch flexibility, and high efficiency-to-cost ratio. With limited defense budgets, small satellites have become a critical component of the US's space cost management strategy. Small satellites have short production cycles and low costs, enabling them to be deployed in large numbers for various military demands. Small satellites allow rapid mass deployment in that multiple satellites can be launched from a single carrier rocket, via air launch, or via on-orbit ejection, lowering the cost of access to space. Small satellites may compete with larger satellites by utilizing satellite networking and lowering orbit heights. According to US government research, for every twofold increase in resolution, the weight and cost of satellites must be increased by $4\frac{1}{2}$–8 times, while lowering the orbital height of small satellites can bring their resolution closer to that of large satellites.

The launch and application of small commercial satellites have accelerated in recent years. O3b, One Web, SpaceX, Google, Facebook, and Boeing, among others, regard low orbits, 200–2,000 kilometers above the Earth, as the orbital Internet's gold mines. They have announced their satellite launch plans in China, from state-owned enterprises such as Aisino Science and Technology and Aerospace Science and Industry to several private space companies, demonstrating their ambition to compete in the orbital Internet race.

On June 26, 2017, "Iron Man" Elon Musk's startup company SpaceX completed two consecutive rocket launches and recoveries within 48 hours, a first in human spaceflight history, while his rival Greg Wyler's One Web had just received a Federal Radio Commission license.

One Web intends to bring Internet service to remote rural mountainous areas and other areas not served by base stations or fiberoptics by connecting hundreds of low-orbiting satellites, allowing 3 billion people to connect to the Internet. According to the US Federal Radio Regulatory Commission, Wyler intends to launch the planned One Web satellite during the 6-year license period. The first-generation low Earth orbit (LEO) constellation designed by One Web consists of 648 operational satellites and 234 backup satellites, for a total of 882 satellites. These satellites will be distributed evenly across 18 orbital planes, with a 1,200 km orbit height, Ku-band capability, a single satellite transmission rate of 6 gigabits per second, a single satellite mass of 150 kg, and a scale flow production cost of less than $1 million. The majority of the bandwidth for One Web satellites has been sold, and Wyler anticipates adding another 2,000 satellites for a total of 2,882. One Web's objective is to provide Internet access to every school that lacks it by 2022 and close the global digital divide by 2027.

SpaceX has been in discussions with the Federal Radio Regulatory Commission for months following One Web's proposal, and Musk has proposed an even larger Starlink satellite Internet plan than One Web. After the program is completed, two massive low-earth orbit satellite constellations will be formed, with 7,518 satellites orbiting at an altitude of 340 kilometers in the inner layer and 4,425 satellites orbiting at an altitude of more than 1,000 kilometers in the outer layer. In October 2019, SpaceX proposed another plan to add 30,000 small satellites, eventually forming a 42,000-satellite space-based Internet, and integrated operations in a future military field based on the "Nebula" are likely to emerge. In terms of economics and cost, the launch, maintenance, and operation of tens of thousands of satellites are prohibitively expensive, far exceeding the costs of ground-based 5G communication, and are difficult to sustain without the assistance of national defense funds.

It is foreseeable that as space science and technology advance, the cost of developing and launching satellites will continue to decline, and the integration of heaven and Earth communications will become the new era's standard and trend. Satellites enable the networking, command, and control of every corner of the globe, and they will become the primary support for global military operations in the future intelligent era.

While military powers are constructing and applying space-based information, special attention should be paid to adjacent space platforms, cognitive network communication systems, and solar-powered unmanned aerial vehicles (UAVs) in order to create a global, multi-level, and diversified networking communications support system.

13.4 Strategic Delivery and Rapid Action

In general, the strategic delivery of global military forces is accomplished through military and civilian vehicles, classified as air, ship, rail, road transport, and other modes. While air transport is quick and efficient, the volume of cargo it transports is small, and the cost is high (Figure 13.1). Ship transport is inefficient in terms of time, but it has a large transport volume and can operate concurrently with the fleet. While road transport provides excellent mobility, it is susceptible to road conditions. Road transport is well suited for small-scale, short-distance delivery due to its low speed. In comparison, railway transport has some obvious advantages, such as speed and volume, but it is also influenced by line accessibility and layout.

Along with overseas bases and a global information network system, the military must establish fast and mobile operation forces equipped with weapons in order to conduct global military operations. For example, in the strategic realm, the US Air Force Global Strike Command is responsible for commanding two major weapons: land-based intercontinental missiles and long-range bombers, carrying out nuclear deterrence and strikes. For many

Figure 13.1 China's large transport aircraft "Yun-20".

years, the US also established the Marine Corps, Army Special Forces, and Quick Reaction Force for global emergency military operations.

The US Marine Corps [4], the only amphibious assault operation force among the five independent branches of the US military, is tasked with the responsibility of utilizing fleet ships under the US Navy to arrive at the scene of a crisis anywhere in the world and conduct combat missions. Additionally, it is a military service capable of conducting operations missions on its own command.

The US Marine Corps, the world's first elite force, is the long-standing main operation force of the US Quick Reaction Force. The US Marine Corps and the US Navy are administratively subordinate to the US Department of Navy within the US Department of Defense. The US Marine Corps, which shares ranks with the US Army and Air Force, is a separate branch of the US military. It is also responsible for the security guards assigned to the US Department of State's embassies in foreign countries. Typically, it serves as an independent military service that conducts training and executes operational missions.

Its missions are divided into three broad categories. To begin, it conducts amphibious offensive operations, and this is its primary mission. Second, the Marines are stationed at various naval bases and major fleets around the world at the request of the president and the Department of Defense to carry out defense missions. Third, it carries out decapitation strikes, hostage rescues, and other specific presidential tasks.

Currently, the US Marine Corps has a total of approximately 194,000 members, including a reserve force of approximately 38,000. The US Marine Corps is composed of ground, aviation, and logistics forces. It is slightly larger than any other service, with four Marine divisions and four aviation companies. The Marine Division was 20% larger in size than the US Army's motorized infantry divisions, with approximately 18,500 personnel. The service support brigade is the basic formation, and its primary mission is to provide a variety of material support, equipment maintenance, and medical assistance to front-line operation units, as well as to ensure that each operation unit maintains a high degree of operational independence.

The US Army Special Forces, commonly referred to as the Green Berets due to their honorable green berets, evolved from several irregular units that fought behind enemy lines during World War II [5].

The Army Special Forces are comprised of 29,000 personnel and include active-duty and reserve forces, as well as the National Guard Special Operations Forces. Its active special operation forces are composed of five battalions: the 75th Ranger Regiment, the 1st Delta Rangers, the 160th Special Operations Aviation Battalion, the 4th Psychological Operations Battalion, and the 96th Civil Affairs Battalion, as well as several communications and logistics support detachments.

The Special Operations Group is the Army's most elite Special Forces formation, and its units are known as Green Berets. Each special operations group is composed of between 800 and 1500 personnel and is organized into three special operations battalions, one direct special operations company, and one support company. The Green Berets are armed with a variety of infantry weapons and transport helicopters, as well as advanced communications equipment, including satellite and light communications with a range of over 3,000 kilometers.

During the Grenada incident, the Green Berets successfully deployed a small number of commandos to seize control of the island's prisons, airfields, and critical military installations prior to the Marines' mass landing, ensuring successful late operations and minimizing war casualties. In the 1990 Gulf War, the Green Berets again served as the lead US force, successfully conducting the operation code-named "Desert Storm" into Kuwait with the cooperation of local armed forces.

Following the "9/11" terrorist attack, the US refocused its war on anti-terrorism on a global scale, emphasizing the role of special forces in critical missions such as emergency rescue, arresting, and decapitating key adversary figures, a case in point being the "Black Hawk Incident" during the Somali War in 1993. Throughout the Afghanistan and Iraq wars, US Army special forces infiltrated and gathered intelligence in advance, arrested the war's primary targets, dubbed the "people on playing cards", and successfully captured 32 individuals, including Saddam Hussein, his family members, and bin Laden's assistants.

The US Rapid Deployment Joint Task Force, colloquially referred to as the US military's Quick Reaction Force, is comprised of the Army's, Navy's, and Air Force's most elite units [6]. The US military experienced the process of growing from a small number of troops to a large number, from a single service to multi-service cooperation. Initially, the US Quick Reaction Force had a strength of slightly more than 30,000 and was composed primarily of the Army's 82nd Airborne Division, Army Special Forces, and five Marine Corps battalions. By 1984, the US Quick Reaction Force had grown to five ground forces, seven Air Force wings, three aircraft carrier groups, and other forces with a combined strength of more than 200,000 personnel.

On March 1, 1980, the US military established the Quick Reaction Force Command to oversee the unified command of the US Quick Reaction Force. The command was transferred to central headquarters on January 1, 1983 and reports directly to the US military's Joint Chiefs of Staff.

The US military's Quick Reaction Force is strategically controlled directly by the highest military authority, and it is the most adaptable military tool available to the country for achieving its political and diplomatic objectives. The US military believes that Quick Reaction Forces, with their rapid deployment, raids, and reinforcements, can "contain enemy aggression and control developments" as well as "conduct emergency and rapid reinforcement operations". It is the military's sharpest knife, capable of filling roles that other large units cannot. It is responsible for three primary functions.

First, swift deployment. The US Quick Reaction Force is generally operationally ready. The military alert system in the US is divided into five levels. Level 1 alert is the highest level of readiness, with troops ready to deploy immediately upon command. For example, ground forces have a standby time of approximately 2 hours, while warships and aircraft have a standby time of approximately 0.5 hours. Within 2–12 hours of a level 2 alert, the commander must enter command posts, warships must depart port and sail, and ground forces must withdraw. Level 3 requires personnel to be off leave for 290 days, with more than one-third of the force on duty, heightened air and maritime vigilance, and ground forces to mobilize within 12–24 hours. Certain combat-readiness units are tasked with the responsibility of enhancing intelligence secrecy measures. Level 5 is normal. Quick Reaction Forces are required by the US military to airlift brigade-sized forces to any crisis area in the world within 48 hours and maneuver division-sized forces to accident areas within 4 days. For example, during the Gulf War, Bush signed the Desert Shield action plan at 2:00 a.m. on August 7, and at 7:35 a.m., the US 82nd Airborne Division's advance forces boarded planes at Fort Bragg, North Carolina, and flew to Saudi Arabia. Later that year, the 101st Air Assault Division, 24th Air Infantry Division, and other Quick Reaction Forces arrived in Saudi Arabia. US Quick Reaction Forces quickly bolstered Saudi Arabia's defense forces and encircled Iraqi forces.

Second, lightning-fast assaults. The US Quick Reaction Forces primarily intervene via airborne and air–land means in order to quickly intervene in emergencies that jeopardize US interests or to conduct rapid counterattacks against the enemy of sudden attacks in order to control the situation's development and seize the enemy's battlefield priority. The 82nd Airborne Division, for example, is the apex of the United States' airborne forces, with 12,800 paratroopers. It was parachuted into France's Notre Dame Church during World War II to complete the Normandy landing. Following the war, the division served as the "sharp edge" of the United States' involvement in various hot spots around the world, participating in the war against Vietnam and playing a key role in the rapid raids on Grenada, Panama, and other hot spots.

Third, rapid reinforcement. The US military maintains a legion of military bases throughout the world, each with a limited number of troops, and if an emergency occurs, these troops are typically outnumbered and unable to respond. A critical mission of the US Quick Reaction Forces is to provide immediate support to the US military in emergency situations by rapidly rushing to the accident site via advanced air transport and conducting emergency support operations.

13.5 Consumption of Overseas Operations

Global military operations, whether in the current information-based war or the future intelligent war, are characterized by "huge consumption" from operational, logistical, and equipment support perspectives.

First, rising costs. Weapons and other military hardware are becoming increasingly informatized, intelligent, and integrated. A sophisticated piece of weaponry frequently incorporates numerous scientific and technological advancements, incurring enormous costs. The US Department of Defense compared the 13 major technical characteristics of new- and old-generation fighters in the early 1970s. The results indicate that for every twofold increase in aircraft performance, the research cost increases by 4.4 times and the manufacturing cost increases by 3.2 times. High-performance weapons require increased development, procurement, and maintenance costs. Tanks used to cost $50,000 to manufacture at the end of World War II, fighter planes cost $100,000, and aircraft carriers cost $7 million. Weapons prices increased dozens, if not hundreds, of times during the Gulf War. For instance, the M1 tank cost $2 million, the equivalent of 40 tanks in World War II; the Patriot missile cost $1.1 million; the F-15 fighter jet cost $50.4 million, the equivalent of 500 planes in World War II; the F-117 stealth fighter bomber cost $106 million; and aircraft carriers cost $3.5 billion, nearly 500 times what they cost previously.

From the Gulf War to around 2020, the cost of weapons and equipment continued to rise, particularly the cost of the new generation of high-tech weapons and equipment, which has increased several or even dozens of times in price.

The cost of the main battle aircraft has increased four- to fivefold since the previous generation: an F-22 fighter aircraft costs more than $200 million per unit.

The bomber's price increased by a factor of 20–30: the B-2 cost more than $600 million.

Transport aircraft are 30–40 times more expensive: a C-17 costs more than $200 million.

The main battle tank costs six to seven times as much as it did previously: $8.5 million for the Japanese type 90.

Aircraft carriers are now three to four times the cost of previous generations: the Nimitz-class carriers cost $13 billion.

While the cost of some future intelligent equipment may decline, the majority of them will continue to rise as a result of mechanization, informatization, and intelligentization.

Second, the increasing consumption of material goods. For example, during World War II, the average daily material consumption per soldier was 20 kg, 90 kg during the Vietnam War, and 200 kg during the Gulf War. The Korean War consumed 18,000 tons of ammunition per month, the Vietnam War consumed 77,000 tons, and the Gulf War consumed 357,000 tons. The exponential growth in the consumption of battlefield supplies has created significant logistical challenges. During the Gulf War, the US established the world's most extensive logistical transportation system to meet critical operational requirements. Ninety percent of the Military Air Transport Command's transport planes were used, and more than 30 airlines were leased, including domestic carriers as well as Korean and German airlines. In maritime transport, the Military Maritime Command deployed 135 transport ships; the Reserve Fleet deployed 170 merchant ships, and rented 78 foreign ships. On the US mainland, 2,400 transport cars from seven states were used for ground transportation, while Saudi Arabia organized 5,000 transport cars. The total amount of materials shipped from the United States to the Middle East during the Gulf War was estimated to be 186 million tons, the equivalent of transporting a medium-sized US city like Atlanta, Georgia into the Gulf of Mexico.

The US military spent $194 million per day in World War II, $230 million per day in the Vietnam War, and an average of $1.4 billion per day in the Gulf War. During ground combat, an armored division of the US Army consumes between 1,900 and 2,800 tons of fuel per day; an aircraft carrier group composed of eight surface ships consumes an average of about 1,800 tons of fuel per day. With a single serving of military fast food costing approximately $6.77, it would cost $6 million per day to feed 300,000 US and British allied troops and $9 billion to return troops and equipment at the end of the war. The Gulf War cost up to $99 billion in total.

In conclusion, cost reduction and efficiency enhancement are critical issues for the world's leading military powers, particularly those conducting global military operations, and these issues have become a focus for future military procurement reform and management innovation.

To adapt to the joint operation requirements of modern warfare and optimize logistics resources, the US military has placed a premium on developing an integrated joint logistics support system and a series of new logistics concepts and models, including "focused logistics," "joint logistics," "supply chain logistics," and "perception and reaction logistics" [7].

As the US military shifts from "forward deployment" and "rapid reinforcement" to "forward presence" and "power projection," a significant amount of US weaponry and equipment, operation supplies, and logistical units and detachments are withdrawn to the homeland. To align the logistics force with the target's actual requirements, the US military has recently streamlined logistics agencies and facilities in order to establish a leaner, more efficient, mobile, and flexible joint logistics support system.

Between 2010 and 2015, the US military logistics system was reduced by one-third, with the closure of 97 large facilities, the restructuring of 55 others, and the elimination of 30,000 Department of Defense positions. The majority of reductions are in management and support staff [7].

According to the US military, quick response capability will be critical in twenty-first-century wars, and the US military must be able to deliver one light brigade in 4 days and five divisions with combat service support forces in 75 days to any trouble spot [7]. Since the US military transitioned from "forward deployment" to "forward presence", large-scale strategic force delivery has become the primary method of resolving various regional crises. Currently, Transportation Command is primarily responsible for the US military's strategic delivery. The Air Force currently operates more than 300 strategic transport aircraft, including C-17s, C-141s, and C-5s, supplemented, when necessary, by more than 400 large civil aviation transport aircraft, providing a day-and-night delivery capability of more than 67 million ton-miles. In a strategic sealift, the US Military Shipping Department maintains a fleet of over 200 ships of various types and hundreds of ships with reserve forces and has bolstered the development of pre-positioned forces at sea to provide timely mobile support to troops conducting emergency operations. Additionally, to ensure uninterrupted strategic delivery, the US military invested heavily in infrastructure, equipment, and construction support equipment, such as renovating military bases, warehouses, airports, and ports, as well as improving transportation conditions through the acquisition of loading and unloading equipment.

The US military believes that a sufficiently sized joint logistics system should be established to facilitate force transfer, material transport, and mission support. The US military has recently invested heavily in a variety of measures to successfully establish a separated logistics structure that locates the logistics command and management organization and the primary support force on the homeland, thus reducing the logistics configuration on the battlefield. Domestic agencies rely on advanced information technology and modern transportation to deliver logistics forces on time, while operation units can claim supplies via forward bases, establishing a two-tiered security link between domestic and overseas operation units. This places a premium on the capability of strategic transportation and delivery.

The US military also attaches great importance to the role of overseas military bases in logistics supply. The US now maintains military bases in the majority of the world's countries to supply logistical support to US forces stationed nearby. For instance, Kuwait's Doha Barracks serves as the US military's primary base and major logistics hub in the Middle East, housing numerous M1 main battle tanks, M2 infantry fighting vehicles, Apache helicopters, and other combat equipment capable of equipping an armored brigade. The US maintains bases in Turkey (Incirlik base) and Qatar (Camp Snoopy, and As Sayliyah Army Base), where it is constructing weapons depots to support US forces conducting combat missions throughout the Middle East, Southwest Asia, and even Africa. The US military base on Diego Garcia in the Indian Ocean is equipped with enough supplies to equip three armored battalions and three mechanized infantry battalions. The US signed its first Military Logistics Cooperation Agreement with India, which allows the Army, Navy, and Air Force to share Indian military bases for logistical resupply, maintenance, and recovery. The US military has established bases in Yokosuka, Japan; Changi, Singapore; and Pusan, South Korea. The majority of these bases are located on the periphery of the US regional containment zone. When the US military employs forces in these areas in the future, it will quickly acquire the various weapons and equipment required to fight the war. With a well-established system of military base networks, the

US is capable of long-range strategic delivery of operation power and precise logistical supplies.

13.6 New Developments in Overseas Forces

In an era of globalization and intelligence, with an increasing number of overseas missions and diverse operational adversaries, the requirements of overseas operations forces are defined by systematization, unmanned, networked, lightweight, remote, and multi-domain integration, cross-domain attack and defense, and virtual–real interaction. Thus, in addition to traditional equipment upgrades, there is a requirement for the development of technical capabilities such as mixed human–machine formations, light-weighted, rapid adaptation passenger loading, long-range three-dimensional delivery, networked battlefield awareness, unmanned three-dimensional protection, manned and swarm airdrop, unmanned amphibious assault, advanced soldier systems, anti-terrorist attacks, and non-lethal weapons.

Army special operations and peacekeeping forces. Army special operations and peacekeeping forces must meet the demands of international counter-terrorism and stability maintenance, emergency rescue and disaster relief, safeguarding interests, security alert, international peacekeeping, and international rescue. They should conduct technical research on equipment systems and lightweight design, human–machine hybrid grouping, future soldier systems, long-range three-dimensional delivery, rapid adaptation of battle vehicles and personnel loading, comprehensive vehicle protection and unmanned escort protection, cyber and magnetic operations, and non-lethal strike, in order to develop a rapid response, efficient disposal, and protection capability in an overseas operational environment.

The Marine Corps. The Marine Corps must bolster new capabilities such as networked perception, ship–ground/submarine-launched precision fire and cruise missiles, unmanned amphibious/multi-dwelling vehicles, low-altitude unmanned reconnaissance, urban operations, special operations, and intelligent man-portable systems to meet the demands of shore–sea integration, cross-domain assault, and multi-domain operations on future complex battlefields.

Airborne troops. Airborne troops must prepare for an uncertain future battlefield environment by strengthening their equipment's lightweight, systematic, and modular design, as well as their rapid loading and adaptability via a series of transport aircraft and air delivery means. Advancements in manned vehicle or equipment airdrops, long-range unmanned equipment airdrops, multi-source networked battlefield perception, high-precision swarm airdrops, and unmanned three-dimensional detection and strike should be made. Additionally, special operations, emergency rescue, natural energy utilization, a dependable food supply chain, and other technical capabilities are required. These advancements and capabilities will assist the airborne troops in conducting operations, defenses, and MOOTWs autonomously behind enemy lines and in unfamiliar environments.

13.7 Intelligent Global Action

Traditional mechanized warfare's primary objective is to annihilate the adversary's forces, attack cities, and seize land. In modern information warfare, high-tech weapons can be used to strike an adversary's economic infrastructure, such as command centers, transportation hubs, power plants, and industrial centers, from outside the defense zone, disrupting or suspending the economy, military operations, and social activities and compelling the adversary to surrender without occupying the adversary's country. The Kosovo war is a textbook example. Simultaneously, cyber, electronic, and information warfare have taken on increasingly significant roles and responsibilities in a number of recent local wars. With the advent of unmanned, intelligent, and hypersonic weapons in future warfare, strategic, operational, and tactical maneuverability will significantly improve, rapidly expanding the battlefield and increasing troop advance speed. As the line between front and rear becomes increasingly blurred, the battlefield will be characterized by multi-domain intersection, rapid integration, and high-speed movement, necessitating an increase in demands for intelligent perception, decision-making, attack, and support.

Individual intelligent equipment and systems must be used to support intelligent global military operations. This includes developing intelligent OSoS, information systems, open mission systems, support systems, as well as distributed, networked, and cross-domain operation systems. The emphasis is on system planning at the highest level, open task requirement analysis, generic capability support architecture, and common technical underpinnings.

13.7.1 Operation Simulations

Global military operations conducted by major powers are diverse in terms of missions, operations, and adversaries. To increase the intelligence level of an OSoS, it is necessary to construct a distributed, virtualized, and cross-domain operation simulation system, where possible future operation objects, adversaries, patterns, equipment strength, and readiness for operations can be analyzed, and confrontation or training under various conditions can be simulated to accumulate data and optimize models, thus increasing the ability to win. The purpose of operation simulations is to conduct experiments for major power confrontations, proxy wars, counter-terrorism, MOOTWs, distributed heterogeneous information intelligence fusion, cross-border/cross-domain mobility, joint operations, diversified mission planning and decision-making, joint tactical operations, precise and rapid logistics, international cooperative peacekeeping, and security. It also includes semi-physical simulation and military exercise training verification when necessary.

13.7.2 Multi-Source Intelligent Perception

It is necessary to establish information intelligence fusion platforms for military networks, open-source information, Internet of Things, and traditional artificial intelligence resources on a global scale, as well as to establish databases and model repositories for countries, regions, and battlefield environments. To begin, structured and unstructured data, radar signals, spectral information, video signals, audio signals, and text files from various detection platforms and information sources must be unified for screening, labeling, sorting, and storage in order to facilitate computing, processing, machine learning, modeling, and intelligent identification. Second, under the unified space–time reference platform, it is necessary to construct the precise location and motion trajectory of multi-sensor and multi-target signals following varied transmission time delays. Third, it is required to develop a correlation link between the detection systems and methodologies used to identify the same target. Fourth, it is vital to classify dangers and provide early warnings about future trends based on precisely identified, monitored, and located data. To summarize, situation assessment, social public opinion analysis, remote command and control, multi-domain and cross-domain maneuver, swarm cooperative attack and defense, fire strike, and damage requirements for various weapons and equipment can be completed using multi-source information fusion and intelligent perception.

13.7.3 Dynamic Mission Planning

In the age of artificial intelligence, global military operations necessitate the development of OSoS-oriented networked systems, adaptive planning systems for different missions, distributed coordination systems, and the formation of necessary databases and model repositories. Notably, intelligent systems based on self-confrontation and dynamic mission planning are critical for strengthening global intelligent decision-making capabilities.

In terms of networked operating systems, Windows + Intel, the Internet era's prevailing operating system and hardware, has been working successfully for many years and has completed multiple versions of upgrades. Mobile operating systems, such as Android and iOS, have also gained popularity and widespread use. Technical systems and private network operating systems have been developed and deployed for the civilian population. However, networked operating systems and self-adaptive mission-planning systems for military usage are either uncommon or in development.

At the moment, the primary methods for dynamic mission planning and operational decision-making are classified into two broad categories: the traditional method, which includes differential decision-making and expert systems, and the intelligent method, which includes reinforcement of learning algorithms, genetic algorithms, particle swarm optimization (PSO), and ant colony optimization (ACO).

Differential game theory is a branch of game theory that examines the optimal aim of two or more decision-makers when their actions are applied simultaneously to a differential equation-based motion system. Solving a differential game is difficult, and the numerical solution is frequently viewed as the optimal solution. The gradient, singular perturbation, polynomial, suboptimal, parameter, functional analysis, and linear feedback methods are the primary approaches for solving numerical solutions to differential games. Expert systems are computer systems that rely on the empirical knowledge of human experts to solve various unstructured problems in specific fields. They are particularly well suited for complex systems and complicated problems that are not fully understood and are inconvenient to solve through partitioning. The shortcomings of traditional decision-making approaches include ambiguous links between rules, ineffective search methodologies, and a lack of ability to learn.

Reinforcement learning theory can be used to address the low present accumulation of real-world data and insufficient trainable samples. The genetic algorithm is a stochastic optimization approach for addressing optimization issues that were created and developed by simulating Darwin's theory of evolution and Mendel's genetics. Its fundamental phases are: (1) coding, which establishes initial populations; and then (2) employing genetic operations such as selection, crossover, and mutation, which facilitates a structured yet random exchange of information. In this way, positive characteristics are inherited, and negative ones are eradicated. It is applied to the successful solution of combinatorial optimization problems, such as travel merchant issues, based on the characteristics of the genetic algorithms (i.e., the shortest route back to the starting point after a traveling salesman visits multiple locations). PSO and ACO are intelligent optimization methods inspired by natural phenomena such as bird flocks and ants foraging for food. At the moment, these new algorithms are widely used in a variety of domains, including combinatorial optimization.

The Defense Information Systems Agency (DISA) upgraded the Joint Global Command and Control System (GCCS-J) in 2015 by modernizing the software and substituting the newly designed Adaptive Planning and Execution (APEX) software for the legacy collaborative force-planning software. Self-adaptive planning and execution software offers dynamic and adaptable operational plans, transforming formerly fragmented planning capabilities into integrated, interoperable, collaborative planning capabilities [8].

Rafael, a defense company of Israel, announced their completely autonomous "Fireweaver" cyber attack system in June 2018. This is a self-contained software system that is used to allocate fire strike tasks to battlefield reconnaissance and firepower units. It can automatically assign appropriate firepower units to rapidly carry out strike operations based on target information provided by the reconnaissance equipment and concurrently strike multiple targets on the battlefield. The system fully exploits the interconnectivity of the modern battlefield to elevate the intelligence level of fire strike mission command, reducing the time interval between discovery and strike and improving the effectiveness of strikes against time-sensitive targets.

In the future, as operation mission-planning systems advance, the degree of intelligence will continue to improve and may even reach a cross-generational leap.

13.7.4　Intelligent Remote Logistics

At the moment, the Internet and the Internet of Things are widely employed in the civil sector for e-commerce, logistics and transportation, telemedicine, etc., resulting in revolutionary changes to the public's job and life. For example, the CaiBird Alliance maintains complete control over the transportation of online purchases, and DiDi's ride-hailing service has significantly improved travel convenience for people. It is reasonable to expect that when these civilian information technologies are fully utilized, they will have a significant impact on the military operational support model. Since the Gulf War, the US military has reduced support response times by 70–80% and increased support efficiency significantly by introducing civil logistics transportation and item management technologies, as demonstrated by local wars, avoiding the situation in which hundreds of thousands of containers were returned to the US without being opened during the Gulf War. In recent years, civilian technologies such as Google's intelligent data analysis, Apple's intelligent voice control system, IBM's smart speaker, and others have been applied to the military, undermining traditional information processing and data analysis methods.

IBM received a US$135 million order from the US Army Logistics Support Bureau in September 2017. The US Army intends to use IBM's cloud service and artificial intelligence product Watson to evaluate and process data from 17 sensors mounted on each Stryker infantry chariot's transmission, after a maintenance warrant officer teaches Watson how to read the data. Watson can mine millions of data points for early-warning indications of engine failure, significantly enhancing the effectiveness of "state-based maintenance". IBM was provided with the maintenance history of 350 Stryker vehicles over the last 15 years, as well as 5 billion sensor readings and data from Stryker program directors, original equipment manufacturers, and Army research centers. The Watson supercomputer can receive sensor data from the vehicle via IBM's cloud technology and, when combined with data from previous maintenance, can identify anomalies and predict potential breakdowns. Watson can also analyze technical literature, such as electronic technical manuals, articles from the journal *Preventive Maintenance*, material repair manuals, and diagrams, to identify components that may have deviations or potential problems and to recommend the most effective solutions for the causes of these problems, as documented in the technical literature. Watson's information can be used by maintenance personnel to identify how to repair and what spare parts are required. Watson can also incorporate other information into the failure analysis process in the future, such as geographic location, weather, and topography, and constantly learn from maintenance personnel's comments to increase forecast accuracy [9].

Global military operations in the future will necessitate the establishment of a distributed overseas security system supported by an intelligent network information system.

13.7.5 Operational Cooperation

Global military operations are a sophisticated and systematic undertaking. It is required to build a standardized system that combines general and specialized functions in order to enable automatic data exchange, seamless information transmission, efficient operation coordination, and precise support for local and international combinations. It is important to achieve integration and compatibility with the country, region, and industry in which it is in order to conduct multi-national joint military operations, multi-national multi-sectoral task collaboration, and military–civilian integration assistance.

Here is a look at the coordinated military strikes by the US, the UK, and France on Syria outside the defensive zone in 2018.

On April 13, 2018, US President Donald Trump announced that he had directed the US military, in coordination with the UK and France's armed forces, to conduct "precision strikes" against Syrian government military facilities in response to the recent "chemical weapons attack" in Syria's Eastern Ghouta region. The US, the UK, and France launched a total of 105 cruise missiles and air-to-surface missiles on April 14, targeting three suspected chemical weapons manufacturing, development, and storage facilities in Syria (Figure 13.2). The coalition missiles were intercepted by Syrian air defense systems. Because Syria employs Soviet/Russian air defense systems, the conflict is being framed as a direct confrontation between US and European precision strike weapons and Russian air defense systems.

Airstrikes are usually conducted outside the defense zone, being multi-directional and dense. The US Tomahawk missile has a range of around 1,600 kilometers, the Joint Air-to-Surface Standoff Missile (JASSM) missile has a standard range of 370 kilometers, and the extended-range variant has a range of 925 kilometers. With an effective range of around 1,000 kilometers for the French MdCN missile and ranges of more than 250 kilometers for the British and French Stormbringer and SCALP-EG missiles, the use of these long-range missiles reflects the features of out-of-area strikes. Additionally, the coalition chose to launch missiles simultaneously from different areas, including the Eastern Mediterranean, northern Red Sea, and the northern Persian Gulf, indicating the strike's multi-directional nature. Seventy-six missiles were launched at the Bazar Research Center, which consists of only three buildings, and 22 missiles were launched at suspected chemical weapons storage facilities near Homs, illustrating the attacks' high density.

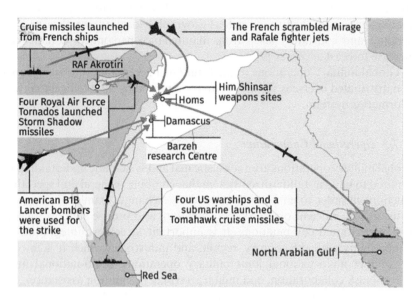

Figure 13.2 Coalition forces choose multiple directions for precision strikes on Syrian targets.

Table 13.1 Interception performance of Syrian Soviet/Russian air defense weapons

Weapon models	R&D era	Range (km)	Launches (pieces)	Interceptions (pieces)	Rate	Ranking
Vega C-200/ SA-5	1960s (inception) 1970s (in service)	250	8	0	0%	7
Beech 9K37/ SA-11	1970s (inception) 1980s (in service)	32	29	24	83%	2
Cube 2K12/ SA-6	1960s (in service)	25	21	11	52%	4
Armor-S1 96K6/ Panzer-S1	1990s (in service)	20	25	23	92%	1
Neva C-125/ SA-3	1950s (inception) 1960s (in service)	15	13	5	38%	6
Wasp 9K33/ SA-8	1960s (inception) 1970s (in service)	12	11	5	45%	5
Arrow 109K35/ SA-13	1970s (in service)	8	5	3	60%	3
Total/average			112	71	63%	

R&D, research and development.

Syrian Air Defense Forces responded forcefully in the aftermath of the bombardment, intercepting incoming missiles with SA-6 Cube, SA-3 Neva, SA-5 Vega, SA-8 Wasp, SA-13 Arrow-10, SA-11 Beech, and Armor-S1 air defense systems, all of which are Soviet/Russian-made. Russian Defense Ministry confirmed the interception of Soviet/Russian-made air defense weapons deployed in Syria on April 14, with 71 missiles intercepted, a 63% interception rate [10]. The Russian military also showed intercepted cruise missiles such as the Tomahawk from the United States, the French SCALP-EG, and the British Storm Shadow. Among them, the Armor-S1 combined bomb and gun air defense system performs admirably in combat (Table 13.1).

References

[1] Ye, Z. C. 叶自成. (2018). *Guanyu luquan de xiangguan yanjiu baogao* 关于陆权的相关研究报告. Beijing: Peking University.

[2] Baidu Baike 百度百科. (2021d, December 13). *Meiguo quanqiu junshi jidi* 美国全球军事基地. Baidu baike 百度百科. https://baike.baidu.com/item/.

[3] Space China 中国航天科技集团. (2020). *Zhongguo hangtian keji huodong lanpishu baogao* 中国航天科技活动蓝皮书报告. Beijing: Space China.

[4] Baidu Baike 百度百科. (2021b, December 13). *Meiguo haijun luzhandui* 美国海军陆战队. Baidu baike 百度百科. https://baike.baidu.com/item/.

[5] Baidu Baike 百度百科. (2021c, December 14). *Meiguo lujun tezhong budui* 美国陆军特种部队. Baidu baike 百度百科. https://baike.baidu.com/item/.

[6] Baidu Baike 百度百科. (2021a, December 12). *Kuaifan budui* 快反部队. Baidu baike 百度百科. https://baike.baidu.com/item/.

[7] Fang, X. Z. 方晓志. (2016, September 29). *Meijun lianhe houqin baozhang tixi gei wojun naxie qishi* 美军联合后勤保障体系给我军哪些启示. Pengpai xinwen wang 澎湃新闻网. www.thepaper.cn/newsDetail_forward_1536427.

[8] China National Defense Science and Technology Information Center 中国国防科技信息中心. (2015). *2015Nian shijie wuqi zhuangbei yu junshi jishu niandu fazhan baogao* 2015年世界武器装备与军事技术年度发展报告. Beijing: National Defense Industry Press.

[9] Chinese Institute of Command and Control. 中国指挥与控制学会. (2019, August 13). *Rengongzhinengjishujiang gaibianmeilunjuncheliangweixiubaozhanm oshi.* 人工智能技术将改变美陆军车辆维修保障模式. www.sohu.com/a/33341 1780_358040.

[10] 105 hits in Syria? Not likely, says Russia & shows fragments of missiles downed in US-led strikes. (2018, April 25). *HomeWorld News.* www.rt.com/news/425120-rus sia-shows-downed-missiles-syria/.

14 Future Urban Operations

Urban operations are the most difficult aspects of military operations, as they require the most intelligent means to overcome tactical and technical impediments. Urban operations in the future will face diversified operational objects, missions, and multi-dimensional domains involving time and space. Traditional urban operations include offensive and defensive operations, counter-terrorism operations, and military operations other than war (MOOTWs), e.g., search and rescue. Future urban operations will include both physical space confrontations, such as combat in streets or buildings, as well as virtual space confrontations, such as cyber attack and defense, electronic confrontation, public opinion control, and psychological warfare. Furthermore, the integration of physical and virtual operations in critical infrastructure management and control, cross-domain attack and defense, and post-war governance will be possible. Future urban operations are the concentrated embodiment and exemplar of intelligent war, combining cognitive confrontation, "intelligence +" and "+ intelligence" operations.

Cities are the gathering places of human civilization; they are densely populated, politically sensitive, economically advanced, and concentrated in science, technology, education, and culture. Cities play a significant role in national and regional development and have become significant targets of various military confrontations. Since the 1980s, the Iran–Iraq War, the Gulf War, the Kosovo War, the Chechen War, the Iraq War, the Afghanistan War, the Libya War, the South Sudanese Civil War, the Syrian Civil War, and the eastern Ukraine War have all been centered on cities. Urban offensive or defensive operations are defined as those conducted in the suburbs and within the urban areas. However, the focus of the research is always on the latter, namely urban operations.

Conducting military operations in urban areas is fundamentally different from conducting large-scale mechanized corps operations in the field. While design and planning are necessary, a more critical task is to specify and manage the diverse and complex issues inherent in urban operations. US military urban operations in Mogadishu, Baghdad, Fallujah, and other cities, as well as Russian military engagements in Grozny, Syria, and cities in eastern

DOI: 10.4324/b22974-16

Ukraine, all involve the interaction of the local and global, the interweaving of offense and defense, and the complementarities of traditional and new methods of warfare. Due to the local advantages in terms of time and space, the powerful sides do not always win. The US military's defeat in Mogadishu was largely due to outnumbering adversaries and a shortage of armor and fire protection. The failure of the Russian military in the first Grozny War was due to a lack of intelligence, insufficient preparation, and inexperience. Confronted with a large number of Chechen militants inside buildings and underground, the Russian armies were attacked from vulnerable areas such as the flanks, tops, and backs, despite being protected by armored firepower, and ultimately failed. The US military's victory in Operation Baghdad occurred as a result of the Iraqi Army's defeat. Saddam had lost command and control of his troops due to the US high-intensity airstrikes and cyber, psychological, and electronic warfare. Iraq's Army was disorganized and distracted, failing to develop a systematic, multi-dimensional, and diverse defense system. Baghdad was essentially in scattered and sporadic resistance, and as a result, the city fell in a short period of time.

In the age of artificial intelligence, future urban operations may be markedly different in terms of environments and adversaries from recent urban operations. As urbanization accelerates, megacities with populations ranging from millions to tens of millions are becoming more prevalent. As a result, it is impossible to attack and destroy targets solely or arbitrarily for military purposes. It is impossible to evacuate a large number of civilians from a city prior to war. Simultaneously, as mobile phones have grown in popularity, the Internet has been upgraded, and the Internet of Things (IoT) has become more pervasive, the battlefields have become increasingly transparent. Once large numbers of civilians are killed, and sensitive locations and facilities such as hospitals, mosques, churches, and cultural relics are destroyed, the media and Internet amplification can easily result in a shift in public opinion, media accusations, and international condemnation. This outcome could have an effect on the war process and leaders' decisions, make it more difficult to accomplish political and military objectives, and ultimately result in failure. When armed forces are confronted with wars against mighty adversaries such as the US and Russia, or their proxies, the situation will become more complicated in terms of targets. As with the battle for eastern Ukraine's cities, there is the possibility of a protracted tug of war, a war of attrition, and a contest of wills because both combatants possess powerful unmanned aerial vehicles (UAVs), long-range precision-guided weapons, fire suppression weapons, and space–air–ground battlefield surveillance systems. Additionally, cyber attack and defense, electronic confrontation, public opinion propaganda, and ideological conflict will be used. As a result, future intelligent urban operations will be defined by a profound fusion of virtual and physical space, as well as complex architectures, communities, and humanistic environments. The orientation will be virtual–real interaction, mind-attack, multi-domain coordination, multi-dimensional offensive and defensive, precision strike, minimal collateral damage, community-oriented, and comprehensive management and control.

Since the 1980s, war practices have demonstrated that each urban operation has its own unique environment and characteristics. Even if the outcome of an operation is different for the same military, they all involve minor physical environment combat issues in streets, buildings, basements, as well as in public facilities, including airports, power plants, and government offices. Simultaneously, there are widespread soft environmental concerns, such as public opinion management, emergency response, and infrastructure management and control. Thus, regardless of the city's size, the focus must be on or narrowed down to community-scale contention and control.

14.1 Introduction: Failure-Related Lessons

14.1.1 Black Hawk Down

In 1993, "Black Hawk Down" in Mogadishu could be considered the US military's Waterloo in overseas operations. Somalia began a war between warlords in 1990. A great famine occurred as a result of wars, natural disasters, and economic collapse. Hundreds of thousands of people died of starvation within 2 years. In 1992, the world learned about scrawny Somalians. At the time, the international community believed that something should be done to help the Somalis, and relief supplies began to arrive in Somalia. However, the warlords were unconcerned about civilian lives and deaths and began looting food on a large scale. As a result, the United Nations (UN) requested that the US send troops to provide protection and transport relief supplies. As the situation in Somalia deteriorated rapidly, the UN was forced to adopt a new resolution authorizing the deployment of additional forces for security maintenance, humanitarian assistance, and peace enforcement. In late 1992, the UN began deploying 38,000 troops to Somalia. About 28,000 of them were US military personnel, who used coercive methods to distribute relief supplies to the victims. While assisting humanitarian relief efforts, US and UN forces dealt with warring tribes. The civilian population welcomed the UN operations, but the warlords, particularly Aidid, the leader of the United Somali Congress Party, which occupied Mogadishu, was opposed. The situation shifted when the rescue operation went smoothly, and the US withdrew 25,000 troops in March 1993. The UN induced the majority of warlords to agree to peace negotiations through mediation, but Aidid refused, and his troops began harassing UN forces. Somalian militias ambushed a Pakistani peacekeeping force on June 5, 1993, killing 24 people and dismembering their bodies. The world was shocked by this incident, and UN forces immediately began searching for the perpetrators. What began as a peacekeeping mission to end famine and restore order in Somalia quickly devolved into combat against Aidid. Aidid, who controlled the Mogadishu news media, incited the populace to believe that the UN would depose him. When combined with the fact that the UN raids were quite disruptive, the peacekeeping forces' status in the eyes of Somali civilians shifted dramatically. They acted as shields for Aidid's militia, allowing them to conceal themselves among the crowd in order to attack UN forces and cover their exit.

The US launched "Operation Gothic Snake" to apprehend Aidid, assembling a 450-member ranger task force. The task force was composed of three army units: the Delta Force's C Squadron, the 75th Ranger Regiment's B Company, and the 160th Special Operations Aviation Regiment. The official operation began on October 3 in the early-morning hours. According to a local intelligence agent, Aidid's senior assistants were scheduled to meet at the Olympic Hotel. General Garrison, the operation's commander, directed deployments and decided to arrest them. The operation involved the use of 19 aircraft and 12 vehicles, with a total of 160 participants [1].

By late spring 1993, the UN military commanders had petitioned the UN headquarters to allow additional armored forces to be provided by other accrediting states. The armored force was required for one simple reason: they could move freely in Mogadishu due to their protection, mobility, and lethality advantages over lightly armed tribes. In Somalia, the senior US military commander agreed to this request. As attacks on UN and US forces intensified in July, US policymakers decided to increase the number of US special operations forces, combine them with US Army Commandos, and use more targeted military forces against key tribes' leaders. Major General Thomas Montgomery, a senior US commander, had applied for a mechanized infantry unit and an armored task force in the event of an emergency. This could result in increased freedom of movement in Mogadishu's streets and the ability to deploy rapid reaction forces when necessary. Regrettably, US politics remained committed to a limited and temporary role in Somalia. As a result, they voted against the request for armored forces.

Despite the lack of heavy armor protection, the special operations task force carried out a raid on October 3, inserting special operations forces into one of Mogadishu's most vital tribal defense areas via helicopters. Although the plan was deliberated, well organized, and included a detailed sequence of operations, the use of light Humvees instead of tanks and armored personnel carriers (APCs) resulted in catastrophic consequences. The task force quickly realized that they were facing what appeared to be a city-wide siege and blockade rather than community-scale resistance.

Special operations forces apprehended several senior tribal leaders quickly. The plan called for US Army commandos to drive a light Humvee across the city to the capture point and quickly transport the captured prisoners back to the commando station. However, the capture point was located in the militia fighters' primary stronghold area. Unarmored Humvees have been vulnerable to small arms, machine guns, and rocket-propelled grenades (RPGs). Somalian tribal militias, along with civilians, conducted swarm attacks and erected roadblocks to prevent vehicles from reaching the capture site. Two US Black Hawk helicopters were eventually shot down, exacerbating the story.

Lieutenant Colonel Danny McKnight, the commanding officer in charge of the commandos being escorted in light Humvees, was tense. There were so many casualties that he was unable to assist at the crash site of one of the downed helicopters. He then alerted and directed helicopters to hover over them, thereby reducing casualties.

Tribal members fought with handguns, RPGs, and, on rare occasions, light mortars and large-caliber machine guns mounted on pickup trucks. Although these tribesmen lacked sophisticated organizational structures and command and control, they were tenacious and exploited their familiarity with the urban terrain of Mogadishu to gain operational advantage.

With the assistance of Pakistan's Armored and Mechanized Forces and Malaysia's Mechanized Infantry Emergency Unit, the task force finally disengaged from serious contact with Somalian tribal personnel after a 24-hour close-combat period. At this point, 18 army commandos had been killed, and over 50 were wounded. This was the bloodiest shootout involving combat troops from the US since the Vietnam War. Although these commandos were highly trained and deployed in well-equipped light infantry combat teams, they sustained significant casualties fighting Somalian tribal members. Due to a lack of armored vehicles, it was difficult to escort the personnel evacuation from the crisscrossed city blocks.

Columbia Pictures adapted the US military's battle in Mogadishu into a war-themed film, *Black Hawk Down*, in 2001, directed by Ridley Scott and starring Josh Harnett, Ivan McGregor, and Tom Sizemore, based on the novel *Black Hawk Down*. The film's script went through numerous revisions. Prior to the start of filming, all of the actors attended a 2-week military training program at the US military's Delta Forces, Ranger units, and "Black Hawk" helicopter base. Additionally, they spoke with relatives and friends of those killed in the Battle of Mogadishu in order to gain a better understanding of the roles they would portray. Due to the impossibility of shooting on location in Somalia, the film was completed in Morocco. Except for those currently serving in the US military and those on special missions, all characters in the film were depicted using their real names.

14.1.2 *Grozny Street Battles*

Two street brawls in Grozny exemplified the failure and success of Russia's Chechen War. In 1994, Russian troops and illegal Chechen armed forces engaged in two large-scale street battles in Grozny, Chechnya's capital. These were the bloodiest battles since World War II and the most heinous urban street battles since the Vietnam War. Russian troops had a clear advantage in personnel and armored tank vehicles during the initial street combat in late 1994 and early 1995, but they were defeated. Between December 1999 and February 2000, the second Grozny street battle took place, and Russia eventually won [2].

Grozny, founded in 1918, was designed as a battle fortress. The city's fortifications are as dense as a spider web, making them easy to defend and difficult to attack. There are underground passageways and air defense facilities that were constructed during the Soviet era for wartime use. The Russian 131st Brigade and 81st Motorized Infantry Regiment, which entered the city center for the first time in late 1994 and early 1995, were attacked by the elite Chechen armed forces from Abkhaz and Muslim battalions. Out of more than 300 people who entered the city center, the 131st Brigade suffered 70 casualties, and the brigade commander was killed in action. In the case of the 81st Motorized Infantry Regiment, only one officer and ten soldiers survived the city center evacuation, 20 of 26 Russian tanks were destroyed, and 102 of 120 armored vehicles were destroyed. The Russian troops' corpses were even used as sandbags by the Chechen armed forces to construct "human bunkers." Between December 31, 1994, and January 2, 1995, Russian troops lost approximately 250 tanks and armored vehicles. The Chechen armed Muslim battalion and the Abkhaz battalion were so powerful because nearly half of their fighters were retired special forces from Western military powers, including US Navy Sea, Air and Land (SEALs), German border guards, French *gendarmerie* troops, and Poland's Thundering Special Forces. These members had extensive combat experience and were battle-hardened. They were divided into several combat teams of three when they engaged the Russian troops, each equipped with bazookas, long-range sniper rifles, and heavy machine guns. These fighters were dexterous and adaptable in their combat against Russia. They used heavy machine guns to attack distant targets, rocket launchers to attack close targets, and sniper rifles to attack available rest targets. This special operation method resulted in significant losses for the Russian Army, which had been out of combat for an extended period of time.

Five years later, in December 1999, the second Grozny street fighting commenced, lasting until February 2000. The Russian soldiers had gained valuable experience from the previous street combat. President Putin gathered about 2,000 men of his special forces; the majority of them were snipers or even sharpshooters. This time, the Russian troops used a decentralized-versus-decentralized, group-versus-group strategy rather than a mass march strategy of deploying conventional forces. Russian special troops were organized in groups of five. Each group consisted of two or more snipers equipped with shoulder-launched guided missiles and the latest long-range rocket launchers and was prepared to call in armed helicopters and fighter jets at any time to assist with the mission. After a hard battle, over 1,173 of the 2,000 Russian special personnel were killed in action. However, when combined with Chechen rebel deaths, the overall number of members of the enemy's special forces slain by Russian forces exceeded 3,000, increasing the total to more than 10,000. This operation was a tremendous success for the Russian forces.

The primary reason the Russian troops were defeated in Grozny's first street battle was that they had oversimplified and underestimated the enemy's will and ability to engage in street warfare. The initial plan called for the deployment of a 6,000-strong tank regiment and infantry, presuming Grozny could be captured unopposed. They underestimated the defensive advantages of the urban environment and widened rather than limited the operational range. The elements of the city's landscape are depicted in Figure 14.1. Enemy soldiers and militias familiar with the city's geography may easily transition from the attack posture in multi-story buildings to the basement and sewage system. These underground locations were the Russian attack range's dead ends, as their tank guns were unable to reach the pitched angles or were being lifted or lowered adequately.

As with the majority of combat vehicles, the Russian combat vehicles were built to fight largely head to head, with relatively minimal protection on the top, sides, and rear. Chechen extremists exploited these flaws. Russian military were heavily reliant on armored tanks and vehicles. The cumbersome mechanized formations, however, were no match for the Chechen guerilla forces in the urban setting. The guerilla troops launched three shells before fleeing the launching point to avoid counterattacks. Grozny's narrow and winding streets enabled snipers to aim at targets and suppress the Russian tank formation's front and rear, causing the tanks in the center to succumb to RPGs. As a result, despite its valiant efforts, the Russian Army ultimately failed to gain control of the outer fringe heading to Grozny's city core.

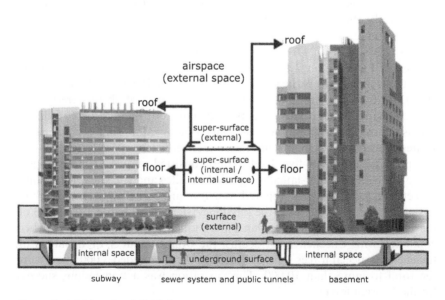

Figure 14.1 Urban battlefield elements.

The Russian army did not conduct thorough intelligence preparations for the battlefield. They were unaware of the drawbacks of urban terrain peculiarities for mechanized forces, as well as the indigenous language and culture.

Additionally, the Russian military's apparent vulnerability was a lack of training in urban operations. Since 1992, the Russian Army had not undertaken divisional or regimental field exercises. Prior training focused mostly on preparing soldiers for war in open terrain throughout Eastern Europe rather than in urban situations. While less than 5% of training was dedicated to urban operations, in the majority of cases, the only preparation for urban missions was a limited-edition street combat manual that had to be shared among service members. The majority of operating troops are formed of inexperienced conscripts, who lacked combined training and were unfamiliar with their respective communication protocols, resulting in the leakage of communication lines, allowing the Chechen separatists to eavesdrop on transmissions and offer false intelligence. One typical example is that it took a Russian tank unit 6 hours to engage a Russian mechanized rifle unit before they discovered they were friendly forces. According to estimates, a sizable proportion of Russian casualties were self-inflicted.

By mid-January 1995, the Russian troop presence in Chechnya had risen to 30,000, with the majority of them stationed around Grozny. Russian forces were unable to blockade and control the city despite seizing and controlling critical infrastructure and destroying the presidential palace. Meanwhile, separatists readily infiltrated civilian homes and committed high-profile kidnappings and terrorist assaults. By August 1996, Russia's demoralized troops had negotiated a ceasefire and relinquished control of the resistance.

In the second Grozny street battle, the Russian troops had learned from their failures and made substantial preparations to strengthen their strategic advantages and tactical capabilities. To be more precise, in contrast to the initial conflict, the Russians did extensive pre-war reconnaissance and made use of linguistic interpretations and intelligence from Chechen loyalists. Russian mission planners meticulously analyzed the streets in key portions of the city, as well as the roads going to specific places and all of the city's municipal institutions. After poring over all the information, the Russian planners discovered several maps and determined the location of sewage pipelines, the route of the heating line, and even discovered a maze that was one person high and two to three meters broad. Both engineers and reconnaissance scouts evaluated these municipal facilities prior to entering the city. This strategy was critical as Russian soldiers marched towards the city center.

The Chechen insurgents had 2 months to prepare for the operations, during which time they established numerous ambush positions. The rebels set up two defensive lines and sent the least skilled soldiers to the front line. Snipers occupied building roofs and upper levels and monitored long-distance passageways at certain intersections. They attempted to herd the Russian soldiers into the roadways. Snipers were also concealing themselves beneath trenches and the concrete slabs that covered the basement. Chechen insurgents were reportedly separated into groups of 25, with each group subdivided into three teams of roughly eight men each, in order to approach near enough to the Russian troops to control them, as they had done in the 1995 operation by reducing the strength of Russian artillery. According to one Russian officer, when fighting in Grozny's urban environment, the combat effectiveness of a single Chechen company was even equivalent to that of an entire Russian brigade.

After receiving thorough information about the Chechen rebels' operational plans, the Russians altered their tactics as well. Their response was to divide Grozny into 15 sectors in order to locate enemy strongholds, underground tunnels, and armament arsenals. This helped them to better understand the distribution and vertical dimensions of high-rise buildings and basements. Furthermore, rather than unleashing major carnage on the city, they besieged it with a force of 50,000 and deployed tiny squads to attack it, attempting to exterminate the city's more than 4,000 militants in a discrete and slow manner. Simultaneously, Russian aircraft targeted dams, weirs, water distribution systems, oil depots, petroleum facilities, telephone lines, and power supply systems with the stated purpose of undermining the Chechen resolve to resist and destroying Chechen internal infrastructure. Grozny devolved into a "free crossfire zone."

Between 1999 and 2000, Russian troops suffered only one tank loss, with a low rate of friendly accidental harm. This demonstrated that they were coordinating their efforts more effectively between tanks and infantry, ground forces, forward artillery, and air power. It is also worth noting that Russia primarily employed forces to besiege the city, with only a few tank attacks. It employed a broad spectrum of conventional operational capabilities, including aviation forces, mortars, howitzers, RPGs, and missiles, with ground firepower accounting for approximately 70% of the total and aviation contributing for 30%.

Throughout the Second Grozny War, the Russians exerted tight control over public opinion and propaganda, reversing the trend in most television reports and newspaper articles indicating sympathy for the rebels during the First Grozny War and sustaining public support for the Chechen war. Russian forces used psychological warfare to persuade people to flee Grozny and to entice Chechen terrorists to surrender. Additionally, they employed electronic warfare to fool the insurgents. There was a well-known incident during the second Grozny street fight. Russia used a phony radio network to convince Chechen defenders to evacuate from the southwest in the nighttime. They purposefully opened the radio network to Chechen troops and publicized the vulnerability. Then they employed landmines and barricaded troops in order to wait and combat Chechen terrorists who fled the city. Russia's triumph in the second Grozny operation brought the operation to an official close.

14.2 Urban Constructions

Cities are intricate assemblages of man-made terrains, dense populations, and enormous infrastructure. Due to the multitude of environmental, social, cultural, and economic conditions, as well as street architecture, transportation and logistics, commerce, and financing, urban areas constitute the most complicated terrains in modern military operations. Complex urban surroundings have a higher impact on operations, particularly in central business districts, manufacturing service centers, and cargo distribution centers.

Cities are not self-contained natural entities. They originate as a result of human-induced changes to the natural environment. Typically, cities serve as hubs for regional political, financial, transportation, industrial, and cultural activities. All cities contain infrastructures, residents, streets, and public utility networks, among other things. The disparities in scale, level, and style of development are reflected across cities. The larger the city, the more regional impact it has. Cities are often classed into metropolises, medium-sized cities, villages, and rural communities, depending on their population. Typically, a metropolis has a population greater than ten million, whereas other categories have populations ranging from thousands to several million. Many cities have expanded into the countryside in recent years, obliterating the distinct demarcation between them. Cities are linked through transportation networks such as highways, canals, and trains, while rural areas are linked by roads.

According to the US National Intelligence Council's report *Global Trends 2030: Alternative Worlds*, nearly 60% of the world's population, or around 4.9 billion people, would live in urban areas by 2030 [3]. According to the report, "the world's urban population will grow by 65 million people per year, equivalent to the population of seven Chicagos or five Londons." Additionally, references predict that the world's megacities with a population of ten million or more will continue to grow, from three in 1980 to 24 in 2014, 30 in 2018, and approximately 37 in 2025. Urban periphery and "urban–rural fringe" areas will develop more rapidly than the urban center, in part because they can provide more affordable housing and manufacturing space. Population expansion, polycentricity, coastalization, and network connectivity are all trends in urban development.

14.2.1 Urban Layouts

According to their layouts, cities can be characterized as central, satellite, network, linear, or sector-shaped, as illustrated in Figure 14.2.

Satellite cities are smaller cities that are connected to larger cities or urban hubs. Between satellite cities and large cities, the topography is typically homogeneous. Satellite cities are cored at urban hubs or primary urban districts, whereas peripheral cities are radial. Urban hubs are usually the principal targets and impediments for attackers.

Natural features or traffic axes impact linear or sector-shaped urban layouts, with some following rivers or beaches, others following narrow valleys, and yet others following land traffic arteries.

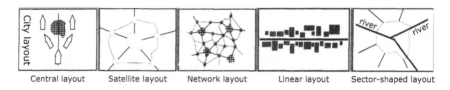

Central layout   Satellite layout   Network layout   Linear layout   Sector-shaped layout

Figure 14.2 Typical urban layouts.

Table 14.1 Operational condition of urban and other terrains

	City	Desert	Jungle	Mountain
Non-combatant population	Multiple	Limited	Limited	Limited
High-value infrastructures	Multiple	Limited	Limited	Limited
Multi-dimensional battlefields	Positive	Negative	Neutral	Positive
Strike/detection/observation limits	Positive	Negative	Negative	Negative
Strike distance	Short	Long	Short	Medium
Approach to the enemy	Multiple	Limited	Limited	Limited
Mobility of mechanized forces	Low	High	Low	Medium
Communication conditions	Weak	Normal	Normal	Weak
Logistics requirements	High	High	Medium	Medium

14.2.2 Urban Terrain

The urban landscape is both complicated and demanding. Urban terrains combine the qualities of natural landscapes with human-created structures, providing a complex environment that has a unique effect on military operations (Table 14.1). Urban terrains are a complex synthesis of horizontal, vertical, internal, and external forms classified into four levels: airspace, near-ground, ground, and underground.

Airspace is a passageway above the Earth that is accessible to aircraft and air munitions. In metropolitan regions, the airspace is divided by buildings of varying heights and densities, as well as varied natural terrains. This results in the formation of an "urban canyon" effect, which can have a detrimental effect on urban operations because it frequently results in higher wind speeds, unexpected wind directions, and turbulence, all of which can cause ammunition to miss its target. The airspace provides a direct channel for forces to use aviation equipment for observation, reconnaissance, strikes, and rapid deployment or withdrawal of troops, materiels, and equipment. Certain surface obstructions have no effect on aviation equipment. However, the profusion of urban structures such as towers and power lines may obstruct flight and ammunition and restrict low-altitude mobility in urban airspace.

Near-ground space includes both natural and human-created structures that can be used for movement, observation, and firing. Certain roofs are configured as helipads, making them ideal for helicopter landings.

Parking lots, airports, highways, streets, walkways, parks, and fields are all examples of ground areas. These surface areas conform to natural topographies but are fragmented as a result of human activity.

Streets serve as the primary avenues for approaching and advancing. In contrast, buildings and other structures frequently obstruct troop advancement. Constructed barriers in urban ground areas are often more conducive to the defensive side than in other open terrains.

Large open spaces such as stadiums, sports fields, school playgrounds, and parking lots are vital in urban operations. They can be used to house displaced civilians, for interrogation centers, facilities for enemy prisoners of war, and detainees. These locations can be used as aviation landing zones, takeoff zones, and artillery positions. Because they are typically placed in strategic locations, they can also provide logistical assistance and air supplies.

Subways, tunnels, sewers, drainage systems, cellars, air defense shelters, public corridors, public utility networks, and other underground spaces comprise the underground areas. These locations may even include ancient hand-dug tunnels and catacombs in older towns. They can be used for concealment, movement, and strike. Both attackers and defenders can stage ambushes in underground regions. However, utilizing underground areas requires a thorough study of their properties and may include consideration of potential paths supplied by rivers that border metropolitan areas and underground sections of bodies of water.

14.2.3 Urban Functions

Cities can be classified according to their functions into industrial districts, high-rise areas, residential areas, commercial districts, and military zones.

14.2.3.1 Industrial Districts

Industrial districts are usually built on the outskirts of cities, near airports, seas, rivers, trains, and highways. Typically, industrial buildings are spaced to accommodate huge cargo trucks, material handling, and associated equipment and facilities.

14.2.3.2 High-Rise Areas

High-rise areas include multi-story residential complexes, commercial office buildings, and corporate buildings that are typically separated by large open spaces such as parking lots, parks, and stadiums, as well as smaller single-story buildings.

14.2.3.3 Residential Areas

Residential areas are typically scattered throughout urban areas and are served by urban infrastructures, including electricity, water, and communications.

14.2.3.4 Commercial Districts

Commercial districts are made up of shopping malls, businesses, hotels, and restaurants that are typically concentrated in urban centers and along arterial highways or in central locations and densely populated blocks.

14.2.3.5 Military Zones

Most major cities have fortifications and infrastructure constructed expressly for military purposes. Permanent fortifications can be constructed above or below ground using mud, wood, rock, brick, or concrete, or a mix of these materials. Countries with lengthy coasts may have also invested in coastal defense projects.

14.2.4 Urban Infrastructures

Urban infrastructures are primarily concerned with the resources and support systems upon which the population depends for survival and productivity. Public infrastructures include bridges, highways, airports, ports, subways, sewers, power plants, networks, communications, and social services. Cities have varying infrastructures. Urban infrastructures in industrialized countries are highly developed and well functioning but are less so in developing countries. The quality of urban infrastructures has a direct impact on how people live and how cities function, as well as on how military operations are conducted and how long-term effective occupation and control are implemented.

In developing countries, communication infrastructures may be rather primitive, and information dissemination may rely more on traditional local media such as broadcast, television, and newspapers. Radio, mobile, and satellite communications may be less utilized. In general, it is critical to understand the communication infrastructures of cities since they ultimately govern information flow into local civilians and adversaries.

Cities cannot function without electricity. Electricity firms provide essential services such as heating, lighting, and power. Due to the fact that electricity cannot be stored in significant quantities, any failure of the power provider will have an immediate effect on the public. As a critical hub of the entire urban service industry, power facilities are always possible targets of urban conflict. Electrical services may be disrupted by riots, military activities, and other forms of warfare. Armed forces may target these facilities in order to erode the adversary's control of the city and popular support.

14.2.5 Urban Population

Economic, political, and cultural priorities in urban areas are influenced by urban residents' expectations, attitudes, and behaviors. The populace can have a positive or negative effect on military operations and missions. Urban residents usually have an intimate knowledge of their surroundings, and their observations can supply intelligence that aids in military comprehension of the urban environment and activities. For example, residents are typically familiar with shortcuts through towns and are able to witness and report on rallies, meetings, and other events as they occur.

Urban populations can also physically impede movement and mobility by altering the width of highways, as well as providing cover for other forces. For example, refugee movements can be used to infiltrate forces covertly. Normal urban life may also have an impact on military operations. For example, during rush hours, military movement in urban locations may be frequently obstructed by traffic.

14.2.6 Urban Information

Urban information is critical for combat and military operations since it serves as the foundation for operational plans and decision-making. Due to the complexity, multi-dimensionality, and rapid change of urban information, it is necessary to conduct continuous monitoring, evaluation, and analysis of the urban environment and key targets. Newspapers, televisions, broadcast stations, computer networks, information technology centers, and postal systems are all examples of information sources. What is worth emphasizing is the importance of analyzing, evaluating, and correlating the information on a global scale as well as at national and local levels.

The preceding six aspects are the fundamental components of a city. For urban offensive forces, the city resembles a massive and impenetrable "black box." Urban operations are fraught with unknowns and uncertainties due to the complexity of the environment, the variety of structures and targets, and the mix of forces and civilians. All of these factors can significantly complicate and obstruct the conduct of combat operations in the following ways.

To begin, targets are difficult to identify. Military targets may be frequently hidden in buildings and underground facilities, making them difficult to detect and locate in a timely manner. It may also be difficult to track significant targets and groups.

Second, options are limited. With military forces mingling with a dense population and military equipment entwined with civilian infrastructure, various traditional weapons are prone to cause collateral damage, making it difficult for them to operate at full capacity.

Third, military deployments are constrained. Since the battlefield capacity is constrained by urban streets, troop and weapon mobility and firing are restricted. Expanding large corps becomes more difficult.

Fourth, command-and-control capabilities are lacking. In an urban environment, battlefield observations and communications are easily disrupted by tall buildings and other electromagnetic signal noises. As a result, information sharing may be difficult for all parties, and command and control of troops' movements and real-time coordination are also challenging.

Fifth, operational risks are substantial. Operational forces stationed near tall buildings or well-developed underground facilities are vulnerable to sniping, ambushes, and sneak attacks from all directions. They will also face close-range attacks by terrorists disguised as civilians during counter-terrorism operations.

Sixth, public opinion is difficult to manipulate. A "butterfly effect" may be easily initiated following an accidental injury or bombing. In cities with complete communication infrastructure and highly developed media, countless eyes will be glued to the battlefield, broadcasting live combat from anywhere and at any time via the Internet. Certain incidents, such as injuring civilians, destroying people's livelihoods, and striking sensitive targets, are extremely susceptible to hype and amplification, which can mislead the public and even spark a global anti-war storm.

14.3 Urban Operations

In comparison to other types of operations, urban operations are characterized by environments that place significant constraints on military deployment, tactics, and equipment usage.

A well-developed urban transportation network serves as the primary conduit and the focal point of control. Highways, railways, and waterways are the three primary components of an urban transportation network. They are the arterial blood vessels of a city's transportation system. Urban roads have varying line fonts, types, and widths. For example, straight roads run east to west and north to south through the city. Circular roads encircle the city, with the city's center serving as the circle's center. Trunk roads connect transportation hubs and public activity centers, while urban branch roads connect residential neighborhoods and alleys. Together with the city's railways and waterways, these roads crisscross and interconnect to form a multi-dimensional urban transportation network. Urban arterial roads are wide and straight, with a high carrying capacity, allowing for a rapid military assault, detour, and retreat, as well as maneuvering of heavy equipment such as armored combat vehicles. Complications do arise due to the limited maneuvering space available, which is obstructed on both sides of the road by the buildings. The troops are limited to movement along the road. Once they are obstructed by fire and obstacles, it will be difficult for them to advance or retreat. As a result, it is critical to seize and control urban roads in urban operations, and the battle for road control is usually fierce.

Complex urban blocks serve as the primary battlefields for street combats. These blocks are also physical impediments to weapon performance. Cities are made up of blocks of varying sizes and purposes. Due to differences in function and construction eras, blocks exhibit a range of characteristics. Central business districts and new urban areas are typically made up of high-rise reinforced concrete structures. Commercial districts are densely packed with buildings, and both sides of the main streets are lined with shops and restaurants. Residential areas are composed of detached houses ranging in height from one to several stories and densely packed high-rise apartments with small open spaces interspersed throughout. Industrial districts are composed of several courtyard-style blocks and clusters of industrial buildings with a maximum height of five stories. Suburbs and historic towns are composed of low-rise densely packed blocks composed of low-rise buildings and bungalows.

Central business districts, commercial districts, and modern residential areas are defined by tall, sturdy, and evenly distributed buildings, which are surrounded by wide, solid streets with direct access, allowing city defenders to quickly deploy troops and organize solid defensive positions. It is advantageous for the offensive side not only for firing and maneuvering tanks and other armored combat vehicles, as well as armed helicopters, but also for organizing a coordinated infantry–tank attack on a certain scale. However, the flaws include the ease with which the action can be exposed and threatened by enemy firepower. Additionally, communication can be easily disrupted, making troop coordination difficult. In the suburbs and old towns, due to the narrow and curved streets with limited visual range and poor visibility, the observation, firing, and maneuvering of military weapons are severely restricted. Armored units may be denied entry, leaving only infantry squads to fight.

Dense urban structures serve as the primary support, focal point, and direct target for urban operations. Urban buildings come in a wide variety and a large number, including office buildings, residential buildings, shopping malls, business buildings, hospitals, gymnasiums, and cinemas. Modern urban architecture is primarily frame-structured or mixed-structured. Mixed-structured buildings are made up of walls and floor slabs that are typically made of bricks, stone masonry, and cast-in-place concrete. The wall structure is robust enough to serve as a support for firing anti-tank missiles or man-portable air defense systems. The ground is dense enough to provide shelter for large equipment such as tanks. The disadvantages include a relatively fragile roof and the fact that the inner walls of some buildings with large internal spans are not load-bearing and thus are easily destroyed. Buildings within urban blocks can provide effective concealment for troops, particularly defenders, limit offensive troops' observation and firing ranges, reduce the range of anti-aircraft weapons, and impede the actions of armored troops. In some ways, urban operations resemble battles for urban structures. Conflicts over key structures pervade the entire urban operation.

Underground facilities are well equipped and serve as vital conduits for troop infiltration and assault, as well as concealment for defense. Underground facilities in cities typically include pipelines, subway tunnels, canals, shopping malls, plants, parking lots, and basements. Underground facilities in modern cities are usually advanced, with long lines, widespread distribution, good concealment, high robustness, strong integrity, and extensive supporting resources.

Underground engineering has facilitated the organization of a multi-faceted defense. The defenders of a city may deploy troops underground, conduct covert maneuvers, attack in time, and collaborate with ground forces to attack offensive forces, eventually forming a complex multi-dimensional defense system. Underground engineering facilities also benefit the offensive side's covert mobile force. Within these facilities, offensive forces can infiltrate sensitive targets in urban areas and launch surprise attacks. However, the underground space is confined in comparison to the surface, and the contact zones between the two parties are small. Observing, shooting, commanding, and coordinating are all impossible. Additionally, because the combat formation is highly dispersed, internal and external communication and support are extremely inconvenient underground, and the combat maintains a high degree of independence.

In short, an urban environment lends itself to "defense" but not to "attack," to the "small" but not to the "large," to "independence" but not to "union," to "nearness" but not to "far," to "controlling" but not to "destroying."

As global urban development continues to innovate, smart cities have become mainstream. Smart cities are connected by ubiquitous intelligent sensors based on cloud computing, and big data are used to enable rapid response and intelligent decision-making for purposes including government affairs, civilian life and productivity, environmental protection, and public safety. Its purpose is to connect the "digital space" with the "physical space" of a real city via the IoT. In the future, in a highly automated and intelligent urban operation space, operational concepts will be distinct from those used in typical urban environments. New operational theories and concepts should be advanced in the following areas: full environmental perception, accurate target positioning, precise strikes, enhanced command-and-control capabilities, operational risk management, collateral damage minimization, public opinion crisis prevention, and social change adaptation. Urban operational capabilities must be strengthened by introducing novel concepts, ideas, and measures.

Intelligent urban operations in the future will be based on cognitive communication and network data. The Internet, the IoT, big data, and cloud computing are all contributors. Physical and virtual space operations will converge to form multi-dimensional, fast, precise, and constrained space operations. Due to non-linear mutations such as universal perception, instantaneous linkage, and blowouts of data, future urban operations may have unintended consequences such as shifts in wartime public opinion, psychological disorder, and social turmoil, all of which will gradually contribute to the birth of virtual space battlefields. Unmanned ground vehicles (UGVs), bionic robots, UAVs, micro-guided weapons, and autonomous swarm systems have emerged in large numbers, stimulating the emergence of new warfare tactics and formations such as unmanned operations, multi-dimensional attack and defense, swarm operations, distributed strikes, and human–machine hybrids in the urban physical domain.

Future urban operations may rely on the Internet, mobile networks, and air–space–ground information systems to ensure multi-dimensional reconnaissance, comprehensively and accurately grasping the target city's geographic, traffic, meteorological, electromagnetic, and social conditions, as well as defensive adversaries. Psychological warfare and opinion-based confrontation can be used to disintegrate an adversary's defenders and win the support of local urban citizens via radio broadcasts, television, and the Internet, among other media. Rapid assault urban operations across land, sea, and air may be performed using unmanned, multi-dimensional, and intelligent mobile vehicles. Additionally, unmanned, precise, and low-collateral-damage air–ground firepower can be used to clear and expel enemies both inside and outside the building or underground space. Information and fire suppression, as well as other control methods, will be used to blockade, isolate, and clear combat areas, as well as to isolate, restrain, and destroy adversaries. This will help to rapidly take over cities, defend critical targets, restore infrastructure, control public opinion, and maintain urban stability.

14.4 Multi-Domain Reconnaissance and Perception

Mastering a city's information in multiple domains is necessary for knowing yourself and your adversary. It is crucial for outwitting the foe. Continuous multi-modal reconnaissance and surveillance of the target city can be conducted using the Internet, big data, satellites, and UAVs. Massive amounts of data should be analyzed to create multi-dimensional city images that include social media, geographic, and electromagnetic elements, among others. This will help to clarify: (1) geographic environments, such as urban traffic networks, building distribution, above- and below-ground layouts, spatial structure, meteorology, and hydrology; (2) electromagnetic environments, such as electromagnetic spectrum, communication links and hubs, and information flow; and (3) social environments, such as ethnic compositions, religious beliefs, and ideology, among other things. In general, a systematic analysis of a city's operational environments should be conducted.

Acquiring abundant, qualified, and accurate data is a significant challenge for urban operations. As cities continue to develop, a variety of infrastructures and social services become available. These diverse software and hardware capabilities, as well as complex urban environments, have resulted in an increasing number of advantages and disadvantages for both offensive and defensive sides. Intelligence is in greater demand in urban operations, as is reliance on the Internet and other non-traditional techniques of intelligence collecting. The method of collecting unidimensional intelligence is becoming less adaptable. It is necessary to analyze the relationship between multi-dimensional and multi-source data, ascertain the information's intent, and integrate network data with reconnaissance operations.

Prior to conducting operations in an urban area, the following questions must be addressed:

- Who are the adversaries?
- How to attack the adversaries?
- How to isolate the adversaries?
- How to create circumstances that expose the adversaries?
- How to decapitate key targets?
- How to locate and obtain critical information?
- How to deal with underground structures?
- How to deal with improvised explosive devices (IEDs) and barricades?

Through multi-dimensional urban reconnaissance, it is possible to track and monitor adversarial forces' deployment and movement, as well as gather pertinent information about key targets. It is possible to obtain the typical urban battlefield environment, target location, and personnel distribution by collecting and analyzing big data from open sources such as the Internet and IoT. For important sites such as government agencies, military garrisons, airports, docks, and transportation hubs, web crawling and modeling should be conducted to collect historical and dynamic real-time data, realizing multi-dimensional acquisition and correlation for the operation and providing intelligence services to operational, tactical situational awareness. Simultaneously, deep neural networks and machine learning should be used to collect, analyze, and model the city's electromagnetic, optical, and spectral characteristics, which may significantly improve intelligent target recognition.

It is also critical to apprehend targeted enemy leaders and significant targets or conduct decapitation operations. The US has used comparable tactics in Afghanistan, Iraq, and Libya, as well as against international terrorist leaders. Russia also employed this method during the Chechen War to assassinate Dudayev. This technique is highly dependent on precise intelligence, positioning, and swift raids against a target.

To comprehend the locations and operations of a city's defenders, it is necessary to understand the troop's size, strength, equipment, garrison and distribution, commander capabilities, as well as the troop's associated blocks, buildings, maneuvering routes social organizations, personnel, and events. Urban operations are more complicated than field operations due to these requirements. Civilians frequently outnumber enemy military personnel, particularly in MOOTWs such as counter-terrorism, stability maintenance, riot control, and post-war security. Familiarizing oneself with the population composition and the distinction between military and civilian targets is a time-consuming and laborious task, but one that is necessary.

The traditional urban operation approach encircles a city's central business district, evacuates non-combatants, and then searches for the enemies block by block, eliminating any stubbornly resisting enemies in the process. Numerous conflicts throughout history and the present day follow this pattern. Ground forces launch attacks after precision airstrikes against critical targets such as the city's government buildings, air defense bases, ammunition depots, and military barracks. Each of these instances involved the offensive side relying heavily on firepower to compensate for a lack of intelligence data. This is a relatively effective but crude method that results in significant collateral damage. Although meticulously planned, each city ended up in a shambles, laying the groundwork for major hidden dangers and complicating post-war recovery, governance, and reconstruction.

This model may become obsolete in the future. Logic dictates that resettling tens of millions of residents in a megacity is nearly impossible. With increasingly transparent globalization, the international community will condemn excessive civilian casualties and operations that are inconsistent with the trend toward a civilization of war. This will dwindle domestic and international support for the war. As a result, there is a need for more information and intelligence to improve accuracy and mitigate the risks mentioned above. To minimize collateral damage in the operational area, it is necessary to distinguish between military targets that must be attacked, civilian targets that should not be attacked, and neutral targets in between. Cultural relics and heritage sites, hospitals, schools, orphanages, churches, and cathedrals must all be designated as restricted areas with red lines separating them.

Clearly, as cities, particularly smart ones, develop rapidly, there will be more potential sources of information for urban operations. Advanced network information technology serves as the foundation for smart-city planning, governance, resource management, infrastructure, architectural design, transportation, security, and disaster relief systems. This technology can be viewed as a critical tool for information disclosure and data mining.

Urban operations in large cities and megacities require multi-dimensional intelligence data, the complexity of which will eventually exceed the "hard perception" capabilities of conventional intelligence, surveillance, and reconnaissance (ISR) sensor systems, necessitating the exploration of new information sources and collection methods. The city's complex natural terrains and large population provide ample opportunities for infiltration, making separating combatants from civilians extremely difficult, let alone identifying enemies and friends. Even identifying who lives in the city has become a challenge. According to research, of the world's 6 billion people today, 4.5 billion are migrants who lack local household registration or home ownership. Urbanization and population growth will exacerbate this problem, much like looking for a needle in a haystack in urban slums.

The military will also face technical challenges in utilizing data collected from smart cities. Among them, two issues must be resolved. The first vexing issue is data standardization, as the Internet contains a variety of systems and sources, including images, sounds, videos, and text. Specifically, texts exist in structured and unstructured forms, and their generation methods and transmission channels vary. This issue must be resolved before data can be recognized uniformly and viewed by the commander. The second issue is data timeliness or the degree to which data is up to date. The abundance and accessibility of data vary by city. Before deciding when and where to launch an attack, the commander must track and analyze intelligence from hot spots and major cities in real time, based on a large amount of dynamic data reflected via networks.

Intelligence gathering via open-source information and big data technology is gaining traction. It helps to screen and correlate massive amounts of data and provide a wealth of valuable information, particularly during the early stages of an operation.

With the increased use of the Internet and mobile phones, the massive amount of open-source data available will provide more information than ever before, while human intelligence will dwindle in quantity and value. According to media reports, the Central Intelligence Agency (CIA) and US military intelligence agencies attempted to establish and verify their informant network in 2003, the early days of Iraq's post-war insurgency. The Army's initial 69 tactical teams provided less than a fourth of the daily intelligence expected. The US commander in Baghdad believed that much of the intelligence gathered was more rumor than fact. The dissatisfactions stem from the difficulty in establishing a human intelligence network prior to the invasion, the fear of potential informants retaliating, and the difficulty of verifying the accuracy of intelligence sources. These stories demonstrate how difficult it is to obtain reliable human intelligence during the initial stages of operations.

Due to the dense population and high volume of communication in megacities, the amount of intelligence collected will soon exceed the analysts' processing capacity. As a result, more automated forms of analysis, such as automated data filtering, are required to reduce complex issues to a manageable level.

Image data acquisition requires the use of non-military channels and methods. When combined with other intelligence sources, image data will become a very effective positioning tool for precision strikes. Law enforcement and intelligence agencies worldwide have adopted security and crime prevention cameras, which have grown in popularity in urban operations. These cameras were useful in the aftermath of terrorist attacks. The best example is the terrorist bombing of the London bus system on July 7, 2005. Following the attack, British authorities conducted the largest closed-circuit television (CCTV) surveillance system inspection in British history, owing to the proliferation of terrorist attacks in multiple locations and the increased number of CCTV cameras in London. The investigators were able to track the bombers' trajectories in London and eventually tracked them to Leeds, about 230 miles from London.

The workload associated with manually processing image data from cameras is enormous. According to a 2004 report by the British newspaper *The Independent* [4], the UK has one CCTV camera for every 14 people, totaling more than 4.9 million cameras. *The Telegraph* stated that the figure could be even higher, implying one CCTV camera for every 11 people. To transform these networks from post-investigation tools to real-time intelligence collection tools, the military must rely on automated prompts. Computers will sift through these massive images, identifying notable activity trails and providing analysts with cues for tracking and investigation. These efforts have been made to a degree. For instance, law enforcement agencies have used audio sensors to detect gunfire and alert officers to take countermeasures. However, when there is no gunfire or distinctive military equipment images to act as cues, the automatic scanning system may struggle to function effectively.

Significant progress has been made in the area of video prompting. Several high-tech firms have already developed video technology capable of detecting changes in the landscape and alerting users. This technology can be tailored for urban environments and is especially advantageous when combined with facial recognition or biometric intelligence, which enables analysts to determine when and how the target object entered the environment under observation.

Biometric technology became an increasingly vital tool of intelligence gathering during the Iraq and Afghanistan wars. This technology necessitates a large database storing the local population's biometric data, most notably their fingerprints, iris, and face traits. A match can be made by extracting fingerprints from captured IEDs or concealed firearms and comparing them to the database. In 2001, the Kosovo War saw the first application of a biometric automation toolbox, followed by the Iraq and Afghanistan Wars. The US Army launched a handheld inter-agency identification device, a camera-sized information collection tool, in the spring of 2007. As of September 2007, this database had around 1.5 million entries, 167 of which matched ambiguous fingerprints discovered on IEDs. Since then, biometric intelligence has found a broader range of applications.

Technically, biometric reports are data-intensive. Cross-checking biometric entries against databases consumes significant bandwidth, posing issues for the military in a hostile environment. If the database expands to include cloud data, these technical impediments are anticipated to grow.

Apart from more well-known biometric techniques such as fingerprinting and facial recognition, visual and speech biometric techniques can also be used to identify potential targets. Visual biometrics, for example, identifies targets based on open-source and image data about ear shapes, hand features, and walking postures. Speech recognition can be used to match voices from a range of sources, including intercepted phone calls and videos uploaded on social media platforms.

New data-processing and integration technologies are crucial for familiarizing oneself with future urban landscapes. Automatic prompting software that identifies abnormal or suspicious activity will become important for resolving the tension between the number of sensors and insufficient human resources. Additionally, new solutions must be developed to store and access collected data in a timely manner. One of the issues may be how to increase the bandwidth while maintaining confidentiality. The fundamental solution is to develop new integrated software that enables confidential and open-source data to coexist in real time, forming a common operational picture.

With regard to intelligence informants, the military may need to train specialized open-source information (e.g., social media data) analysts, just as it does with human intelligence analysts and electronic signal analysts, as the open-source/social media field continues to grow in importance. The military must overcome another obstacle as well. That is, to ensure that there are sufficient foreign-language experts at all levels of the force, as the location of the next battle is unpredictable.

14.5 Multi-Dimensional Block Control and Precision Operations

Urban central area operations refer to offensive or defensive operations conducted to compete for central city blocks and buildings after obtaining the initiative of the city periphery. Traditional urban operations may result in indiscriminate bombardment and excessive collateral damage. In the future, as precision and multi-dimensionality improve, urban operations should evolve toward a more rational and controllable state, guided by unmanned or intelligent technology.

In terms of future offensive urban operations, there are two possible operational patterns. The first, based on a precise reconnaissance of the air/space and ground, as well as the association and military and civilian data, is to conduct distributed precision strikes and multi-dimensional control of critical fixed targets, mobile targets, and key figures throughout the city, primarily using reconnaissance and strike (R/S) UAVs and loitering munitions. The second option is to conduct large-scale special operations such as decapitation strikes or raids on urban core areas and key targets.

In terms of urban operation tactics, future combat is likely to be unmanned and precise operations aimed at the urban core, following the patterns of strike, control, defense, and stability maintenance. Consider an offensive operation. To begin, troops can use low- and medium-altitude UAVs to detect, track, and strike urban blocks based on previously acquired target information. Then, troops can use UGVs to conduct front-end obstacle removal and target guidance while relying on manned vehicles to conduct rear command, or even conduct precision strikes against armored ground targets, hidden fire points on the streets, and snipers inside buildings and roofs, using front and rear accompanying firepower systems in conjunction with aerial R/S. Meanwhile, troops can command human–machine hybrid special operations teams and individual soldiers, using the "machines in the front, humans in the rear" model to conduct manned–unmanned joint operations in buildings, underground parking lots, tunnels, and subways, among other environments, to clear the city's dead ends and blind spots.

In the future, supporting information systems for urban operations will revolve around space-based information systems, UAV reconnaissance and precision strike systems, ground-associated precision fire support systems, manned–unmanned cooperative assault systems, human–machine mixed formation soldier systems, and area blockade and isolation systems combining software and hardware.

Cognitive communication networks and mobile tactical clouds are critical components of the supporting systems. Four objectives must be taken into account when constructing cognitive communication networks. The first is to maximize the use of space-based information and satellite communication to facilitate over-the-horizon communication and situational awareness in the face of obstacles in the city. They are typically used to establish long-distance communication between the front and rear. The second is to equip autonomous communication systems that can automatically hop and jump in response to building occlusion and interference signals. These systems can be used typically in front-end distributed reconnaissance, swarm attacks, and coordinated manned–unmanned communications. The third is to ensure the accessibility of communication between medium- and long-distance ground forces based on medium- and low-altitude communication relays. The fourth is to resolve communication, navigation, and detection problems encountered indoors and outdoors and ensure situational awareness and intelligent human–machine interaction between individual soldiers and squads.

The mobile tactical cloud is a mobile processing center equipped with an integrated information service platform that supports troops and detachment units during urban operations. It leverages a robust cloud platform in conjunction with front-end processing capabilities to provide real-time data and information services. It is constructed around an intelligent perception recognition module at the front end and a decision-making algorithm and model at the back end. The former ensures limited autonomous decision-making while offline, while the latter ensures rapid, mixed, and optimized decision-making while online. To enable networked sharing, control, and collaboration, the mobile tactical cloud's interface services should be standardized and networked.

Urban unmanned systems are composed of ground, aerial, and underground unmanned vehicles. In coastal cities and river-rich areas, the systems may also include unmanned surface vehicles (USVs) and unmanned underwater vehicles (UUVs). These systems provide real-time, accurate, and comprehensive battlefield information to support back-end data modeling, situation analysis, mission planning, and command decision-making. Additionally, these systems can execute commands issued by cloud, assault, and individual soldier systems and coordinate operations with manned operation vehicles (Figure 14.3).

The urban assault system is based on a manned–unmanned hybrid armored mobile vehicle used for mobile and multi-dimensional operations in urban areas. The system is capable of transporting small UAVs and UGVs. It is also equipped with powerful information-processing systems and firing systems for the capability of rapidly planning and executing multiple tasks, coordinating front-end and back-end firepower, conducting active and passive protection,

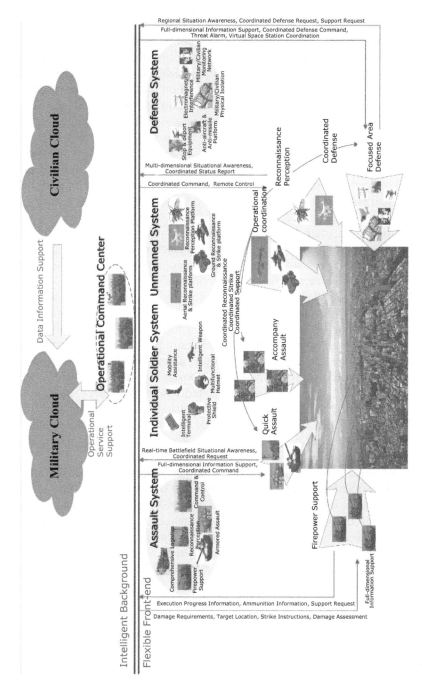

Figure 14.3 Equipment systems in urban central area operations.

striking targets precisely and adaptably, assaulting and maneuvering flexibly, communicating fluently in real time, and providing logistics accurately and rapidly.

The urban soldier system, based on networked perception and multifunctional wearable devices, exemplifies the operational capability of the assault and unmanned systems. It is capable of receiving battlefield environment and adversary intelligence from the back-end cloud and providing real-time feedback to the back end. This system can be used to precisely target remnant concealed enemies. It has superior mobility over urban obstacles, making it ideal for all-weather target search and perception, precise positioning and striking, as well as clearing and controlling local battlefield areas.

The urban defense system protects vital urban targets from attack by cruise missiles, armed UAVs, squads, and terrorists. It possesses capabilities of detecting, tracking, intercepting, and striking low-altitude targets; defending against cyber attacks; reconnaissance, surveillance, and rejection of enemy armed personnel or terrorists; situation control and emergency repair following an attack on critical targets.

Front-end and back-end integrated operation systems of systems (OSoS) can be used to complete a new operational mode by providing diverse front-end functions and back-end cloud support, including front-end target reconnaissance, back-end information fusion, front-end threat monitoring, back-end mission planning, front-end target guidance, back-end precision strikes, front-end assault operations, back-end situational awareness, front-end maneuverability, and back-end logistics.

In urban environments, cooperative manned–unmanned operations typically include coordinated reconnaissance, strikes, protection, and support. For example, UGVs, UAVs, and other unmanned vehicles can conduct coordinated reconnaissance and surveillance in tunnels, underground spaces, and inside buildings, as well as in core areas of engagement, damaged buildings, and biochemical suspect areas. Unmanned vehicles on the ground and at low and medium altitudes can be used to target enemy forces concealed in underground passageways, underground caverns, and high-rise structures. These vehicles are capable of operating in a multi-dimensional attack mode against armored targets, command-and-control facilities, and communication centers. They can detect, clear, or detonate roadside explosives during coordinated protection operations, as well as intercept incoming ammunition and conduct field rescue missions in hazardous situations. Additionally, they can be used in conjunction with manned equipment or individual soldiers to deliver materials and munitions and provide communication relay support.

Urban operation forces in physical space will become unmanned, swarmed, miniaturized, modularized, and mixed, with networked perception, coordinated reconnaissance and strike, multi-dimensional assault, precise damage, and diverse defense functions.

14.6 Virtual and Cross-Domain Operations

The conventional physical space-based attack technique will be abandoned in favor of a novel one in future urban operations. Virtual and physical locations will be dynamically combined to undermine the adversary's morale while causing the least amount of collateral damage possible. Virtual space and cross-domain operations are becoming more prevalent throughout the pre-war, in-war, post-war, and peacetime phases. Examples include perception and cognitive confrontation, public opinion control and intervention, key targets and crowd surveillance, electronic deception and interference, regional hard/soft blockade and isolation, post-war defense and counter-terrorism, crowd emergency evacuation and management, and key infrastructure management and control.

Sufficient information resources and online media are required to consolidate victories in public opinion, cyber psychological, and policy propaganda warfare. The city's information resources have exploded in size as a result of urban informatization, digitization, and intelligentization. Diversified media outlets such as radio, television, communication, and the Internet can act as a conduit for citizens' massive information requests while also serving as a guide for public opinion. Public opinion mobilization and psychological warfare techniques can be used to destabilize an adversary. During the Iraq War, the US military launched a cyber offensive against Iraqi military and government employees in January 2003, primarily through text messages and emails delivered via mobile phones. "We know who you are; lay down your weapons now; there is no way out," Iraqi soldiers were told. As a result of the excellent psychological propaganda campaign, many Iraqi soldiers crossed borders and surrendered to the US Army, and the majority of Iraqi troops abandoned resistance. US troops advanced swiftly into Baghdad from the Iraq–Kuwait border.

Throughout the Libyan war, which began in February 2011, coalition forces used news outlets such as the Internet, radio, and television to condemn Libya's bureaucratic dignitaries for monopolizing various resources, the family rule system for diverging from democratic politics, and Gaddafi for being the executioner who slaughtered the populace while winning their hearts. Simultaneously, coalition forces sent emails and text messages to Libyan military officers stating, "We have the GPS coordinates for your command post and can also lock your mobile phone," and "The cruise missiles have set up programs for these coordinates …," deterring senior Libyan officials significantly. The campaign effectively demoralized the Libyan Army and facilitated the siege of cities.

People congregating in a city can choose to be defensive supports, offensive allies, retreating onlookers, or neutral, depending on public opinion. The perceptions of urban people on war and its results may vary depending on their religious beliefs, cultural education, social standing, and occupational income. Psychological studies on varied groups and the adoption of diverse public opinion orientations are required for urban people to sustain emotional stability and a sense of order. Cyber, public opinion, and psychological warfare are thus critical components of urban operation victory.

Urban offensive operations necessitate constant monitoring, tracking, and location of the city's political officials, military commanders, tribal elders, and social elites. Several are apparent targets for decapitation, while others are necessary for disintegration, popular support, and unification. It is vital for counter-terrorism operations to track high-profile individuals. In addition, early detection of terrorist activity enables early warnings to be made to avert potential disasters.

It is vital to understand the social and cultural history of a local city, as well as the ethnic composition of its inhabitants, in order to follow key targets and groups. In the case of Karachi, Pakistan, a city of over 24 million people, identifying who or what is "local" is challenging due to the huge amount of commodities and people entering and exiting. The city's wide ethnic composition reflects both the various languages spoken and the multiple political parties in the city. Ramadi, Iraq, on the other hand, is one of the province's final border towns which, despite a decline in total population over the last decade, is estimated to have a population of around 500,000 people, virtually completely of the same tribe. In brief, tracking key targets and groups requires various approaches, just as cities differ in their social, cultural, and ethnic composition.

Cyber attacks, electronic deception, electronic jamming, and even capturing control of the defensive side's command-and-control system are all vital strategies for urban offensive operations. These means may help to secure a decisive victory during the early stages of urban operations. The following methods may also assist in subduing an adversary without fighting: to attack the adversary from virtual cyberspace, paralyze its command system, destroy its communication networks, combine political offensiveness and military deterrence, and conduct precise psychological warfare and public opinion confrontation. These can wreak havoc on an adversary's morale, reduce its desire to resist, and cause neutral citizens to escape. At the time, military powers are effectively employing cyber attacks, electronic interference, and deceptions in real-world conflict on several occasions, with remarkable outcomes.

In the latter stages of urban operations, situational awareness, trend analysis, early warning, and population movement guidance in urban combat zones should be conducted based on big data, artificial intelligence, and navigation technology. These will help troops to control crowd behavior and respond to emergencies during military operations such as regional seizure, blockade isolation, key target defense, and counter-terrorism. Multiple wireless-positioning resources should be leveraged along with multi-channel user network information, user behavior analysis, user identification, and dynamic tracking in order to acquire real-time population and crowd flow characteristics. Dynamic identification of regional population movements, analysis of real-time flow trends, alerting to aberrant flows, and fast emergency actions are also necessary. Finally, wireless communications, mobile networks, digital television broadcasting, and IoT technologies should be used to provide accurate crowd directing and emergency management.

Various technical means and weaponry can be used to protect the city's government buildings, transportation hubs, airports, docks, radio and television stations, mobile communications and data centers, power plants, and water plants from enemy destruction and harassment in the aftermath of the war. The defensive side should integrate military and civilian reconnaissance and surveillance capabilities, as well as fire strike, electromagnetic interference (EMI), information isolation, crisis management, and emergency repair capabilities, in order to detect and track low-altitude targets, intercept fire strikes, and defend against armed ground adversaries. Defenders should anticipate, notify, and identify covert terrorists and extremists, as well as discriminate against, seduce, and eliminate them through military operations that leverage big data search, identification, and storage.

Smart cities will make it possible to attack, control, defend, and restore an entire city's information flow, energy network, food chain, water resources, public transportation system, and financial services. As a result, the entire process of future urban operations should be designed holistically, including urban networks, communications, electricity, water conservation, finance, transportation, and logistics. The early stages of combat are focused on disrupting the enemy's network and causing manageable damage, which means eliminating resistance as quickly as possible without causing excessive incidental damage so that in the later stages of combat, the offensive side can focus on restoring social services, supply, and production.

14.7 Theoretical and Technical Support

In terms of urban operation theory, it is critical to consider both operational objectives and the city's development status in order to occupy and govern the city effectively and humanely. To overcome the major constraints on urban operations, it is also necessary to conduct research in the following areas: future urban development, urban operational concepts, operational capabilities, operation system design, equipment image depiction, operational experiments and pre-evaluations, and key equipment and technology verification. Additionally, it is necessary to research the following areas to overcome the major constraints on urban operations: future urban development, urban operational concepts, capabilities, system design, equipment image depiction, experiments and pre-evaluations, and key equipment and technology verification. Advancements should be made in multi-dimensional urban environmental detection and situational awareness, intelligent clouds and cognitive communication networks, operation area isolation and blockade techniques, intelligent weapons and equipment that adapt to urban structures and street combat, public opinion and psychological control, crowd diversion and emergency management, and urban infrastructure management and control. These may contribute to the development of an urban OSoS that combines virtual practices and operation trials, forming a hybrid of virtual and real operation capabilities and advantages. Among them, three key technological concerns demand special attention.

The first is simulating future urban operation environments. It will be possible to construct a physical and virtual simulation environment that encompasses urban social environments, computer-generated forces, and operational evaluation and analysis based on data mining for future urban operations. This may help to verify and evaluate urban operational patterns, capabilities, tactics, coordination, and effectiveness, as well as weapons and equipment adaptability.

The second is isolating and blocking urban operation areas. The future complicated urban operation environment necessitates regional identification, physical isolation, fire blockade, information cutoff, and EMI to quickly isolate geographical, network, and electromagnetic operational zones. Isolation allows for the division, encirclement, and suppression of enemy forces, as well as the quick handling and containment of public disturbances and riots. Among the research areas are crowd recognition and rapid positioning based on multi-source data acquisition and big data analysis, the physical identification of key urban targets, geographic environment isolation, network information-blocking measures, and electromagnetic control, blockade, and isolation.

The third is managing public opinion, which includes psychological manipulation, crowd diversion, and emergency management. Urban operation technology research will focus on the challenges of effective urban administration and control, which may include: (1) emergence effect-based information transmission and psychological behavior correlation; (2) data mining and situational awareness-based public opinion analysis; (3) machine learning-based crowd action analysis and identity simulation; (4) Internet/ mobile communications and digital TV-based public opinion control and psychological impact; (5) virtual reality-based crowd positioning and feature extraction; (6) fuzzy match technology-based risk assessment and early warning; (7) wireless communication and digital TV broadcasting-based precision guidance; and (8) multi-system urban emergency management.

References

[1] Baidu Baike 百度百科. (2020b, March 4). *Heiying zhuiluo* 黑鹰坠落. Baidu baike 百度百科. https://baike.baidu.com/item/.
[2] Baidu Baike 百度百科. (2020a, May 6). *Geluozini xiangzhan* 格罗兹尼巷战. Baidu baike 百度百科. https://baike.baidu.com/item/.
[3] National Intelligence Council. (2012). *Global trends 2030: Alternative worlds.* Washington, DC: National Intelligence Council.
[4] Frith, M. (2004). Big Brother Britain, 2004. *The Independent.* www.independent. co.uk/news/uk/this-britain/big-brother-britain-2004-73167.html

15 Gray Zone Operations

Gray zone operations, or those aimed at gaining control of social infrastructures and systems such as power, networks, food, transportation, and banking, are a critical component of hybrid warfare. Gray zone operations can be interpreted in two distinct ways. One refers to operations that fall between conventional operations and military operations other than war (MOOTWs). The other is a legal term that refers to operations in ambiguous areas. Controlling social institutions that are inextricably related to human productivity and subsistence raises moral and legal issues, particularly under international law.

War is a violent act. In the age of artificial intelligence (AI), the nature of conflict, violence, and coercion may stay identical, even though the manifestations, procedures, and outcomes may change. As science and technology advance and global connectivity improves, future conflicts will entail new social systems and forms. Controlling social systems becomes a necessary precondition for occupying and controlling cities and countries, as well as an unavoidable outcome of the war's progress. Intelligent warfare against social systems is a general trend that must be closely monitored and planned for. A critical strategic issue that remains to be studied and resolved is how to protect the social systems upon which people rely for survival and productivity, how to defend against adversary attacks, and how to combat terrorist infiltration and attacks against routers, servers, local area networks, industrial control networks, and private networks.

15.1 Network Infrastructure Protection

Network infrastructure has become one of humanity's means of survival and spiritual dependence in the age of AI, closely linked in production, life, and the virtual world. Internet, mobile communications, radio, television, and instant messaging apps (e.g., WeChat, Facebook), and online shopping, which are important sectors of the social system and irreplaceable media for interpersonal communications in peacetime, may become the barometer reflecting public opinion in wartime. They may also act as a catalyst for human psychological disturbance, a breeding ground for social turbulence, and a source of collective emergent behavior that both sides in an operation can use effectively.

DOI: 10.4324/b22974-17

The most typical example is Turkey's 2016 coup. President Erdoğan was able to put an end to the riots and significantly alter the situation not through his army, force, or oppression, but by delivering messages to his supporters via Facebook and encouraging them to fight the rebels.

Network infrastructure protection involves maintaining normal social systems of living and productivity, as well as defending against and destroying operational opponents and terrorists. Defense and protection involve the following key points: the first is to prevent network communication interference and hijacking. Encryption technologies are decipherable via open-frequency bands and channels of television, radio, mobile, and satellite communications. Therefore, countermeasures should be taken against decryptions that interfere with and even take over public network communication signals in public space. The second is to defend against cyber attacks. Opponents and hackers may target systems involving telecommunication, spectrum resource allocation, sale and after-sale service, mobile Android/iOS, satellite navigation and communication management, radio/TV production and broadcasting systems via online or offline, wired or wireless attacks. The third is to prevent deep intrusion into radiofrequency (RF) and sensor systems. For example, attackers may interfere with network communications or damage key components of transceiver systems using electromagnetic interference (EMI) or electromagnetic pulse weapons. Vulnerable targets may include core circuits such as base stations, transmission towers, receiving terminals, antennas, mixers, and amplifiers of operation systems. The fourth is to prevent hardware damage. It is necessary to defend against enemy attacks (e.g., explosions and line cuttings) on TV and radio stations, postal and telecommunication centers, underground pipelines, community servers, optical fiber networks, etc. The fifth is to prevent physical destruction and soft killings of network infrastructures' power supply systems.

15.2 Power Protection

Protection of power plants and networks is essential for social security and energy supply. Modern society uses more electrical equipment, with energy supply distributed globally, whether in industrial production, urban infrastructures, or daily life. In the event of a power outage, all electrical equipment, including networks, water supply, electric vehicles, subways, elevators, and household appliances, will cease to function. Finance, retail, and manufacturing will all be suspended as well. This situation could cause social unrest and the collapse of an urban management system if it lasts longer than 3 days.

Strategic grid protection requires better protection and management of the state grid center and key network facilities. It is required to prevent both cyber and physical attacks. An example is the cyber attack on the Ukrainian power system, which resulted in widespread blackouts. Similarly, during the Kosovo War, the US military used graphite conductive fiber bombs to completely disable Kosovo's national power grid, rendering life unsustainable for the populace and ultimately forcing the rebels to surrender.

On March 7, 2019, Venezuela experienced the country's worst power outage in history. Businesses and government agencies across the country were shut down for nearly a week. Companies folded. Hospitals and public services were paralyzed, and society collapsed. On March 15, Venezuelan President Nicolás Maduro claimed the blackout was an "act of war in cyberspace" carried out by the US government's "Cyber Attack System." On March 25, another massive power outage hit 16 states in Venezuela, leaving blocks in Caracas dark. The subways were shut down, and dialysis patients were unable to receive treatment. The Venezuelan government announced a 24-hour national work and school suspension on March 26. The large-scale blackout had already put Venezuela in a gray zone, and it was uncertain whether there would be more blackouts. This incident exemplifies how military operations will be presented in the future era of AI.

Tactical management and protection of power systems like urban power plants are critical. They should always be heavily guarded against physical, terrorist sneak, and cyber attacks. Power supply in buildings and communities is also necessary. This includes solar energy equipment, emergency power supply and generation equipment, and electric meter protection systems. Smart electric meters are used widely in online payment, bank fund transfer, power management systems, etc. Particularly, smart meter controllers are monopolized by a few global suppliers. Once exploited, the whole system may fail or be paralyzed.

Future energy supply will gradually diversify. Nuclear, solar, wind, biofuel, and thermoelectric power generation will all be connected to the grid separately. The energy network structure will become more complex, and risks will increase. For example, aside from natural disasters like earthquakes, nuclear power plants are also vulnerable to cyber attacks like the "Stuxnet" virus attack.

The power grid's topology is both uniform and random. Energy can come from remote sources like hydropower and coal mines, or from local power plants that generate, transmit, transform, and use electricity. In most cases, electricity is transmitted as high-voltage alternating current (AC), although most power applications require low-voltage AC, such as 110 or 220 volts. Therefore, transformers capable of converting high-voltage AC into low-voltage AC are required. Also, power will be delivered to customers, where transformers and substations will be concentrated. Thus, transformers and substations are vital nodes, and their protection is vital.

Nowadays, power transmission equipment and management networks are more secure, and power accidents caused by lightning, fires, short circuits, and line aging will automatically close the corresponding electric switches. However, it may trigger chain reactions that result in widespread blackouts.

In 2003, for example, approximately 50 million people in the northeastern United States and Canada lost power for an extended period [1]. On August

14, 2003, the weather was scorching, and air conditioners were operating at maximum capacity, nearly approaching their previous peaks. After 3 p.m., a power outage began in Cleveland, Ohio. The high-voltage transmission line touched overgrown trees. In order to prevent further damage and fire, the self-regulating electricity system disconnected the lines, causing a series of faults. Automatic safety relays throughout the northeastern United States shut down circuits and generators that exceeded preset operating load limits, beginning at 4:10:37 p.m. and lasting less than 8 seconds. The failure of these safety relays cut the power grid connection, causing widespread power outages.

In mid-February 2021, a strong cold-air mass pushed southward, resulting in a record low temperature in the southern United States. According to the US Weather Service's snow forecasts for the entire US territory, snowfall in the southern region would be 150–200% of normal, and at least 150 million people, or about half of the US population, would face severe cold weather.

The southern US began experiencing winter storms on February 15. The storms resulted in a grid failure in the southwest, affecting 14 states, most notably Texas, which rarely sees snow. The massive power outage in Texas occurred because everyone needed to heat up, and as electricity consumption climbed dramatically, so did the load on the Texas power infrastructure. Simultaneously, due to the extreme cold weather that rendered the entire energy infrastructure inoperable (Texas energy system had never encountered such extreme cold weather), the entire system was "frozen." According to incomplete estimates, over 3 million Texans were left without power during the exceedingly rare cold storms. Meanwhile, it was unanticipated that this exceptional snowstorm would directly result in the freezing of most Texas wind turbines.

Authorities deployed 26 fire trucks, 80 police cars, and 13 ambulances to reclaim control of the situation following a series of collisions involving 133 cars on the highway leading to Dallas, resulting in the deaths of six people and 65 injuries. Roads were frozen in many areas, tap water was cut off, and residents lacked adequate heating equipment. Individuals were required to go out and obtain fuel, firewood, food, and drinking water. Everyone donned their warmest clothing and waited for the power to be restored.

The price of electricity was the most notable example. During peak demand, electricity prices increased by over 200 times. Previously, 1,000 kilowatt-hours were $40 wholesale. However, the outage's peak electricity price was estimated to be $8,750.

15.3 Petroleum and Natural Gas Security

Protecting oil and gas facilities is a critical component of gray zone operations. One of the war's strategic objectives is to compete for and control oil and gas resources. Management and control of oil, gas, and hazardous chemicals are critical in peacetime and require special attention during times of war. The Iraq War exemplifies the importance of fighting for and protecting oil fields during a war. ISIS had also grown rapidly, owing primarily to its occupation of northern Iraq's and eastern Syria's oil fields.

On May 12, 2019, four massive oil tankers were attacked near the largest oil port in the United Arab Emirates (UAE). The propellers on the ship's side and tail rudders were blown through, creating a 2-meter-square hole and rendering the ship unusable. Two days later, seven Houthi unmanned aerial vehicles (UAVs) attacked Saudi Arabia's east–west oil pipelines, causing damage to pipelines and facilities. The daily transport of 5 million barrels of crude oil was halted, resulting in enormous economic losses.

Five months later, on September 14, 2019, the Houthis used 18 UAVs and seven cruise missiles to attack and set fire to two Saudi Aramco oil facilities. According to US satellite images, Saudi crude oil production fell by 5.7 million barrels per day due to accurate penetration of large storage tanks (Figure 15.1). This figure equated to approximately 5% of the world's daily oil production.

In the age of AI, oil, gas, and hazardous chemical safety protection includes the following:

1. Preventing hackers and adversaries from launching cyber attacks against energy and chemical management systems. An example of these kinds

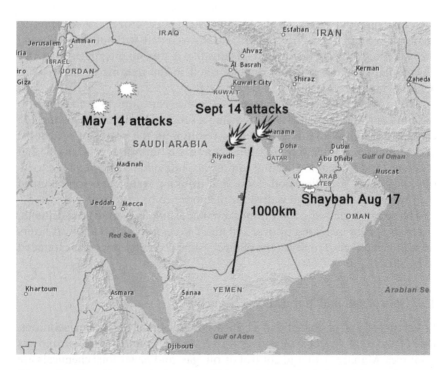

Figure 15.1 Houthi unmanned aerial vehicles (UAVs) and missiles with a strike range of thousands of kilometers.

of attacks is the US military's Stuxnet virus attack on Iran's nuclear centrifuges.

2. Early-warning and management control via global positioning system (GPS)/Beidou + mobile communications. Distributed pressure sensors and toxic chemical sensors, such as laser detection equipment, Raman sensors, and quantum dot micro-spectrometers, can monitor oil and gas leaks and dangerous explosions in critical production pipelines, storage, and transportation systems in real time and alert command-and-control centers to anomalies. Then, risky facilities can be immediately shut down, and emergency teams can be dispatched to treat and control the situation.

3. Developing an integrated emergency command-and-control system that links the networked sensor system to energy and chemical facilities and sites. Public safety, armed police officers, fire control, chemical defense, and other forces should be mobilized quickly in the event of a dangerous substance explosion.

4. Establishing a security system for the whole supplier-to-user chain, including oil and gas fields, refineries, storage sites, pipeline facilities, community stations, supply stations, buildings, and households.

During peacetime, oil and gas are frequently involved in risks of fires and explosions. In wartime, oil and gas facilities are also susceptible targets for adversary mission planning. Given current trends, physical attacks on oil fields, oil depots, and gas pipelines are likely to elicit public outrage and condemnation. However, in the future, the potential for cyber attacks on energy systems will increase as AI makes them more networked, precise, and intelligent. As a result, it is critical to propose comprehensive countermeasures and solutions for the protection and defense of oil and gas systems, as well as hazardous chemicals, during peacetime and throughout the entire wartime operation process, both technically and administratively. Among them should be militarized defense methods and programs.

15.4 Food Chain Management

During and after the war, food and water scarcity may become a major social issue. As with blackouts, 2 days without food or water can result in social panic and even a massive refugee wave. In a war, it is much easier to destroy than to rebuild and sustain. When basic necessities such as food and water are not guaranteed, people flee for their lives, which is why recent Syrian wars and chaos have resulted in waves of refugees in Europe. The safety of the food chain entails preventing the chain from being interrupted or contaminated.

Interruptions can be avoided through the management of the entire food chain. To begin, food production should be safeguarded, particularly grain and vegetable production. Second, smooth transportation between producers and consumers, such as supermarkets and private residences, should be ensured. Third, food storage, preparation, and processing should follow industrial standards. Fourth, the food currency's sale and payment systems should be stable.

In the age of artificial intelligence, e-commerce and multi-dimensional logistics are also important means to ensure food chain security as they connect production sites, suppliers, and consumers. E-commerce and logistics management systems must be safe to ensure food supply. These systems are integral to people's daily lives, involving financial transactions, warehousing, and express delivery. If the e-commerce and logistics systems are attacked and destroyed, people may revert to cash transactions and a barter economy that they have long been unaccustomed to.

Preventing food contamination requires ensuring the hygiene and quality of the food chain, as well as avoiding damage from enemies and terrorists. According to *The Jerusalem Post*, ISIS was plotting and agitating for "Lone Wolf" supporters to carry out terrorist toxic attacks in crowded shopping malls in September 2017 [2]. As was the case with the sarin attack on the Tokyo subway, these toxic chemical weapons were directed at unsuspecting crowds. Toxic chemicals or contaminants may be mixed into food during production, transportation, and processing, spreading the horror. The preceding threat necessitates the development of ultra-small Raman sensors and quantum dot spectrometers that can be installed on mobile phones to continuously monitor food safety concerns.

A space-based hyper-network management system may be formed in the future to ensure the security of e-commerce, logistics, food production, processing, and retail management. This has the potential to improve the quantity, quality, and hygiene of food supplies. The planting, growing, and harvesting of crops and their supply bases in domestic and even international regions can be monitored in real time using futuristic small satellite networks. In the event of a war, e-commerce systems such as Cainiao Logistics, Hema Fresh Food, and unmanned supermarkets can quickly restore urban food supply and social order. These systems are typically based on a recoverable network and are modular in design and simple to assemble. Additionally, advanced Internet of Things and shared devices can aid in the efficient transportation of food. Expedited delivery services are capable of providing efficient and on-time transportation services at any time and from any location.

15.5 Competition for Traffic Lines

Human and cargo transportation are intertwined with roads, railways, waterways, and aviation. Future modes of transportation will include autonomous vehicles, drone delivery, and even space travel. Three factors should be taken into account to ensure smooth and efficient transportation lines. The first is the energy supply. Whether conventional (for airplanes, cars, trains, ships) or novel (for new energy automobiles, drones, and unmanned ships), energy is required. As a result, energy infrastructure security has become the primary pillar of transportation security. The second is network management.

Network information systems underpin air traffic control, railway network operation, highway dispatching, and logistics distribution. Therefore, securing and protecting these networks is vital. The third is communication and navigation services. All vehicles, particularly airplanes, unmanned ground vehicles (UGVs), UAVs, and unmanned surface vehicles (USVs), are all supported by communication and navigation services, including positioning, self-organizing networks, intelligent perception, and intelligent control.

Three measures are notable in terms of protecting and competing for transportation networks. The first is vehicle protection. During peacetime, security and safety inspections can help to prevent terrorist attacks and the transportation of flammable or explosive materials. For security and defense in times of war, armed escort or accompanying protection, and temporary installation of protective devices such as active defense systems and electronic jamming equipment are advised. The second is transportation hub protection. In order to protect critical transportation infrastructures (e.g., bridges, railways, airports, ports, docks, railway stations, and bus stations) from terrorist attacks, air defense and ground security systems must be established and improved. The third is cyber defenses. Cyber attacks are inevitable as vehicles and management systems become more automated, unmanned, and intelligent. To support automatic piloting and data transmission of high-speed railways, UGVs, UAVs, USVs, as well as modern manned cargo or civil aviation aircraft, satellite networks, and back-end clouds will be used. In spite of firewalls and security software, the systems will always be vulnerable to sophisticated hackers and specialized adversaries, posing significant security risks.

For example, modern people rely on civil aviation to travel, study abroad, and do business as globalization progresses. International flight capacity and passenger flow continue to grow. Airspace planning, air traffic control, flight management, and passenger services are all intricate. Major airport air traffic control, airline operations management, and aircraft avionics systems are all complex. Cyber attacks on airlines will result in stranded passengers, a paralyzed aviation system, unexplained connection loss, and even aircraft collisions. In 2014, Malaysia Airlines Flight MH370 jumped to 46,000 feet before free-falling to a height of fewer than 3,000 feet. One possible explanation for this accident was an unanticipated oxygen supply system failure, resulting in an automatic flight behavior that rendered all passengers unconscious due to oxygen deficiency. It is unclear whether this was due to the captain's initiative or being attacked and infiltrated by unknown signals. Nothing is impossible.

Aircraft systems safety is vital. An aircraft's avionics system is vulnerable. At least one of the most common ways terrorists and hackers attack aircraft is through infecting and occupying the ground radar communication channel with computer viruses. Then, via a data bus connected to a processing unit, they can enter the avionics center or the maintenance module, searching for system vulnerabilities or initiating operation instructions, with fatal consequences for the aircraft. The aircraft may lose control or be taken over if the ground tower communication, oxygen generation, and safety control systems are disabled. Additionally, a new destination address may be entered in autopilot mode to change the flight route. This could result in civil aircraft loss. These cyber attacks are both theoretically and technically possible.

Modern civil aviation engines are controlled by a few global suppliers. Data are sent live via satellite to the monopolies' databases. Once the communication channel is exploited and the data are out of control, hackers may issue false commands (e.g., fire detected) that forcefully shut down the engine or the fuel supply system, causing the aircraft to malfunction. Manufacturers of aircraft engines are expected to take all technical precautions to ensure passenger and aircraft safety, reducing or eliminating vulnerabilities and attack possibilities. However, because no problem can be solved thoroughly, theoretically any possibility exists.

Aviation management, railway passenger and freight transportation, and urban subway management systems are all examples of complex transportation management systems. Network communication and machine learning software gradually upgrade the systems into automated and intelligent systems. While the security systems are improving, cyber and physical attacks and destruction during wartime are still possible. The widespread use of small and medium-sized UAVs and UGVs will likely impact human survival and productivity in the future. A single cyber attack may cause catastrophic damage, so immediate security and protection measures are required.

15.6 Financial System Risks

Finance is the lifeblood of an economy, including currency issuance, circulation and withdrawal, loan issuance and repayment, fund deposit and withdrawal, and exchange dealing. Financial management has entered the age of AI, reflected in full network interconnectivity, global settlement and mobile, automatic, and facial recognition payment. Future financial system security risks will be divided into two categories: strategic and tactical.

Attacks on central banks and corporate data centers and information management systems are strategic threats. These systems are interconnected nationally and globally, and most use GPS for timing. An attack on these GPS-based data management systems would be catastrophic. Military powers are currently testing GPS interference in the airspace near major financial systems by wirelessly injecting fake satellite timing signals or viruses over long distances to disrupt the central controller's time–space benchmarks. The effect has been confirmed. Technically, these tests pave the way for long-distance wireless interference and destruction. Financial security requires detecting, warning, protecting, and backing up time–space benchmarks, as well as risk prevention, control, and management. For example, it is vital to prevent a computer virus from entering through insiders, especially high-level management personnel.

While banking systems are fortified with varying degrees of security, they remain connected to public networks and mobile payment. As a result, leaks exist everywhere, which hackers and adversaries can exploit.

Tactical threats mainly manifest themselves as telecommunications fraud. As with drugs, which breed international crimes and global issues, telecommunications fraud is a domestic and international social cancer. For instance, criminals may steal personal identity information or impersonate public security and judicial officers via phony base stations and phones in order to commit psychological intimidation and deception.

Moreover, during times of war, it is critical to emphasize the security of the central bank's headquarters, underground vaults, data centers, and core outlets. Backup and restoration of data following physical attacks or war destruction are also necessary, as the consequences of war are unpredictable in the present and future.

15.7 Military Industrial Security

The military industry, or national defense industry, is comprised of businesses engaged in national defense science and technology research and development, weaponry production, and associated support services. Strategic deterrence, technological advancement, manufacturing complexity, autonomous controllability, and civil–military fusion are all characteristics of the global military industry. The military industry is a critical component of national strategic deterrence, a vital source of support for national armed forces, a guarantee of victory in future warfare, and a strategic high ground for technological and economic competition. During peacetime, operational adversaries and hostile countries may conduct reconnaissance and surveillance on the military industry. In times of war, the military industry is likely to remain a primary target. The military industry is critical both now and in the future age of AI. Thus, the protection of the military industrial base is critical for national and social security.

To achieve technical superiority in peacetime as a critical component of deterrence, the US proposed the "New Three-Pronged" strategic deterrence system in 2001, which includes nuclear and non-nuclear strike systems, active and passive defense systems, adaptable defense infrastructure, and the military industry. Following the 2008 financial crisis, the US bolstered its New Three-Pronged strategic deterrence system (Figure 15.2). As a result, more diverse deterrent forces at national level have been developed. The military industry is a vast arsenal of sophisticated and intelligent weapons, as well as an effective deterrent against potential and future adversaries.

During both peacetime and wartime, the military industrial complex's layout, secrecy, and capabilities must be safeguarded. Cyber security and physical defense are also necessary.

In an era of mechanized warfare, the military industries of major powers are typically spread across multiple fronts. This organizational structure may

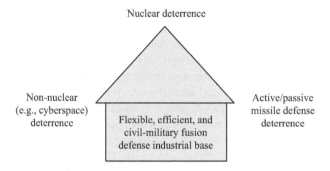

Figure 15.2 The new US strategic deterrent system.

Table 15.1 Hierarchical diagram of systematic equipment supply capability

---

Organization layer →top-level design, operational elements integration
System layer →overall model and weapon system
Framework layer →key subsystems and supporting components
Basic layer→materials, devices, power, and software
Support layer→digitized design, manufacturing, and assembly

---

allow for greater strategic mobility for contact-based operations. Military factories on the third front that are hidden in remote mountainous areas or caves may have a better chance of surviving an initial wave of enemy attacks. However, as the process of military transformation has accelerated, the importance of a detailed strategic layout has gradually waned. The military industry has evolved gradually toward intensive operation and management, industrial chain convergence, and regionalized support.

Military manufacturers of nuclear weapons, strategic weapons and equipment, core devices, and critical materials hold some of the most vital secrets, capabilities, and data in the national defense field. Therefore, they have been subjected to espionage and terrorist attacks. Additionally, these manufacturers are a military industrial security priority. Thus, confidentiality should be maintained during peacetime to avoid leakage. In times of war, data backup, concealment, and physical defense should be bolstered.

Intelligent equipment design and manufacturing (Table 15.1) will necessitate the evolution of scientific research and manufacturing capabilities toward those that are more networked, digitalized, multi-dimensional, coordinated, visualized, flexible, and intelligent. These advanced capabilities will enable cross-industry collaboration in product design, development, and production, ensuring product reliability and consistency of quality.

Network security concerns will inevitably arise as weapons and equipment acquire networked intelligence in the future. Even internal and private networks, as well as those that are exclusive, may have unexpected flaws. Therefore, data security should be prioritized throughout the value chain of the military industry, from design to trial production, manufacturing, assembly, inspection, acceptance, and after-sales service and support.

It is also critical to safeguard critical scientific and technical military personnel. Their minds are imprinted with technical products' design concepts, experience, craftsmanship, and know-how. Thus, along with increased confidentiality and legal education, a set of effective incentive mechanisms is required. Many secret leaks are motivated by economic considerations rather than political or ideological ones. As a Chinese proverb puts it, "A man without remote consideration may experience near sorrow." For technical employees and teams, short-term incentives such as annual salary, wages, and benefits for technical employees and teams must be addressed, as well as long-term incentives such as long-term patents, dividends, and equity.

In conclusion, military industrial security involves personnel, finances, resources, scientific and technological advancements, education, ideological and confidential information, cyber defense, and physical defense. Most wars occur beyond the borders of military powers, where security and defense issues are rarely encountered during hostilities. Rather than that, network security is more critical during peacetime. It is critical to remember that for small countries, the military industry is frequently the first target.

References

[1]   Kramer, F. D., Starr, S. H., & Wentz, L. K. 弗兰金·D.克拉默, 斯图尔特·H.斯塔尔, 拉里·K.温茨. (2017). *Saibo liliang yu guojia anquan* 赛博力量与国家安全 (G. Zhao 赵刚, X. H. Kuang 况晓辉, L. Fang 方兰, D. X. Wang王东霞, F. Xu 许飞, & J. Tang 唐剑, Trans.). Beijing: National Defense Industry Press.

[2]   Yonah. J. B. (2017). Israeli experts point at a new ISIS trend: Poison in malls. *The Jerusalem Post*.

16 Intelligent Construction and Evaluation

What is the crux of an intelligent military? How can the intelligent capabilities of an operation system of systems (OSoS) be quantified? Which metrics and indicators are most critical? How to evaluate strategies and tactics, or how to conduct qualitative and quantitative analyses? These are the remaining questions in developing an intelligent military and assessing its effectiveness. Nine factors should be considered to accelerate the development of an intelligent military: cyber contribution, parallel intelligence, autonomy, swarming, quick chains, emergence effects, controllability, economy, and side effects.

16.1 Cyber Contribution

Since the turn of the century, cyberspace has been established and officially integrated into battlefields due to the widespread use of the Internet and mobile phones. Cyberspace, an artificial world, allows human activities to expand from the physical to the virtual space. Simultaneously, the Internet has created a new channel for spiritual communication. Over time, cyberspace has exerted an increasingly strategic, global, profound, and long-term influence on human society and military domains. Cyberspace lays the technical groundwork for a global military network by enabling humans to enter a parallel world populated by virtual forces and military systems. The evolution of cyberspace features information linking, data exchange, model algorithm development, and advancement of human–computer interaction, which underlie intelligent military and operations. The exponential growth of cyberspace has also sparked artificial intelligence (AI).

The Internet, mobile communications, space-based information, sensors, embedded systems, cloud computing, and integrated electronic information systems all contribute to an intelligent military. The three main indicators are network, data, and computing power.

16.1.1 Network Power

The network communication system serves as the prerequisite for cyberspace that includes civilian and military networks as well as their interaction and

DOI: 10.4324/b22974-18

complementarity. External metrics such as operational efficiency and effectiveness can be used to quantify network power.

Effectiveness-wise, network power is defined by military and civilian networks' popularity, sophistication, and interactive efficiency. Indicators include civilian mobile Internet uses and military Internet protocol (IP) nodes. Metcalfe's Law states that the Internet's value is proportional to the square of its user base. The efficiency of a network of 100 users can be increased 10,000-fold by connecting them in a way that allows everyone to see each other's content, and 10,000 users may result in a 10^8-fold increase in efficiency. Despite the impossibility of everyone viewing the information of everyone else, Metcalfe's Law emphasizes the potential and possibility. According to Metcalfe's Law, the military value of a country's network power is worth the following:

$$W = Rlg(S^2_m/T_1)+lg(S^2_j/T_2)$$

In the formula, W stands for network value (number of connected users per unit of time). R stands for civil–military fusion degree (the highest number is 100%). S_m stands for civilian mobile Internet users, while S_j stands for military IP nodes. Information propagation in civilian and military networks requires T_1 and T_2, respectively, periods and average time lags. As network users, groups, and channels grow in size and complexity, the network value W will grow exponentially. Thus, the logarithmic approach to expressing value gain in decibels is more scientific.

For security and other reasons, military systems are not regularly connected to civilian networks. However, as cyber systems improve, open-source intelligence gains value, and military and civilian systems will become more interconnected and intertwined. Unlike high-level strategic operations, low-level tactical operations (e.g., emergency rescue, disaster relief, counterterrorism, stability maintenance, and military operations other than war (MOOTWs)) rely more on civilian networks and information. As can be seen, the link between military and civilian systems weakens with operational level. Therefore, if $R = 0$, military networks are completely isolated from civilian networks and are only used to assist operations.

In contrast, if $R = 1$, the civilian network is ready to assist military operations at any time and from any location, and the value gain may represent both networks' combined support. If $S_j = 0$, civilian networks are only used to support non-military activities, which is rare. So R is always between 0 and 1, never zero or one.

16.1.2 Data Power

Data power can be accessed by collecting and preserving all historical military data:

$$B = \{B_z, \triangle B/\triangle T\}$$

In the formula, B denotes data power. B_z denotes the total amount of military data (as measured by the number of accessible bits). $\triangle B / \triangle T$ denotes the growth rate of military data, or the update rate, which indicates the data's degree of freshness. Military data can be generated by military systems or mined from open-source information systems operated by civilians.

16.1.3 Computing Power

A computer's computing power can be quantified in four ways: processing speed, storage capacity, physical size, and energy consumption.

$$J = \Sigma(S \times C)/(V \times P)$$

In the formula, J denotes computing power, S denotes processing speed, C denotes storage capacity, V denotes the volume of computing and storage devices, and P denotes energy consumption. When volume and energy consumption are kept constant, the faster the processor and the larger the storage capacity, the higher the computing power.

16.2 Parallel Intelligence

Intelligent operations require virtual space battlefield brain systems that can predict, plan, guide, learn, optimize, and self-evolve. "Parallel intelligence" refers to the level of intelligence required to perform the above functions. It also refers to the intelligence that these systems' virtual and physical forces share. Parallel intelligence demonstrates the superiority of battlefield brains (e.g., models and algorithms) in virtual space after simulating various environmental conditions. Before engaging in a real-world conflict, a troop can assess their chances of victory through virtual combat, semi-physical simulation, and physical training. Based on sufficient and precise virtual models of both red and blue forces, more realistic simulation systems of confrontations and operations, as well as advanced battlefield brain systems, can be developed with minimal investment, high efficiency, and rapid evolution. The battlefield brains, or the parallel systems, are a distributed collection of engagement models and strategy libraries within the virtual domain. These include the super brains of virtual staff officers and commanders for decision-making, the brains of weapons and equipment vehicles, and the brains of swarm units. Command-and-control systems, sensing terminals, primary battle vehicles, strike weapons, and support vehicles will all be equipped with these brains. The degree to which virtual and physical forces interact and the extent to which virtual forces aid physical forces in their victory define the strength of virtual forces.

Parallel intelligence consists of four components: network power (previously discussed), operational data accumulation and timeliness, tactical model and algorithm libraries, and advanced human–computer interaction and iteration. Model and algorithm libraries are crucial, and the other three are supportive.

Parallel intelligence may be quantified in three ways: data capability, model and algorithm count, and model and algorithm update rate:

$$P_Z = \{B, M, \triangle M / \triangle T\}$$

In the formula, P_Z stands for the level of parallel intelligence, B for data capability, M for the number of models and algorithms, and $\triangle M / \triangle T$ for the update rate of the models and algorithms (reflecting the advanced nature of the models).

The accumulation and update rate (timeliness or freshness) reflect the mapping and interaction frequency between virtual and real forces. Therefore, more data mean more frequent virtual–real interactions, and more real-time data. The data can support continuous model and algorithm optimization and increase update efficiency.

Models and algorithms based on various missions and tactics are core to parallel intelligence. A library of models and algorithms may include intelligent perception, self-adaptive mission planning, autonomous decision-making, intelligent cyber attack and defense, intelligent swarm attack and defense, intelligent air defense, anti-mischief, intelligent mobility and strike, intelligent operational support, battlefield environment simulation, virtual staff and commander, computer-generated effectiveness, and operational training exercise models and algorithms that work in both physical and virtual space.

16.3 Autonomy

Autonomy is the ability of unmanned vehicles to operate without human intervention. Intelligent operations seek the fewest possible frontier combatants, the least possible injury to people and civilian infrastructure, and the most precise strikes against targets. The intelligence depends largely on unmanned vehicle autonomy, intelligent perception and identification of complex battlefields and targets, self-adaptive mission execution, and manned–unmanned collaboration.

Mechanized vehicles place humans in the physical ring. Humans are closely connected to vehicles and actively involved in the entire operation mission. Humans directly operate, control, and support the vehicles. As a result, a mechanized vehicle usually has at least two crew members. Tanks and armored vehicles, for example, have three distinct roles: commander, gunner, and driver. Warships require more crew. In the age of AI, vehicle automation and mission intelligence have reduced the number of combat aircraft operators from multiples to one. Intelligent munitions have also emerged due to the deep integration of advanced navigation systems, mission systems, and network information.

Humans and vehicles will become physically separated as technology advances. Humans will be removed from the physical ring and remain responsible for remote vehicle control, command, control, and execution of operational tasks. The relationships between humans and vehicles are typically divided into three categories: "humans within the ring," "humans above the ring," and "humans beyond the ring." The vehicle's autonomy is first demonstrated by its ability to make autonomous decisions within the observe–orient–decide–act (OODA) loop and to automatically interact with and correlate operation tasks, referred to as task chain AI capability. A second reflection of autonomy is the ability to cooperate on multi-vehicle operations with manned or unmanned vehicles.

Quantifying vehicle autonomy is difficult. Each military service's distinct missions, operational space, and styles necessitate the use of distinct criteria for determining autonomy. Nonetheless, common evaluation methods from the perspective of human–machine interaction can be used.

The average amount of time spent by the vehicle AI (after learning missions and tasks) and the average amount of time spent by humans controlling the vehicles help determine the value of vehicle autonomy:

$$Z = T_{AI}/(T_{AI} + T_r)$$

In the formula, Z denotes the vehicle's autonomy, T_r denotes the average time spent by humans controlling the vehicle, and T_{AI} denotes the average time spent by the vehicle's machine AI controlling the vehicle. A higher Z value indicates a greater degree of autonomy and unmanned function, and vice versa. Fully autonomous unmanned vehicles have Z values close to 1, while remotely operated unmanned vehicles have Z values less than 0.3, implying that they require close human supervision and control most of the time.

When evaluating a vehicle's autonomy, the comparison to adversary vehicles and manned vehicles, the difficulty of manned control, and the complexity of control nodes must also be considered. To the extent that AI is capable of self-learning and self-evolution, all of these factors can be left to machine learning. Other factors affecting autonomy may include security, reliability, logistics, and economy, but these are also indicators of conventional vehicles and may be overlooked.

Unmanned vehicles will begin to demonstrate value if they exceed 1% of the total fleet and achieve the same combat effectiveness as manned vehicles. If they exceed 50%, the value will increase significantly, as half of the labor costs will be eliminated, and combatant safety will improve.

Military superpowers like the US and Russia are increasing the proportion of unmanned vehicles in their total military. While it is impossible to completely replace manned vehicles, increasing the proportion of unmanned vehicles or improving the degree of vehicle autonomy will increase the economic benefits of military operations while decreasing costs. This is undeniable.

Manned and unmanned controls are interdependent. While most vehicles will be unmanned, human intelligence will be heavily incorporated into their design, development, manufacturing, application, test, and role planning.

All operations are customized. In a war, when missions with varying weights are involved, the analytic hierarchy process can be used to determine the autonomy of multi-functional vehicles:

$$Z = \Sigma Q_i Z_i$$

where Q_i denotes the task weight, and Z_i denotes vehicle autonomy during specific operations. Accordingly, obtaining unmanned and autonomous data requires in-depth research and evaluation methods for various services, arms, and operational environments. However, the autonomy of operational vehicles is a critical criterion for evaluating intelligent warfare.

16.4 Swarming

Swarming is based on the quantity and quality of operational vehicles. The Lanchester square rate states that doubling the number of operation units with direct-point weapons quadruples the combat effectiveness. The numerically superior side (designated as *A*) can outnumber the adversary (*B*), whereas the numerically inferior side (*B*) can only confront *A* with half or less combat power. *B* can improve its chances of victory by increasing its combat effectiveness, or quality; however, if the quantity of *A* doubles that of *B*, *B*'s only option is to quadruple its quality. Therefore, quality advantages can only compensate for a limited quantity shortfall. It is clear from the US military's defeat in Mogadishu and Russia's disastrous failure in the first Chechen street battle that qualitative superiority cannot always trump quantitative superiority. In certain time and geographic windows, even inferior equipment can outperform strong opponents. This is also backed up by Mao Zedong's famous military maxim: "Concentrate numerically superior forces to wage an annihilation war."

Augustine's Law shows how improving equipment quality increases costs and possibly reduces quantity. Norman Augustine observed an exponential rise in US military aircraft spending and a linear rise in overall defense spending in 1984. "By 2054, the US defense budget will be sufficient to purchase only one tactical aircraft, which will be used alternately by the Air Force and Navy three and a half days per week," he predicted, "The Marines can only use the aircraft on leap years' extra day" [1]. From the US F-16 to the later F-22 and F-35, the cost of a single aircraft has risen from around $30 million to $160 million within 25 years. As a result, between 2001 and 2008, the US Navy and Air Force's basic budgets increased by 22% and 27%, respectively (after inflation), while the number of warships and aircraft on the equipment list decreased by 10–20%, respectively. Despite an increase in total budget, less equipment is purchased. Overemphasizing quality over quantity will also result in a decline in operational capabilities.

The cost of personnel deployment and life support systems will decrease as unmanned equipment becomes more common. Many low-cost unmanned vehicles and ammunition will be available. At this point, the benefits of scale and quantity become apparent. The numerical scale encourages non-linear amplification of combat effectiveness. Intelligent coordinated detection, strike, and defense may be equipped on low-cost swarm operational vehicles and systems, such as unmanned ground and aerial vehicles (UGVs and UAVs), missile, loitering munition, terminal-sensitive bomb, and bionic robot swarms. The vehicles and systems may contribute to the tactical value of "overcoming intelligence with swarm stupidity," increasing efficiency with quantity, the emergence effect, and the many-to-one and many-to-many strike capabilities. Calculating intelligence operation capabilities requires determining both sides' proportions of autonomous swarms. Swarm units have the following value in the Lanchester square law:

$$Z_j = (J - D)^2$$

where J denotes the number of friendly swarm units, D denotes the number of adversary equivalent vehicles, and Z_j denotes the swarm advantage.

The preceding formula involves adversarial unmanned units. When comparison to friendly manned units is required, another algorithm applies:

$$Z_j = (J - Y)^2$$

where J denotes the number of friendly swarm units, Y denotes the number of friendly manned equivalent vehicles, and Z_j denotes the swarm advantage.

The same formula applies to manned–unmanned swarm operations, though the number of equivalent vehicles used in manned and unmanned operations is difficult to determine.

16.5　Quick Chain

The efficiency in transmitting information between reconnaissance, control, strike, evaluation, and support stages is a key indicator of operation system of systems (OSoS) intelligence. Speed is the magical weapon that ensures humanity's victory in war. Detecting, attacking, defending, and occupying areas ahead of adversaries' expectations all depend on speed. Due to varying historical periods, the meaning and scope of speed in war vary greatly. Speed in ancient warfare refers to the speed of manpower, horses, bows, arrows, and chariots. The speed in mechanized warfare is exemplified by the rapid mobility and firepower of tanks, aircraft, warships, and artillery. The speed in information warfare relies on rapid detection, perception, and integrated command and control. Speed in intelligent warfare is revealed in the processing speed of operation information and the efficiency of kill chains. In the early stages of intelligent warfare, the speed with which humans analyze, judge, and make decisions is critical. Later on, the speed of autonomous perception, decision-making, action, and support becomes more critical. All of the preceding indicates intelligence.

Two factors determine the value of a quick chain. One criterion is how far machines can outperform humans in terms of analysis, judgment, decision-making, perception, and vehicle control abilities throughout the OODA loop. In other words, the question is whether machines can accomplish the same task faster than humans. The second criterion is whether the time spent transmitting information and executing the kill chain (reconnaissance, control, strike, evaluation, and support) is shorter when compared to adversaries. These two dimensions are intertwined. Outperforming the adversary may become more likely, as AI becomes more intelligent and applicable, and the kill chain becomes shorter.

Military operations entail a variety of missions, each with its own set of variables that must be managed separately. While tactical and operational machine intelligence is more straightforward to achieve, strategic machine intelligence entails considerations of politics, economics, society, and the military. As a result, relying entirely on AI to analyze, judge, and even make decisions is impractical in the foreseeable future. Thus, the quick chain's evaluation is primarily operational, if not tactical.

During the early stages of a war, the quick chain of any mission may be quantified in terms of time spent on perception, analysis, decision-making, and action, denoted by T_1, T_2, T_3, and T_4, respectively.

The intelligence of the quick chain can be measured in two ways: total time of machine control (ΣT_{AI}, the shorter, the better) and ΣT_{AI} as a percentage of total operation time (the higher, the better). Fully autonomous operation (ΣT_{AI}) should take less time than fully human operation (ΣT_r). When $\Sigma T_r - \Sigma T_{AI} > 0$, machine intelligence triumphs because automating tasks saves time. In terms of total time spent on an operational chain loop by humans and AI, the quick chain intelligence, or value, equals the ratio of:

$$V_z = \Sigma T_{AI} / (\Sigma T_{AI} + \Sigma T_r)$$

where V_z denotes the value of a quick chain associated with specific operational tasks, ΣT_r denotes the average time required for human control, and ΣT_{AI} denotes the average time required for machine AI control. ΣT_{AI} and ΣT_r are the sums of T_1, T_2, T_3, and T_4 respectively. Elements other than T_1, T_2, T_3, and T_4 may be added in later stages if necessary (e.g., $\Sigma T_i = T_1 + T_2 + \ldots + T_n$).

A higher V_z value indicates a mission that is more intelligent and efficient, and vice versa. A fully autonomous OSoS has a V_z value close to 1, whereas a fully manned OSoS has a V_z value of zero or a value infinitely close to zero.

A troop can theoretically win by striking first and acting first. Therefore, the quick chain value can also be quantified by comparing the quick chain duration on both sides. The formula is as follows:

$$V_d = \Sigma T_d - \Sigma T_j$$

where V_d denotes the difference between our and the adversary's quick chain duration, ΣT_d denotes the duration of the adversary's quick chain, while ΣT_j denotes our quick chain duration. When V_d exceeds zero, an operation may be executed faster than the adversary, and vice versa.

The outcome is contingent upon the adversary's strength, whether superior, inferior, or evenly matched. Given an adversary whose weaponry performance is comparable to ours, the duration of the quick chain is directly proportional to the amount of time that a human or AI controls the weapons and equipment. If both sides' strength, weaponry, strategy, and tactics are significantly different, the formula above may need to be adjusted slightly, but the results may only be slightly changed. Regardless of military power on either side, the critical factor is the time required to complete the OODA loop, determined by the time spent on sensor perception, analysis, and decision-making, as well as the primary operational vehicles' range and penetration efficiency. The quick chain is a closed loop, where a single efficient element may not guarantee the loop's short duration. However, it is not rare for a single front-end link to fail completely, thus terminating the whole loop. For example, if something goes wrong during intelligent perception, none of the loop's subsequent pieces will occur.

The range of the vehicle's mobility and strike, the speed of flight, the defensive penetration capabilities, the ability to endure electronic interference, the accuracy and controllability of destruction, and real-time damage evaluation are all variables of the quick chain. These variables are affected by the degree of mechanization, informatization, and intelligentization. In particular, speed, precision, and accuracy may be improved by intelligent applications and also present the issue of intelligent support. Intelligent support or logistics is concerned with timely delivery of weapons and ammunition, intelligent diagnostics, automatic fault tolerance, and timely energy supply. All of these factors add up to a time-sensitive cumulative effect. The quick chain will be prolonged or even halted if the strike distance is too far, or if the logistics are delayed, or if the hit is insufficiently precise and must be repeated.

A matrix index system based on the four "fasts" (fast perception, analysis, decision-making, and action) and three levels (strategy, campaign, and tactics) may be developed to evaluate the quick chain under different operations. The operational process may be described as five-, six-, or N- "fasts" in the future as operational space and missions develop. The command level, however, may become progressively flattened. The multi-level command-and-control tree structure may evolve into one composed of two layers: a manned decision-making and intervention layer, and an unmanned AI intelligent decision-making layer.

In many cases, the quick chain value is the first indicator of victory in intelligent warfare. The quick chain value reflects the level of OSoS intelligence in perception, decision-making, autonomous strike, and logistics capabilities, as well as operational efficiency (especially the V_d value compared with the adversary).

To summarize, troops must strive to be faster than their adversaries and attempt to break their quick chain loop. Once the loop is broken, subsequent hostile operations will be terminated, and adversaries will become passive. While defenses at the early stages of the loop may be effective, defenses at the loop terminal are commonly the last resort. The recent return of the US military to the Asia-Pacific region, for example, included the deployment of radars as close to China as possible in order to disrupt Chinese operational loops as quickly as possible in the event of a war.

16.6 Emergence Effects

The term "emergence" may be defined as "a significant number of unexpected events occurring in a short time span." According to systems science, emergence effects are caused by non-linear interactions between scale and structural influences.

Emergence effects are increased and converged in networked, swarm, or intelligent operations. The emergence effect can be studied in two ways: swarm attack and defense effects in the physical domain, and non-linear dissemination effects of human social mentality and behavior in the psychological domain. Studies on the emergence effect are mainly concerned with the latter. In an era of global interconnectivity, operation and war outcomes may be influenced by the dramatic effects of social mentality and online public opinion, the deterrent and operational effects of military powers, and the unanticipated effects of decapitation strikes. In general, the emergence effect can present itself in a variety of physical and psychological ways, including quick military collapse, victory reversal, populist eruption, and national marches and protests.

Nowadays, the international community can monitor an operation's processes and stakeholders closely, apart from the combatants. A single event may trigger a chain reaction in social psychology and public opinion, resulting in a monster wave. For instance, a successful decapitation strike may have a rapid effect on the adversary. The Internet era has resulted in information being disseminated in near-real time. It has been estimated that an event requires no more than six retweets to reach every person on the planet.

The use of big data search engines and associative computing enables the examination of the Internet's impact on human society. To begin, algorithms such as TOP-K can be used to rank hot spot popularity in both traditional and digital media. Second, positive and negative comments on Internet events can be assessed using knowledge-mapping and machine-learning techniques.

Third, rapid shifts in public opinion can be monitored in order to forecast trends associated with hot spot events. Fourth, developing an intelligent evaluation and early-warning system may aid in the correlation and verification of public opinion observations with operational progress, combat gains/losses, and public demonstrations.

$$Y = \{T_{OP\text{-}K}, \pm B, \pm \triangle B / \triangle T\}$$

In the formula, Y denotes the public opinion early-warning index, which is classified according to its severity as blue, yellow, red, or purple, $\pm B$ denotes data on public opinion, either positive or negative. $\pm B$ denotes the increase in either positive or negative public opinion, and $\triangle T$ denotes the rate of update.

In public opinion analysis, statistical analysis of positive and negative views expressed on online social media and self-media may more accurately reflect true public opinions than news reporting and analysis on traditional broadcast television and mainstream media. In addition, charts, data, and colors can be used to visually express public opinion.

Based on prior peacetime and wartime experiences, targeted and systematic public opinion analysis and countermeasures may be proposed for various major events. The offensive–defensive split necessitates a more targeted public opinion early-warning system, research analysis, and intervention system, segmented by country, region, city, and industry.

16.7 Controllability

Controllability is the ability to precisely control weapons and equipment, combatants, vehicles, maneuvers, and logistics support at any time and location. Controllability in physical space (e.g., precision strike and destruction) and virtual space (e.g., precise tracking and influence of critical targets and crowds, precise control of critical infrastructure, risk management for unmanned systems) both fall under this umbrella.

16.7.1 Precision Strike and Controllable Damage

Precision strikes and controllable damage represent two of the weapon development trends in the future. Precision strike technology is nearing maturity at the moment, and the accuracy of strikes is typically determined using the circular error probable (CEP) method. CEP is used to determine a weapon's ballistic accuracy. If a weapon has a 50% chance of hitting the target, a circle with a radius equal to the CEP can be drawn around it. For example, a missile with a CEP of 90 meters has a 50% chance of either hitting the target or deviating from its intended path by no more than 90 meters. In contrast to a precise strike, precise destruction control remains in its infancy. As effective destruction improves, excessive, ineffective, and accidental destruction of civilian infrastructure will decline. Further research will need to be conducted to determine whether an effective destruction rate of greater than 90% should be used as a criterion for intelligent warfare. The explosive power of ammunition, the control of explosives, and the target's vulnerability all contribute to the effectiveness of destruction. The effectiveness also connects to various modes of delivery and strike. For example, the attacking position and speed may also modify the effect. Due to the diversity of strike objectives, there is no simple or uniform standard for evaluating the effectiveness of destruction. Therefore, future precision-guided weapons should be capable of determining the geometry, nature, and vulnerability of targets using machine learning, image recognition, sound recognition, and, most importantly, multi-spectral spectrum recognition.

It is possible to evaluate the precision and effectiveness of a strike in two dimensions:

$$K = R + CEP$$

where K denotes the comprehensive precision of a precise strike or effective destruction, R denotes the radius of a weapon's or explosive's effective destruction sphere, And CEP refers to the precision of the weapons. When CEP equals zero, no controllable damage occurs because it equals explosive destruction.

16.7.2 Key Target Tracking and Influencing

Public figures (e.g., government heads and senior military personnel) may leave digital traces, images, and sounds that can be accessed via big data. An association search can provide an accurate picture of the key figures in the target population for precise tracking and influence of their social relationship and daily lives. To begin, data from mobile phones and news reports can be used to create a dynamic behavioral model of the target population. Then, real-time positioning and research can be used to ascertain permanent, alternative, or temporary residences, office building and private office locations, military command posts, mobile command posts, frequently visited venues, traffic routes, and personal preferences. A static behavioral model of the target population can then be established using network big data correlation. Third, the target group's social relationships can be mapped through their family members, relatives, and friends, as well as through their social circles. Knowledge mapping and reasoning can be used to ascertain the target group's values and ideologies based on their comments, articles, thoughts, and opinions. Fifth, precise and effective intervention or interference can be accomplished through psychological, physical, chemical, or other methods.

As Edward Snowden, a former Central Intelligence Agency (CIA) employee, revealed, tracking and stealing the mobile phone signals and personal preferences of the key figures may have become possible. Then, in

peacetime, it is also possible to develop a real-time behavior model and correlate it with intelligent image reconnaissance.

The social, professional, or personal endeavors of someone within a period
of time can be considered a full probability event. According to the full probability formula, suppose the event group is a division of the sample space Ω,
and $P(B_i) > 0$ ($I = 1, 2, ..., n$), then for any event A, there is:

$$P(A) = \sum_{i=1}^{n} P(B_i) P(A \mid B_i)$$

Assume $P(B_i)$ is the probability of the target person engaging in a particular activity in a location. The probability $P(A)$ of various activities in a
given location can then be calculated using statistical analysis. Additionally,
the Bayesian formula and a model of a large data structure can infer and
optimize correlations and probabilities between the locations and activities
of target personnel. For instance, tracking mobile phones or staff, family,
drivers, and secretaries enables the conditional probability of the key target
staying in a fixed apartment at night to be calculated. Attacks similar to the
one that resulted in bin Laden's capture may be launched. These attacks are
typically planned for late at night or early the next morning.

16.7.3 Precise Control of Infrastructures

Infrastructures are diverse, as are their precise control methods. Precise control of infrastructures is more likely to employ a cyber attack and defense
model rather than a large-scale attack and destruction model that would be
detrimental to post-war recovery and social stability. For example, to precisely
control network infrastructure, one must first grasp its structure, software
and hardware vulnerabilities, and real-time control methods. At the moment,
this level of precise control is not quantifiable. It can, however, aid in the
assessment of special forces' combat effectiveness.

16.7.4 Risk Management for Unmanned Systems

Military intelligence studies necessitated a greater emphasis on ethics and
morality, the nature of war, its social ramifications, and risk management.
Unmanned autonomous systems require a technical, procedural, ethical,
and legal framework capable of preventing and restricting their harm to
civilians. For unmanned systems, a self-destructive function or an effective
control should be designed. Fire suppression systems must be removed
from prohibited target areas such as hospitals, religious institutions, cultural relics, and schools. To rein in hackers and illegal organizations, mandatory restrictions, certifications, and strict control measures similar to those
used in drug management are required. Analyzing the effectiveness of risk

management and control strategies for unmanned systems requires both technical and legal analysis.

16.8 Economy

Each business system and area of expertise must consider and evaluate the economy, or economic efficiency. Future operation and equipment systems, likewise, involve economic efficiency with a focus on three factors.

The first focus is the cost analysis and pricing of human resources, money, and materials for electronic and mechanical systems, which encompass processing, manufacturing, assembly, testing, and integration stages.

The second focus is the pricing and investment of data collection, software development, and intellectual property. Software, data, web traffic, and intellectual property rights are already priced and evaluated in the commercial sector, but not yet in the military.

The third focus is the cost analysis and value evaluation of developing models and algorithms, particularly those used in AI systems, which requires a large number of talents, materials, and financial investments, as well as civil–military fusion. Furthermore, it requires extensive data use, intellectual property transfer, and complex issues like incremental value and non-linear input–output relationships.

Economic standards for intelligent systems must be established independently of mechanized information products. Cost analyses, pricing models, and evaluation criteria for various types of equipment should be proposed, including intelligence-intensive, hardware-intensive, conventional, and commercial equipment. For example, cloud platforms, command-and-control centers, and cognitive confrontation systems are intelligence-intensive; power transmission and fire destruction systems are hardware-intensive; electronic confrontation, avionics, and vehicle-electric systems are intelligence-intensive with partial hardware focus; and tanks, armored vehicles, artillery, ships, and airplanes are hardware-intensive with partial intelligence focus.

16.9 Side Effects

While the intelligence trend has a plethora of benefits, it also has a number of obvious drawbacks and even side effects, as detailed below.

16.9.1 Data Redundancy

Intelligent systems are built on numerous data centers. According to Internet Data Center (IDC), 8.4 million data centers would have been built globally in 2017, primarily in the US, EU, Japan, and China. After peaking at 8.55 million in 2015, the global data center count began to decline in 2016 and reached 7.2 million by 2021, a 15% decrease from 2015. While global data centers are dwindling, the average center size is increasing, indicating that data volume is increasing. The latest data from Synergy Research Group show that the total number of hyperscale data centers reached 597 by the end of 2020, more than doubling the 2015 number. The US retained nearly 40% of cloud and Internet data center locations geographically. Following the US, China, Japan, Germany, the UK, and Australia accounted for a combined 29% of the total [2]. Large amounts of data are required to train AI systems in machine learning, pattern recognition, image processing, audio recognition, simulation countermeasures, simulation training, and self-adaptive mission planning. Costs associated with data storage and management may continue to rise as a result of their continuously optimized models and algorithms, with new data generated and old data not yet discarded.

16.9.2 Energy Consumption

At the moment, Internet behemoths construct dozens of large-scale data centers throughout the world to ensure data integrity and freshness. Data centers house hundreds of thousands, if not millions, of servers and storage devices, which require enormous amounts of energy to operate and cool. Continuous backups and full-speed operation for security reasons can generate extreme heat in data centers that operate 24 hours a day. Numerous pieces of cooling equipment are required to maintain the temperature of these devices. A data center that contains servers, storage devices, and refrigeration systems consumes a significant amount of energy.

A large data center consumes enough energy to power the residents of a small town. Global data center electricity consumption accounts for 3% of total global electricity consumption. According to some industry analysts, global data center electricity consumption will reach tens of billions of dollars by 2025, growing at a compound annual growth rate of nearly 6% [3]. Data centers have become one of the most rapidly growing industries in the United States, consuming 91 billion kWh of electricity in 2013. This figure reached 1.38 billion kWh by 2020. Despite efforts to minimize energy waste, it is an unavoidable fact of life. "Data centers have an important role to play in making the economy more efficient," said Pierre Delforge, Natural Resources Defense Council (NRDC)'s director of high-tech energy efficiency [4]. "We see great leaders in the cloud space, but we need the whole industry to participate". Delforge believes small, medium, corporate, and multi-tenant data centers are still squandering huge amounts of energy and that social responsibility extends to all data centers.

Google claims 36 data centers globally, while Microsoft asserts that it operates over 100 data centers, both with over one million servers. Tencent, Alibaba, and Baidu are just a few of the companies accelerating their efforts to build data centers in China. Broadband providers are the Internet's nerve centers and also the primary builders of data centers. China Unicom now owns a total of one million servers spread across 31 provinces. However, most servers in large data centers are idle for most of the time, accounting for approximately 80% of computing power. For example, during normal business hours, major bank servers are used exclusively for data backup and security. Alibaba's servers only run at full speed on "Double Eleven," an online shopping carnival.

16.9.3 Privacy Leakage

In the Internet or AI eras, influential individuals and user groups leaking private data will be a serious issue. Verizon, the largest US telecommunications company, admitted in July 2017 that a cloud server used to store user data had been misconfigured and made publicly available. Verizon customers' names, phone numbers, and account passwords were compromised. Uber revealed in November 2017 that it was hacked in 2016, resulting in a massive data breach.

According to a *Guardian* and *New York Times* investigation [5], Cambridge Analytica collected the profiles of over 50 million Facebook users without their consent in mid-March 2018. Christopher Wylie, the co-founder of Cambridge Analytica, confirmed the occurrence. Following an analysis of their voting patterns, personality traits, values, and upbringing, the Facebook users were sent targeted messages and campaign advertisements. Under Armor announced the largest data breach in history in March 2018, affecting 150 million MyFitnessPal accounts (a diet and fitness app). The Identity Theft Resource Center and other sources estimate that the ten largest data breaches of 2018 compromised nearly 182 million records, including sensitive information from government agencies, private companies, and financial, medical, and educational institutions.

16.9.4 Security Risks

Intelligent technologies have fundamentally altered operational procedures and have been extensively used to increase the flexibility and accuracy of military operations. Automated labor has reduced the dangers faced by human soldiers. Unmanned operations represent the way war will be fought in the future. However, the security flaws and vulnerabilities of unmanned or autonomous systems may be exploited by cyber intruders and strikes, posing a threat to human life. Unmanned misfires are likely to be frequent and difficult to avoid during the early stages of development. If UAVs and robots are not properly regulated, they will cause significant harm to human society.

There are currently no effective controls in place to prevent autonomous systems from making errors. If a program malfunctions or is maliciously tampered with, a bloody tragedy will ensue. For instance, on July 3, 1988, the USS *Vincennes*, a cruiser operating in the Persian Gulf, detected the approach of an Iranian civilian A300 airliner. The cruiser's automatic air defense system and radar identified the airliner as an Iranian Air Force F-14 fighter and launched a standard air-defense missile on it, killing all 290 passengers and crew. The US deployed three armed "SWORDS" ground robots into

Iraq in 2008. These remote-controlled robots were quickly removed from the battlefield after they turned their weapons on their human commanders. In conclusion, if no constraints on the development of intelligent killer robots are imposed, they will enter the battlefield and arbitrarily determine human life and death. Using AI, specifically facial, image, and audio recognition, it is technically possible to avoid the issue of a gun pointed at the commander. Targets that must be hit and those that cannot be hit can be marked on high-definition maps. Consequences must be minimized, even more so when operational targets coexist with civilians.

16.9.5 Social Ethics

Inevitably, unmanned and intelligent weapons will have a detrimental effect on war ethics and accountability. The concept of using military robots appeals to both those concerned with soldier safety and those concerned with war costs. Military robots reduce costs associated with military salaries, housing, pensions, and medical care by operating at a faster and more precise rate than humans. They will not experience post-traumatic stress in the same way that human combatants do. They are emotionally stable and therefore do not need to be mobilized. They can access areas that are inaccessible to humans and perform difficult tasks. They are tenacious and fearless in their pursuits regardless of intelligence. Additionally, robots do not require human intervention or life support. Nobody will perish as a result of their destruction. Simply replace a robot with a new one. This is unavoidably economical.

On second thoughts, who should be held accountable if military robots kill innocent civilians? "It is possible for robots to fire at their computers and go insane," according to Noel Sharkey, a British robotics professor. "Whoever is to blame, it is obvious that the robot is not to blame. As a result, war laws must establish accountability" [6].

When used effectively on the battlefield, military robot development should maximize expected effects while minimizing collateral damage from a humanitarian standpoint.

Robotic advancements will have an effect on the military. Robots deserve human eyes, whether it's Alibaba Cave or Pandora's Box. Only when technology is effectively and safely used for human development can the world's Alibaba Cave be built.

These are significant issues that may have emerged. As intelligence advances, new issues will arise. They are critical issues in the development and evaluation of intelligence. In the future, the by-products must be addressed through technological, management, and legal means.

References

[1]   Work, R. O., Brimley, S., & Scharry, P. 罗伯特·O.沃克, 肖恩·布瑞姆利, 保罗·斯查瑞. (2016). *20YY: jiqiren shidai de zhanzheng* 20YY: 机器人时代的战争 (H. Zou 邹辉, Trans.). Beijing: National Defense Industry Press.

[2]   Data Center Operation and Maintenance Management. 数据中心运维管理 (2021). Zui xin: Quanqiu daxing shujuzhongxin zongshu zeng zhi 597 ge, shi 2015 nian shuju zhongxin shuliang de liangbei 最新：全球大型数据中心总数增至597个，是2015年数据中心数量的两倍. Sohu搜狐. www.sohu.com/a/44812 8462_610463

[3]   Gu, A. T. 骨傲天. (2018, August 20). *Shuju zhongxin haodianliang taixiaren yongdian zongliang zhan quanqiu yongdianliang 3%* 数据中心耗电量太吓人用电总量占全球用电量3%. Tianji wang 天极网. https://server.yesky.com/datacenter/405/1676472405.shtml.

[4]   NRDC. (2014). America's data centers consuming massive and growing amounts of electricity. www.nrdc.org/media/2014/140826

[5]   Farr, C. (2018). Whistleblower Christopher Wylie says he's now been blocked by Facebook. www.cnbc.com/2018/03/18/whistleblower-christopher-wylie-says-hes-now-been-blocked-by-facebook.html

[6]   Sharkey, N. (2011). Automating warfare: Lessons learned from the drones. *Journal of Law, Information and Science*, *21*(2), 140–154.

Concluding Remarks
Is There a Bright Future?

The train of intelligent warfare is speeding down the tunnel of the twenty-first century. Should we abandon it to the darkness of avarice and uncontrollability, or should we lead it into the light of civilization and illumination?

Intelligentization represents one way for the future, but it is not the full story. While intelligence manifests itself in various military duties, it is not omnipotent. Faced with acrimonious clashes between civilizations, faiths, countries, and classes, as well as extreme terrorists or events such as armed mobs, suicide bombers, and mass riots, intelligence's function remains limited. Global politics, power, and trade are inherently unbalanced and unjust, resulting in a variety of social paradoxes that make wars and conflicts unavoidable.

After all, world dominance is defined by the strength of nations, which includes their scientific, economic, and military capabilities. While the military cannot impose its will on politics or the economy, it may ensure economic prosperity. Stronger intelligence operational capabilities increase the likelihood of deterring and preventing wars, implying a greater likelihood of world peace. As with nuclear deterrence, public fear of trauma and disasters plays a critical role in averting large-scale wars.

In wars, the degree of intelligence may reflect the degree of civilization. The history of human battle is replete with slaughter, violence, and repression, progressing from tribal conflicts over food and shelter to land occupation, resource robbery, political power expansion, and spiritual world dominion. War is the ultimate means of resolving human society's insoluble problems. The ideal goal of future warfare would be a civilized one, which would entail either winning without fighting or resolving the conflict with the least amount of resource consumption, casualties, and social damage possible. However, political instability, ethnic conflict, divergent economic interests, and lethal technologies frequently cause things to backfire – battles wreak havoc on our lands and homes.

Fortunately, with the advancement of intelligence, dreams previously unattainable in conventional warfare may come true in the future. Increased technological transparency and economic reciprocity will also make it possible for robots to eventually take the place of humans on future battlefields.

Future battles would result in fewer deaths, reduced material consumption, and other negative consequences. As everyone hopes, "civilized warfare" is extremely likely to become a reality in this sense. It is hoped that future warfare will shift away from intra-human combat toward operational deterrence and balancing in the physical world, as well as toward a highly simulated war game between artificial intelligence and robots in the virtual world. War consumption would be limited to a manageable number of unmanned systems, simulated clashes, bionic experiments, and even the resources required to launch a computer war game. Humans will transition from planners, designers, participants, oppressors, and victims to logical thinkers, organizers, controllers, bystanders, and adjudicators throughout the war. They will be spared the physical agony, emotional anguish, and loss of personal belongings that accompany wars.

There may be a chasm between our hopes and realities, but we pray for a speedy resolution. This may represent the final stage in intelligent warfare, the book's ultimate goal, the author's dream, and a grand vision for humanity.

Postscript

I was initially inspired to write the book by the 2016 Human vs Machine Go Match between AlphaGo, Google's artificial intelligence based on deep neural network technology, and Lee Sedol, a professional Go player with a 9 dan rank. That game reminded me of a course I took in my master's program on mission planning and decision-making for complex systems, where the professor taught us how to model and compute with neural networks. At the time, I was awestruck and fascinated by the fact that humans could build decision-making models by mimicking the neural system of the human brain. I never imagined that this technology would be applied to human vs machine Go battles after two decades of development. Shortly after the battle, as the associated research progressed, I began to consider how artificial intelligence would affect future warfare. In August 2016, I was invited to deliver a speech at a high-level forum on future intelligent warfare. The draft of this book was completed based on that speech and after 3 years of extensive literature study.

The book has gone through ten printings since it was officially published and released in January 2020. In the English version of this book, I have restructured it into theoretical and application sections, revised several chapters, added new material on the 2020 drone war on the Turkish–Syrian border, the Nagorno-Karabakh conflict, the assassination of Iranian nuclear scientists, the US military's "joint all-domain operations", and the large-scale Palestinian–Israeli conflict, as well as updated China's gross domestic product (GDP) and global trade data.

I am indebted to the Chinese and international scholars who worked to provide references and feedback for this book. I can list only a portion of their names here due to space constraints and concerns about confidentiality.

Many thanks to Jian Yang, who provided most of the images and illustrations for the book and contributed to the content on military technological development and brain–computer technology. I would also like to thank colleagues from the National Defense Industry Press who provided invaluable assistance in editing, organizing, illustrating, proofreading, and publishing this book.

I would also like to convey my gratefulness to Prof. Qichao Zhu and his translation team, including Prof. Chaowei Pang, Dr. Rong Xiao, Dr. Zhen Deng, Dr. Xiaoyu Zhang, Dr. Kun Long, Mr. Nie'ao Chen, and Miss Jiaxi Liu, for their commitment and professionalism.

I hope this book can serve as a source of information and inspiration for readers. I would appreciate readers' constructive feedback, suggestions, and comments.

Index

For Product Safety Concerns and Information please contact our EU
representative GPSR@taylorandfrancis.com Taylor & Francis Verlag GmbH,
Kaufingerstraße 24, 80331 München, Germany

Printed and bound by CPI Group (UK) Ltd, Croydon, CR0 4YY
01/05/2025
01858492-0001